86

Fortschritte der Chemie
organischer Naturstoffe

Progress in the
Chemistry of Organic
Natural Products

Founded by
L. Zechmeister

Edited by
W. Herz, H. Falk,
and G. W. Kirby

Author:
A. Gossauer

SpringerWienNewYork

Prof. W. Herz, Department of Chemistry,
The Florida State University, Tallahassee, Florida, U.S.A.

Prof. H. Falk, Institut für Chemie,
Johannes-Kepler-Universität, Linz, Austria

Prof. G. W. Kirby, Chemistry Department,
The University of Glasgow, Glasgow, Scotland

This work is subject to copyright.
All rights are reserved, whether the whole or part of the material is concerned, specifically those of translation, reprinting, re-use of illustrations, broadcasting, reproduction by photocopying machines or similar means, and storage in data banks.

© 2003 by Springer-Verlag Wien
Printed in Austria

Product Liability: The publisher can give no guarantee for all the information contained in this book. This does also refer to information about drug dosage and application thereof. In every individual case the respective user must check its accuracy by consulting other pharmaceutical literature. The use of registered names, trademarks, etc. in this publication does not imply, even in the absence of a specific statement, that such names are exempt from the relevant protective laws and regulations and therefore free for general use.

Library of Congress Catalog Card Number AC 39-1015

Typesetting: Thomson Press (India) Ltd., Chennai
Printing: Druckerei Theiss GmbH, A-9400 Wolfsberg
Printed on acid-free and chlorine-free bleached paper

SPIN: 10896068

With 1 Figure

ISSN 0071-7886
ISBN-3-211-83889-9 Springer-Verlag Wien New York

Contents

List of Contributor VII

Monopyrrolic Natural Compounds Including Tetramic Acid Derivatives
A. Gossauer

1. Introduction	2
2. Pyrroles from Vertebrates	9
3. Pyrroles from Invertebrates	12
3.1. Pyrroles from Insects	12
3.1.1. Pyrroles from Ants	12
3.1.2. Pyrroles from Beetles	15
3.1.3. Pyrrole Pheromones from Butterflies	16
3.2. Pyrroles from Sponges	16
3.2.1. Pyrrole-2-Carboxylates from Sponges	17
3.2.2. Pyrrole-2-Carboxamides from Sponges	20
3.2.2.1. Pyrrole-2-Carboxamides of the Oroidin Type	23
3.2.2.2. Cyclized Oroidin Metabolites	28
3.2.2.3. Dimeric Oroidin Metabolites	40
3.2.3. Alkyl Pyrroles from Sponges	44
3.2.4. Aryl Pyrroles from Sponges	48
3.2.5. Hydroxy Pyrroles from Sponges	52
3.2.6. Tetramic Acid Derivatives from Sponges	53
3.3. Pyrroles from Other Invertebrates	58
3.3.1. Aryl Pyrroles	58
3.3.2. Other Pyrrole Derivatives	64
3.4. Pyrroles from Protozoa	66
4. Pyrroles from Plants	67
4.1. Dihydropyrrolizine Derivatives	72
4.2. Pyrrole-2-Carboxylic Acid Derivatives from Plants	76
4.3. Pyrrolinone Derivatives in Plants	80
5. Pyrroles from Fungi	81
5.1. Pyrroles from Basidiomycetes	81
5.2. Pyrroles from Deuteromycetes	82
5.3. Pyrroles from Myxomycetes	85
5.4. Pyrrolinone Derivatives in Fungi	86
5.5. Tetramic Acid Derivatives in Fungi	86

Contents

6. Pyrroles from Bacteria 93
   - 6.1. Halogenated Monopyrroles from Bacteria 94
     - 6.1.1. Halogenated Benzylpyrroles 94
     - 6.1.2. Halogenated Benzoylpyrroles 95
     - 6.1.3. Halogenated α-Arylpyrroles 97
     - 6.1.4. Halogenated β-Arylpyrroles 98
   - 6.2. Pyrrole-2-carboxylates from Bacteria 102
   - 6.3. Pyrrole-2-carboxamides from Bacteria 108
     - 6.3.1. The Distamycin Group 108
   - 6.4. Pyrrol-2-carbacyl Derivatives from Bacteria 114
   - 6.5. α-Pyridylpyrroles from Bacteria 117
   - 6.6. Other Monopyrrole Derivatives from Bacteria 120
   - 6.7. Prodigiosins 122
     - 6.7.1. Prodigiosins from Eubacteria 122
     - 6.7.2. Prodiginines from Actinomycetes 126
   - 6.8. Hydroxy Pyrroles from Bacteria 129
   - 6.9. Tetramic Acid Derivatives from Prokaryotes 131
   - 6.10. Derivatives of 3-Acetyltetramic Acid from Actinomycetes 134

References 140

Author Index 189

Subject Index 213

List of Contributor

Gossauer, Prof. A., Department of Chemistry, University of Fribourg, Pérolles, Ch. du Musée 9, CH-1700 Fribourg, Switzerland

Monopyrrolic Natural Compounds Including Tetramic Acid Derivatives

Albert Gossauer

Department of Chemistry, University of Fribourg, Fribourg, Switzerland

Contents

1. Introduction	2
2. Pyrroles from Vertebrates	9
3. Pyrroles from Invertebrates	12
3.1. Pyrroles from Insects	12
3.1.1. Pyrroles from Ants	12
3.1.2. Pyrroles from Beetles	15
3.1.3. Pyrrole Pheromones from Butterflies	16
3.2. Pyrroles from Sponges	16
3.2.1. Pyrrole-2-Carboxylates from Sponges	17
3.2.2. Pyrrole-2-Carboxamides from Sponges	20
3.2.2.1. Pyrrole-2-Carboxamides of the Oroidin Type	23
3.2.2.2. Cyclized Oroidin Metabolites	28
3.2.2.3. Dimeric Oroidin Metabolites	40
3.2.3. Alkyl Pyrroles from Sponges	44
3.2.4. Aryl Pyrroles from Sponges	48
3.2.5. Hydroxy Pyrroles from Sponges	52
3.2.6. Tetramic Acid Derivatives from Sponges	53
3.3. Pyrroles from Other Invertebrates	58
3.3.1. Aryl Pyrroles	58
3.3.2. Other Pyrrole Derivatives	64
3.4. Pyrroles from Protozoa	66
4. Pyrroles from Plants	67
4.1. Dihydropyrrolizine Derivatives	72
4.2. Pyrrole-2-Carboxylic Acid Derivatives from Plants	76
4.3. Pyrrolinone Derivatives in Plants	80
5. Pyrroles from Fungi	81
5.1. Pyrroles from Basidiomycetes	81
5.2. Pyrroles from Deuteromycetes	82
5.3. Pyrroles from Myxomycetes	85
5.4. Pyrrolinone Derivatives in Fungi	86
5.5. Tetramic Acid Derivatives in Fungi	86

6. Pyrroles from Bacteria 93
  6.1. Halogenated Monopyrroles from Bacteria 94
    6.1.1. Halogenated Benzylpyrroles 94
    6.1.2. Halogenated Benzoylpyrroles 95
    6.1.3. Halogenated α-Arylpyrroles 97
    6.1.4. Halogenated β-Arylpyrroles 98
  6.2. Pyrrole-2-carboxylates from Bacteria 102
  6.3. Pyrrole-2-carboxamides from Bacteria 108
    6.3.1. The Distamycin Group 108
  6.4. Pyrrol-2-carbacyl Derivatives from Bacteria 114
  6.5. α-Pyridylpyrroles from Bacteria 117
  6.6. Other Monopyrrole Derivatives from Bacteria 120
  6.7. Prodigiosins 122
    6.7.1. Prodigiosins from Eubacteria 122
    6.7.2. Prodiginines from Actinomycetes 126
  6.8. Hydroxy Pyrroles from Bacteria 129
  6.9. Tetramic Acid Derivatives from Prokaryotes 131
  6.10. Derivatives of 3-Acetyltetramic Acid from Actinomycetes . 134
References 140

1. Introduction

Pyrrole was discovered in the coal tar by Friedlieb Ferdinand Runge (1794–1867) in May 1834. Only 24 years later – in 1858 – it was obtained in a pure state by Thomas Anderson (1819–1874) by distillation of bone oil. The actual structure of pyrrole, which was suggested by Adolf von Baeyer (1835–1917) in 1870 on the basis of his work on the elucidation of the structure of indigo (*1*), was proved later (1877) by synthesis by Chichester A. Bell (*2*). The later development of synthetic methods for the preparation of pyrrole derivatives by Ludwig Knorr in 1884 and Carl Ludwig Paal in 1885 paved the way for Hans Fischer's overwhelming work on the synthesis of naturally occurring pyrrolic pigments, which culminated in 1929 with the total synthesis of haemin. The relationship of the latter with pyrrole had been already established, at the beginning of the 20th century, by the Russian physician Marceli Nencki (1847–1901) and by William Küster (1863–1929). On the other hand, Leon Pavel Teodor Marchlewski (1869–1946), in collaboration with Nencki, proved in 1901 that haemin and chlorophyll are structurally related. The enormous biological significance of the tetrapyrrolic pigments – the so-called pigments of life (*3*) – delayed manifestly the search for more simple derivatives of pyrrole in nature. Of great significance, therefore, was the observation made by Sachs in 1931 that the urine of patients affected by acute porphyria gives a positive

References, pp. 140–188

Ehrlich reaction, which is characteristic for pyrrole and its derivatives (*4*). Only in 1952 Westall (*5*) was able to isolate the pure crystalline component responsible for this reaction, which was characterized two years later by Cookson and Rimington (*6*) as porphobilinogen (**1**), the biogenetic precursor of all tetrapyrrolic pigments known until now. Before 1954 (the year in which the structure of porphobilinogen was elucidated) only four natural compounds – 2-acetylpyrrole (**139**), prodigiosin (**267**), nicotyrine (**141**), and ryanodine (**178**) – had been characterized as pyrrole derivatives. After the discovery of the first antibiotic derived from pyrrole, netropsin (**246**), the structure of which was elucidated in 1957, the number of simple pyrrole derivatives isolated from natural sources increased rapidly as a consequence of the systematic search of new antibiotics, mainly produced by microorganisms.

The present review deals with monopyrrolic derivatives which occur in nature as products of the secondary metabolism (*7*). It does not consider, therefore, tetrapyrrolic pigments such as porphyrins, chlorophylls, and bile pigments, the chemistry and biology of which are object of excellent recent monographs on this field. For the sake of completeness, natural occurring pyrrole derivatives – *e.g.* methoxatin (**260**) – in which the pyrrole ring is a cleraly distinct part of the molecular structure, as well as a few polypyrrolic compounds, such as bipyrrole and pyrromethene derivatives, are included in this work. However, derivatives of 2*H*- or 3*H*-pyrrole as well as natural compounds in which the pyrrole ring is constituent of a heteroaromatic system (*e.g.* indole, indolizine, etc.) and metabolites derived therefrom are considered to be beyond the scope of this review. At the borderline of the definition of a pyrrole derivative are compounds, the actual structure of which is tautomeric with that containing the conjugated system of two C=C-bonds and the lone pair of electrons on the nitrogen atom, which is characteristic for pyrrole. Therefore, naturally occurring α-hydroxypyrroles – including derivatives of tetramic acid (*cf.* Section 3.2.6.) – have been considered as pyrrole derivatives, despite the fact that they occur usually as lactame tautomers, whereas derivatives of succinimide, the tautomeric structure of which is that of 2,5-dihydroxypyrole, have been generally excluded from the present review.

A classification of the hitherto known naturally occurring monopyrroles is difficult not only because the extreme diversity of their structures but also because the lack of knowledge, in most cases, of their biosynthetic origin. Actually, depending on the substituents present on the pyrrole nucleus, the biogenetic origin of the latter may be very varied (*cf.* Table 1.1.). On the other hand, a large number of natural compounds

Table 1.1. Origin of the Carbon Atoms of the Pyrrole Ring in Natural Occurring Pyrrole Derivatives Other than Pyrrole-2-Carboxylic Acid

Biogenetic precursor	Pyrrole derivative
Acetate	Chilocorines A and B (**21** and **22**, respectively)
Acetate	Exochomine (**20**)
Acetate (*via* a polyketide)	Myrmicarins (**11–19**)
Acetate (*via* a terpinoid)	Molliorins (**101a–d**, **102**)
Acetate, serine or alanine	Prodigiosins, ring B and C, respectively (Section 6.7.)
Acetate (*via* a polyketide), glycine	Prodiginins, ring C (Section 6.7.)
Acetate	Verrucarin E (**191**)
δ-Aminolaevulic acid	Porphobilinogen (**1**)
2,5-Dioxogluconic acid	Funebrine (**151**)
Isopentenylpyrophosphate	6-(3-Methylpyrrol-1-yl)adenosine (**156**)
Leucine?	Methyl 4-methylpyrrole-2-carboxylate (**9**)
Leucine	Glycerinopyrin (**244**)
Ornithine	Loroquin (**161**)
Ornithine	Dihydropyrrolizine alkaloids (Section 4.1.)
Ornithine	β-Nicotyrine (**141**)
Ornithine	Danaidone (**23**) and related pheromones
Proline	Pyrrole-2-carboxylic acid and its derivatives (*cf.* Table 1.2.)
Proline	Dioxapyrrolomycin (**222**) and other pyrrolomycins(?)
Proline	Prodigiosins, ring A (Section 6. 7.)
Proline	Pyoluteorin (**225**)
Proline, serine and acetate.	Tambjamins (**133–135**)
Tryptophan (side chain)	Lycogalic acids (**198a–c**)
Tryptophan (C2, C3 and side chain)	Pyrrolnitrin (**229**)
Tryptophan (side chain)	Rhazinilam (**157a**)
Tyrosin (side chain)	Lamellarins (Sections 3.2.4. and 3.3.1.)
Tyrosin (side chain)	Storniamides (**106a–d**)
Tyrosin (side chain)	Lukianols (**123a,b**)
Tyrosin (side chain)	Ningalins A and B (**125** and **126**, respectively)
Tyrosin (side chain)	Polycitone A (**129a**)
Tyrosin?	Rigidin (**131**)

containing pyrrolic substructures are merely esters or amides of 2-pyrrolecarboxylic acid and some of its halogenated or methylated derivatives, which represent, in the most cases, only a minor part of the molecule (*cf.* Table 1.2.). For these compounds it is obvious that the

References, pp. 140–188

Table 1.2. Pyrrole-2-Carboxylic Acids Which Commonly Occur in Natural Compounds

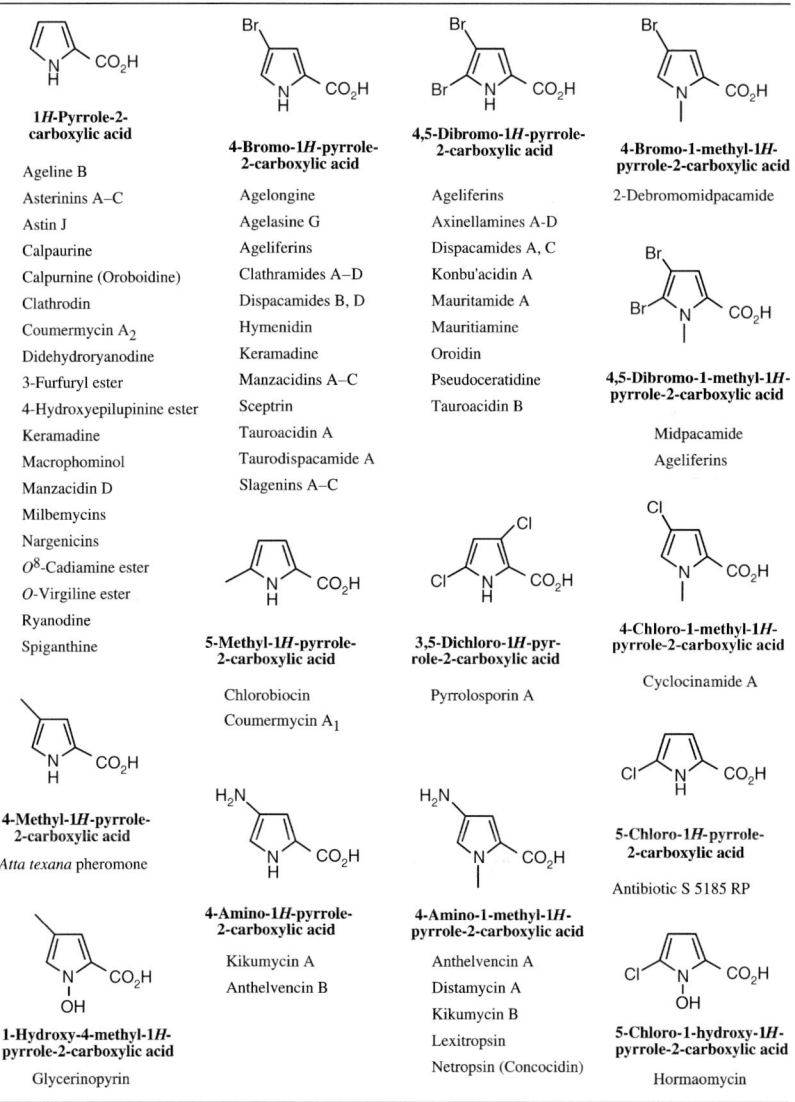

biogenetic precursor of the pyrrole moiety may be L-proline, although the origin of the pyrrole ring has not be elucidated in all cases. Pyrrole-2-carboxylic acid itself, which has been isolated along with other low-molecular weight metabolites from shake cultures of the fungus

Marasmiellus ramealis (*8*), is supposed to be accumulated extracellulary in other bacterial cultures (*9*).

The most common pyrrole derivative in nature is, however, Porphobilinogen (**1**). It is an ubiquitous metabolite which occurs in all aerobic and most anaerobic living organisms as a biogenetic precursor of porphyrins (haeme and cytochromes), corrinoids (*i.e.* vitamin B_{12}), coenzyme F-430, chlorophylls, bacteriochlorophylls, and bile pigments, including the phycobilins which act as photoreceptors in photosynthetic bacteria and some algae.

Porphobilinogen (PBG) (**1**) ⇌ (Ac₂O / pyridine, OH⁻) Porphobilinogen lactam (**2**)

Owing to the presence of three ionizable functional groups in the molecule, **1** is soluble in water, but insoluble in apolar solvents. At its isoelectric point, at pH = 4.3 (*10*), porphobilinogen can be precipitated from the aqueous solution as the corresponding mercuric salt, which on treatment with hydrogen sulfide regenerates the starting material. *In vitro*, **1** is a readily polymerisable compound (*5, 11*). In the presence of acids it is transformed, in high yield, to a random mixture of the four isomeric *uroporphyrinogens* (Scheme 1.1.), in which uroporphyrinogen III (50%) statistically predominates (*12*).

Porphobilinogen is accessible by different chemical syntheses which have been from time to time critically reviewed (*13, 14*). As a synthetic intermediate in most synthetic methods, porphobilinogen lactam (**2**), which is more stable than **1** itself, plays an important role, as it can be easily transformed into porphobilinogen by alkaline hydrolysis (*15*). The synthesis of **2** with the best overall yield has been reported by Battersby *et al.* in 1975 (*16*). In this approach (*cf.* Scheme 1.2.), 4-methyl-5-nitro-2[1*H*]-pyridone was reacted with dimethylformamide in the presence of phosphorus oxychloride to give a dimethylimminium salt, which was hydrolyzed by sodium hydroxide in aqueous acetone and subsequently reacted with sodium benzyloxide in benzyl alcohol to give the benzyl ether of an α-(4-pyridyl)acrolein derivative. Reduction of the latter with zinc dust and aqueous acetic acid yielded an azaindole, which reacted with monobenzyl malonate in a dry pyridine-piperidine mixture to form the corresponding acrylic acid derivative. Finally,

References, pp. 140–188

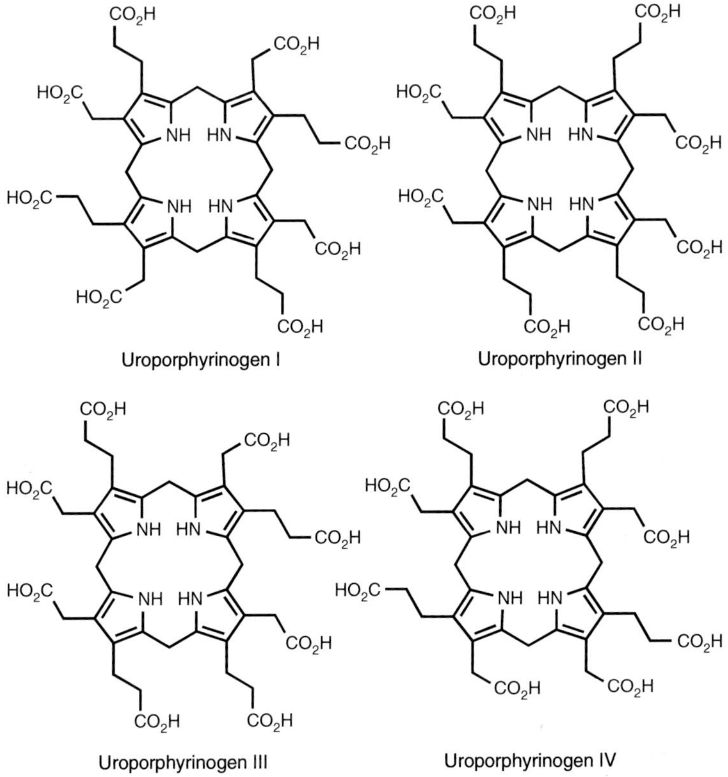

Scheme 1.1. The four isomers of uroporphyrinogen in which each pyrrole ring is substituted at the β-positions with an acetic and a propionic acid chain

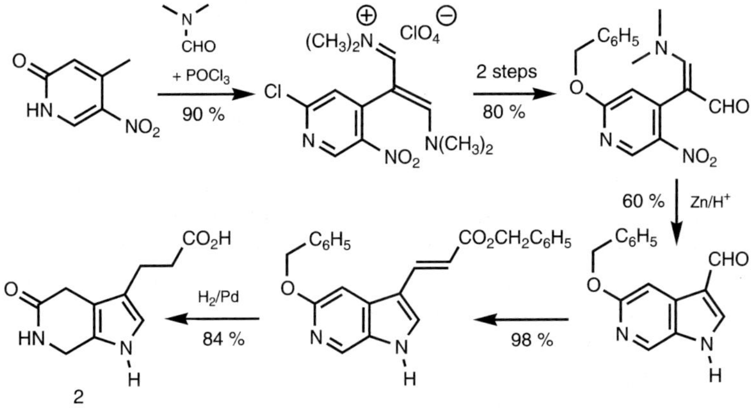

Scheme 1.2. A short synthesis of porphobilinogen lactam

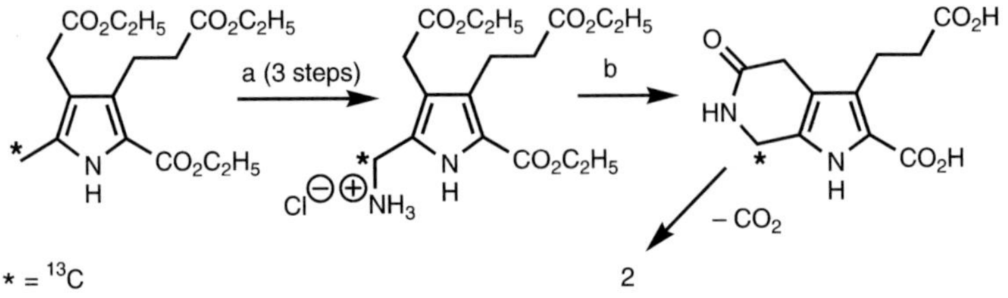

Scheme 1.3. The conventional synthesis of porphobilinogen (*17*): **a**) *i*: SO$_2$Cl$_2$ in diethylether, *ii*: NaN$_3$ in acetone/H$_2$O; *iii*: H$_2$/Pd(C) in ethanol; **b**) 2N NaOH

catalytic reduction of the latter cleaved the benzyl groups, saturated the exocyclic C=C-bond and reduced the pyridone ring to give porphobilinogen lactam in 35% overall yield from the starting pyridone. More conventional porphobilinogen syntheses may be, however, advantageous for the synthesis of isotopically labeled derivatives (*cf.* Scheme 1.3.).

From the practical point of view, however, the preparation of porphobilinogen from 5-aminolaevulinic acid either using extracts of beef liver (yield > 60%) (*18*) or from *Propionibacterium shermanii* cultures (yield up to 54%) (*19*) may be more appropriate than chemical procedures. Still higher yields have been obtained attaching 5-aminolaevulinate dehydratase to a Sephadex column in which the enzyme remains active for several weeks while 5-aminolaevulinic acid is continuously passed through the column. Between 50 and 94% of the substrate was dimerized to porphobilinogen by this procedure (*20*). Contrarily, the chemical condensation of 5-aminolaevulinic acid under basic conditions yields only 3% porphobilinogen (*21*).

The biosynthesis of porphobilinogen is broadly similar in all living systems, the first common intermediate being the highly reactive 5-aminolaevulinic acid (δ-ALA) (*22*). There are, however, two totally distinct pathways by which 5-aminolaevulinic acid is produced, one utilizing the carbon skeleton of glutamate and the other involving glycine and succinylcoenzyme A as precursors (Scheme 1.4). The glutamate route is found in many anaerobic bacteria and in plants, whereas the pathway using glycine and succinyl-coenzyme A, although occurring in some bacteria, is mainly confined to animals, fungi and other eukaryotes. There is, however, a growing amount of evidence to suggest that some photosynthetic eukaryotes may operate both glycine and glutamate pathways, the former being employed for the synthesis of haeme and the latter for chlorophyll production.

The condensation of succinyl-coenzyme A with glycine is catalyzed by the 5-aminolaevulinic acid synthase, a pyridoxal 5′-phosphate-dependent enzyme, which activates glycine by formation of the

References, pp. 140–188

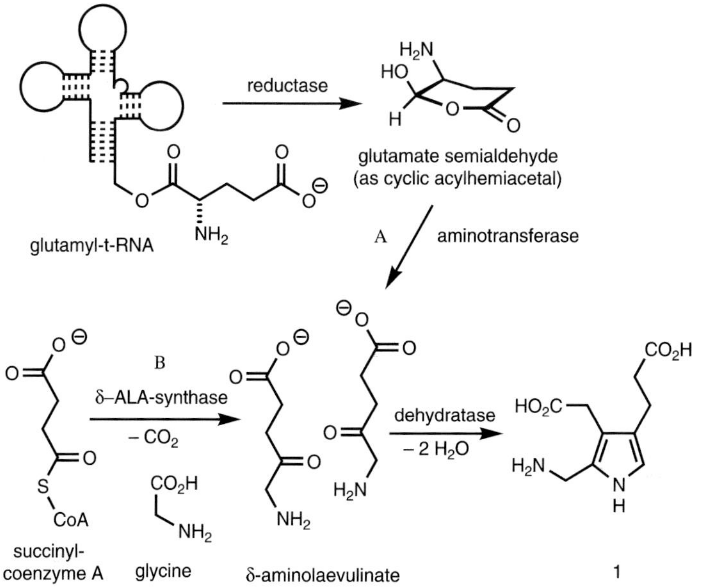

Scheme 1.4. Biosynthetic pathways to porphobilinogen in plants (**A**) and animals (**B**)

corresponding imine. Thereon, reaction of succinyl-coenzyme A with the deprotonated imine precedes decarboxylation. From the stereochemical point of view the overall reaction of 5-aminolaevulinic acid synthase takes place under retention of the *pro*-(*S*)-H-atom of glycine, which is found finally in the *pro*-(*S*)-position of the final product. In the glutamate pathway, glutamate is bound to the corresponding *transfer*-RNA by a ligase before the α-carboxyl group is reduced by a NADPH-dependent reductase (dehydrogenase). The resulting glutamate 1-semialdehyde, which most likely adopts the more stable cyclic acyl hemiacetal structure represented in Scheme. 1.4., is transformed subsequently into 5-aminolaevulinic acid by a aminotransferase enzyme, which presumably catalyzes the ring opening of the hemiacetal as the first step of the reaction. Once formed, two molecules of 5-aminolaevulinic are transformed into porphobilinogen by a dehydratase, a reaction which is formally analogous to the Knorr's pyrrole synthesis.

2. Pyrroles from Vertebrates

With the exception of porphobilinogen (**1**), which is biosynthesized by all eukaryotes, monopyrroles are rare in vertebrates. Porphobilinogen itself, however, is a characteristic abnormal component in the urine of patients affected by disorders of haeme metabolism, so that its detection is of diagnostic significance. On the other hand, the occurrence of some other pyrrole derivatives in urine of humans has been recognized to be significant for other pathological disorders. Thus, the so-called "mauve factor", which was identified with kryptopyrrole = 3-ethyl-2.4-dimethylpyrrole (**3**) (*23*), and pyrrole-1,2-dicarboxamide (**4**) (*24*) have been found in the urine of patients suffering from psychoses or chronic polyarthritis, respectively.

Kryptopyrrole (**3**)   Pyrrole-1,2-dicarboximide (**4**)

Contrarily, some pyrrole derivatives which have been detected in vertebrates must be considered as xenobiotics (*25*). Thus, the report of the presence of several halogenated bipyrroles in the eggs of Pacific and Atlantic Ocean seabirds and in bald eagle liver samples reveals that natural organohalogen compounds are more pervasive in the environment than previously believed (*26*). Although sufficient material was not available for full characterization of these compounds, the probable structures of two of them (**5** and **6**) have been established by synthesis (*27*). In the same work, the structure of the synthetic bipyrrole **5** was confirmed by X-ray diffraction analysis. The presumed dietary origin of these halogenated pyrroles is unknown, although it is relevant that 3,3′4,4′.5,5′-hexabromo-2.2′-bipyrrole (**219**), along with 2,3,4,5-tetrabromopyrrole (**218**), have been isolated from a marine *Chromobacterium* sp.

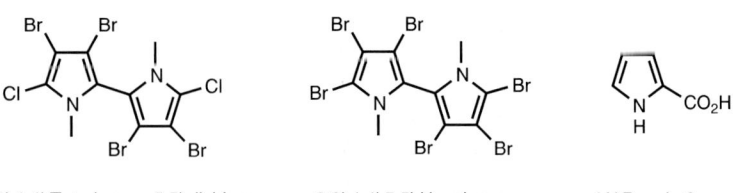

3,3',4,4'-Tetrabromo-5,5'-dichloro-   3,3',4,4',5,5'-Hexabromo-   1*H*-Pyrrole-2-
1,1'-dimethyl-2,2'-bipyrrole (**5**)   1,1'-dimethyl-2,2'-bipyrrole (**6**)   carboxylic acid (**7**)

The most common monopyrrole present in vertebrates is pyrrole-2-carboxylic acid (**7**) which is excreted with urine in mammals as the principal metabolite of both hydroxy-D-proline (*28*) and hydroxy-L-proline (*29*).

References, pp. 140–188

Two derivatives of pyrrole-3-carboxylic acid, namely 2,4-dimethyl- and 2-ethyl-4-methyl-3-carboxylic acid, are constituent parts of batrachotoxin (**8a**) and homobatrachotoxin – formerly called isobatrachotoxin (**8b**) – respectively. They were isolated, along with two other pyrrole-free alkaloids (batrachotoxinin A and pseudobatrachotoxin), from 5000 Colombian arrow poison frogs (*Phyllobates aurotaenia*) (*30–32*). Later on, two additional batrachotoxin class alkaloids, which were characterized as 4β-hydroxybatrachotoxin and 4β-hydroxyhomobatrachotoxin on the basis of their mass and ^{13}C-NMR spectra, were isolated as minor constituents from skin extracts of *Phyllobates terribilis* (*33*). Interestingly enough, homobatrachotoxin is found also in the feathers and skin of the rubbish bird (*Pitohui*), which is endemic to the New Guinea subregion (*34*). The occurrence of homobatrachotoxin in some *Pitohui* sp. suggests that birds and frogs independently evolved this class of alkaloids as chemical defense.

R = CH$_3$: Batrachotoxin (**8a**)
R = C$_2$H$_5$: Homobatrachotoxin (**8b**)

Hydrolysis of the pyrrole moiety of batrachotoxin yielded batrachotoxinin A, which can be transformed back into batrachotoxin on esterification with the mixed anhydride of ethyl chloroformate and 2,4-dimethylpyrrole-3-carboxylate under Schotten-Baumann conditions. The structure of batrachotoxinin A was elucidated by X-ray diffraction analysis (*35, 36*) and provided the key to the structures of the corresponding 2,4-dimethyl- and 2-ethyl-4-methyl-3-pyrrolecarboxylates, batrachotoxin and homobatrachotoxin, respectively. A partial synthesis of batrachotoxinin A was achieved in Wehrli's laboratories at the ETH in Zurich (*37*). For a comprehensive review on the chemistry, spectroscopic data, and biological activity of the batrachotoxins see Ref. *38*.

Batrachotoxin is among the most toxic substances known to man. On subcutaneous injection, the lethal dose in mouse is about 100 ng and it has been estimated that in man a lethal dose would be much less than 200 μg. It is noteworthy that batrachotoxinin A is remarkably less toxic (LD$_{50}$ = 1000 μg/kg) than batrachotoxin (LD$_{50}$ = 2 μg/kg) and homobatrachotoxin (LD$_{50}$ = 3 μg/kg), thus pointing out that the pyrrole carboxylate moiety has an influence on the physiological properties of these alkaloids. Batrachotoxins depolarize neurons and muscle cells *via* a specific interaction with voltage dependent sodium channels in plasma membranes (*39*). Binding of batrachotoxin to sites associated with sodium channels appears to prevent the physiological inactivation of the channels. A resultant massive influx of sodium ions leads to membrane depolarization. The effects of batrachotoxin in nerve and muscle preparations are often relatively irreversible. This apparent irreversibility reflects a slow removal of the alkaloid from tissues because of lipid solubility and because only a small percentage of sodium channels ($<5\%$) need to be activated to cause and maintain complete depolarization in most electrogenic membranes. Thus, batrachotoxin has proven to be an invaluable tool for the mechanistic study of voltage-dependent sodium channels and for investigation of effects of depolarization and/or influx of sodium ions on physiological functions.

3. Pyrroles from Invertebrates

3.1. Pyrroles from Insects

3.1.1. Pyrroles from Ants

Methyl 4-Methylpyrrole-2-carboxylate (**9**) is an insect pheromone which has been isolated from the secret used by the workers of the leaf-cutting ants *Atta texana* (*40*), *A. cephalotes* (*41*), and *Acromyrmex subterranous* (*42*) to mark their tracks. The sensitivity of the insects for this pheromone is enormous. As few as 0.33 mg would be enough to mark a track around the earth, which would be recognized by *A. texana*. Methyl 4-methylpyrrole-2-carboxylate was a known synthetic pyrrole derivative before the discovery of the pheromone (*43*). Several syntheses of this compound have been reported since then (*44*). Presumably, the biosynthetic pathway of this insect pheromone resembles that of glycerinopyrin (**244**), as described in Section 6.2,

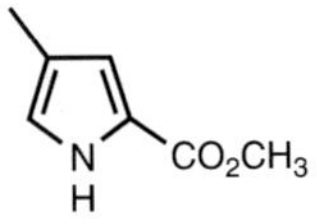

9

Also peculiar is the supposed biosynthesis of the myrmicarins which are contained in the venom of ants of the genus *Myrmicaria* (Myrmicinae).

References, pp. 140–188

The carbon skeleton of these polycyclic alkaloids, some of which contain a pyrrole ring as a substructure element, consists of one, two, or three unbranched chains of 15 carbon atoms which are derived from the acetate pool. In the oligocyclic systems, each of the carbon chains is joined at two or three sites to a nitrogen atom forming indolizine, or more frequently, hexahydropyrrolo[2.1.5-*cd*]indolizine systems which had not been described as natural compounds before. Since all the C_{15}-chains in the oligocyclic alkaloids are functionalized in a similar fashion, it appears that piperidine derivatives may represent common biosynthetic precursors of the *Myrmicaria* alkaloids. Thus, all these compounds can be postulated to derive biochemically from the structurally more simple myrmicarins 237A (**10a**) and 237B (**10b**), which theirselves are supposedly biosynthesized from acetate through the polyketide pathway. Myrmicarins 237A and 237B have been identified as major constituents of the poison gland secretion of the African ant *Myrmicaria eumenoides* (*45*).

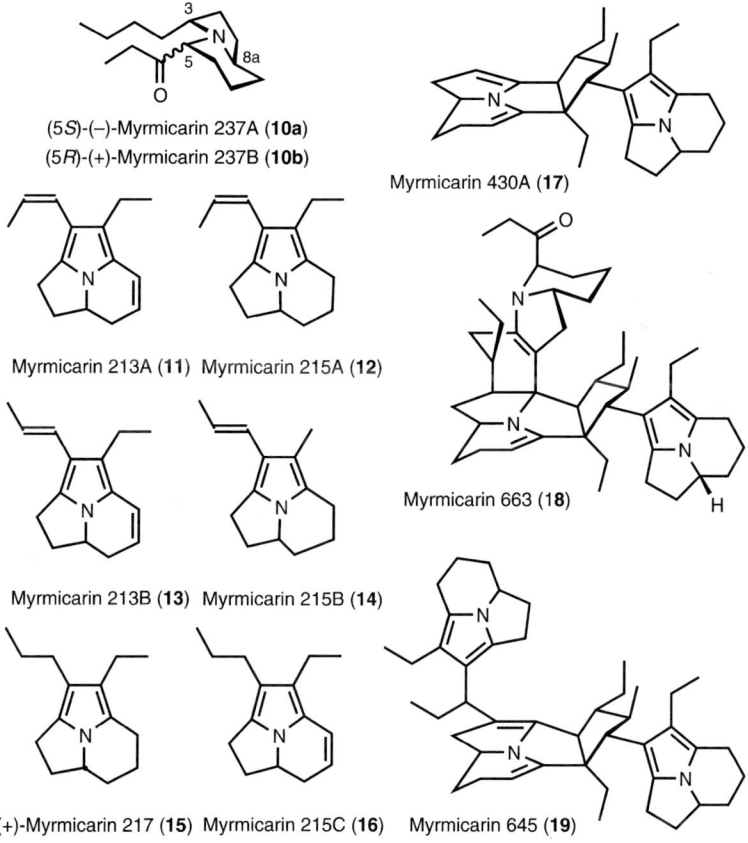

(5*S*)-(−)-Myrmicarin 237A (**10a**)
(5*R*)-(+)-Myrmicarin 237B (**10b**)

Myrmicarin 430A (**17**)

Myrmicarin 213A (**11**) Myrmicarin 215A (**12**)

Myrmicarin 663 (**18**)

Myrmicarin 213B (**13**) Myrmicarin 215B (**14**)

(+)-Myrmicarin 217 (**15**) Myrmicarin 215C (**16**) Myrmicarin 645 (**19**)

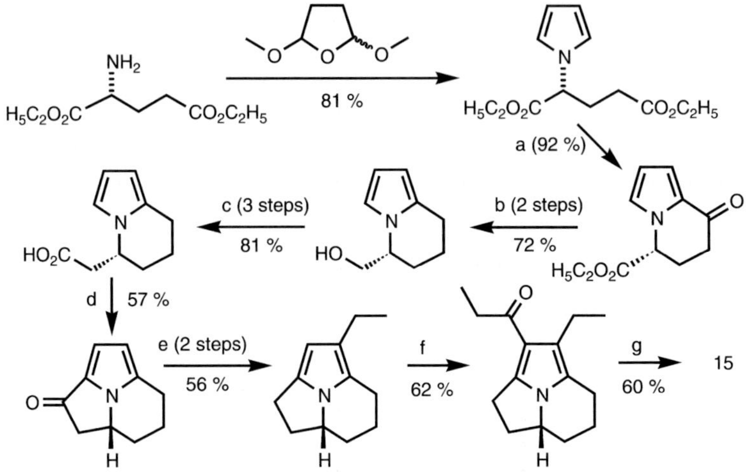

Scheme 3.1. Synthesis of *R*-(+)-myrmicarin 217 (**15**) (*48*): **a**) BBr$_3$ in CH$_2$Cl$_2$. **b**) *i*: NaBH$_3$CN/ZnI$_2$ in CH$_2$Cl$_2$; *ii*: LiAlH$_4$ in tetrahydrofuran. **c**) *i*: H$_3$CSO$_2$Cl/pyridine in CH$_2$Cl$_2$; *ii*: NaCN in dimethylformamide; *iii*: aqueous NaOH in methanol. **d**) ethyl chloroformate/triethylamine in tetrahydrofuran. **e**) *i*: H$_3$C–COCl/AlCl$_3$ in CH$_2$Cl$_2$; *ii*: LiAlH$_4$ in dioxane. **f**) *N,N*-dimethylpropionamide/POCl$_3$ in toluene. **g**) LiAlH$_4$ in dioxane

From the African ants *Myrmicaria opaciventris* three groups of alkaloids have been isolated, containing 15, 30, and 45 carbon atoms, respectively (*46*). The C$_{15}$N alkaloids are pyrrolo[2.1.5-*cd*]indolizines while the two other groups can be considered as dimers (C$_{30}$N$_2$) and trimers (C$_{45}$N$_3$) of the C$_{15}$N basic skeleton. Myrmicarin 215A (**12**), 215B (**14**), and 217 (**15**), the names of which refer to the molecular weight of the corresponding compounds, are the main alkaloids of the venom of colonies from Kenya. The determination of their structures was based mainly on their spectral properties. The structure of myrmicarin 217 has been confirmed by the synthesis of both the racemate (*47*) and the optical active (*R*)-(+)- and (*S*)-(−)-enantiomers (*cf.* Scheme 3.1.) (*48*).

Myrmicarin 213A (**11**), 213B (**13**) and 215C (**16**) are less abundant but a secretion that had been exposed to air showed higher amounts of these compounds. This suggests that they may simply be products of non enzymatic oxidation during storage and/or isolation (*46*). The main component of the C$_{30}$N$_2$ family is myrmicarin 430A (**17**), which can be regarded as a dimer of myrmicarin 215. Its structure was established by extensive two-dimensional NMR experiments (*46*, *49*). This compound is very sensitive to air, showing more than 90% decomposition after only one hour at ambient temperature. Myrmicarin 663 (**18**) is the major alkaloid in colonies of *Myrmicaria opaciventris* from Cameroun and

M. striata (*46*, *50*). It was shown by two-dimensional NMR that it is a decacyclic compound representing a new class of alkaloids. It is by far the most complex alkaloid isolated until now from insects. Like myrmicarin 430A, myrmicarin 663 is also sensitive to air. Another compound, myrmicarin 645 (**19**), has been tentatively assigned a trimeric structure containing two pyrroloindolizidine moieties.

3.1.2. Pyrroles from Beetles

Structurally related to the myrmicarins is the alkaloid exochomine (**20**), which affords the European ladybird *Exochomus quadripustulatus* chemical protection against predators (*51*). Its structure and absolute configuration have been determined by single crystal X-ray diffraction analysis on the hydrochloride. This dimeric alkaloid belongs, together with the chilocorines A (**21**) and B (**22**), to a series of ladybird alkaloids possessing a 2-methyl perhydro-9b-azaphenalene skeleton of the hyppodamine type linked to a 3,4-dimethyloctahydro-8b-azaacenaphtylene ring system. Presumably, originates the latter, as hyppodamine itself, from acetate, trough the polyketide pathway.

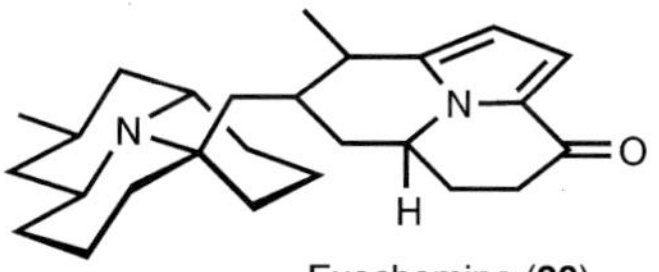

Exochomine (**20**)

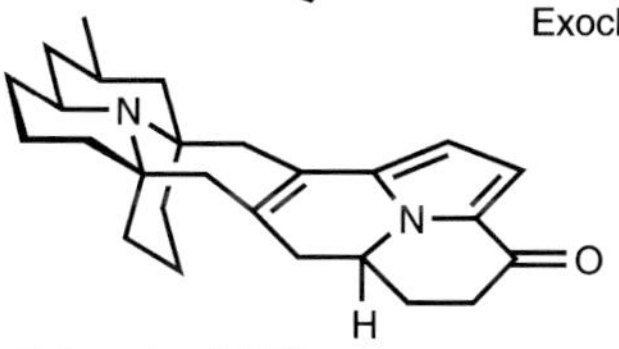

Chilocorine A (**21**)

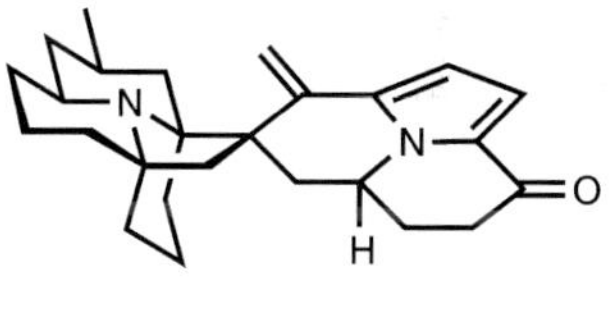

Chilocorine B (**22**)

Chilocorine A (originally named chilocorine) (*52*) and chilocorine B (*53*) are heptacyclic dimeric alkaloids, which have been isolated from the coccinellid beetle, *Cholocorus cacti*. Like many other ladybird beetles, *C. cacti* excretes droplets of blood when disturbed. Such reflex-bleeding is defensive and protects the beetles against such predators as ants. The unique structures of the chilocorines are made up of two tricyclic substructures: hippodamine (2-methylperhydro-9b-azaphenalene), which is frequently encountered among alkaloids from coccinellid beetles and 3,4-dimethyloctahydro-8b-azaacenaphthylene. Both structures are based on mass spectrometric, as well as ultraviolet and NMR spectroscopic evidence.

3.1.3. Pyrrole Pheromones from Butterflies

Danaidone (**23**)

X = H : Danaidal (**24a**)
X = OH : (–)-Hydroxydanaidal (**24b**)

The hair-pencil secretions used to disseminate pheromonal substances during courtship by male butterflies of the family Nymphalidae, subfamily *Danainae*, which occurs in tropical regions, contain danaidone (**23**) (*54–60*). Its structure was confirmed by synthesis (*55, 61*). Danaidone and related pyrrolizine derivatives possessing pheromonal activity are structurally similar to the aminoalcohols of the hepatotoxic pyrrolizidine alkaloids (*cf.* Section 4.1.) and even more closely similar to the dihydropyrolizidine derivatives which are produced metabolically in rats treated with these alkaloids. This supports the possibility that the butterflies obtain the hair-pencil dihydropyrrolizines by transforming precursors found in their food plants (*62*). Actually, studies on the arctiid moths *Utetheisa ornatrix* have proved such kind of plant-insect relationship (*63*). Larvae of both sexes feed on *Crotalaria* plants, which produce pyrrolizidine alkaloids. These alkaloids are in part sequestered by the insects, and they protect adults from predators such as spiders. In addition, the males convert some of their alkaloid into simple pyrrolizidines, *e.g.* danaidal (**24a**) or hydroxydanaidal (**24b**), which serve as sex pheromones (*59*). Moreover, the alkaloid is transmitted from male to female during mating and the females are able to put this alkaloid into their eggs, thereby rendering them distasteful to predators (*63*). Analogous results were obtained from detailed studies of alkaloid transfer in the Florida queen butterfly *Danaus gilippus* and in two East Asian arctiid moths, *Creatonotos transiens* and *C. gangis*, whose males also produce hydroxydanaidal (*64*). It seems therefore likely that the male pheromone, since it can be produced only from acquired alkaloid, provides a female with unambiguous chemical evidence that the male can contribute to the defense of their offspring.

3.2. Pyrroles from Sponges

The most of over 250 pyrrole-containing compounds, which are known from marine organisms have been isolated from sponges. It is noteworthy, however, that whereas a large number of these compounds

References, pp. 140–188

are brominated derivatives, chlorine-containing metabolites are much more common in terrestrial bacteria (*cf.* Section 6.1.) than in marine organisms. A short compilation of the different halogenated monopyrroles encountered in nature can be found in Ref. 26.

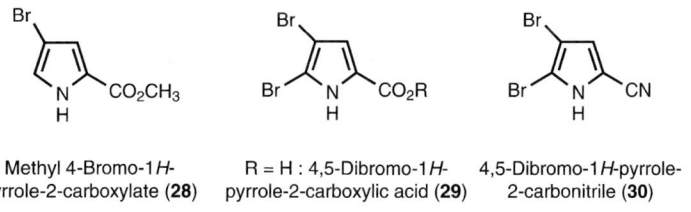

2,3-Dibromo-
1*H*-pyrrole (**25**)

2,3-Dibromo-5-methoxy-
methyl-1*H*-pyrrole (**26**)

2,3,4-Tribromo-
1*H*-pyrrole (**27**)

2,3-Dibromopyrrole (**25**), a putative decarboxylation product of 4,5-dibromopyrrole-2-carboxylic acid (see below), and its 5-methoxymethylderivative (**26**) have been characterized, among 4,5-dibromopyrrole-2-carboxylic itself and four previously known derivatives of the latter (see below), by NMR spectroscopy and found to be very unstable sponge metabolites of *Agelas* sp. (*65*). The more halogenated analog, 2,3,4-tribromopyrole (**27**), on the other hand, has been isolated from the polychaete (a marine worm) *Polyphysia crassa* (*66*). It is also a very unstable pyrrole derivative which slowly disproportionates, in dilute solution above $-18°C$, forming di- and tetrabromopyrrole. Most bromine-containing pyrrole derivatives encountered in marine sponges are, however, esters or amides of 4-bromo- and 4,5-dibromopyrrole-2-carboxylic acid.

3.2.1. Pyrrole-2-Carboxylates from Sponges

Methyl 4-Bromo-1*H*-
pyrrole-2-carboxylate (**28**)

R = H : 4,5-Dibromo-1*H*-
pyrrole-2-carboxylic acid (**29**)

4,5-Dibromo-1*H*-pyrrole-
2-carbonitrile (**30**)

Both Methyl 4-bromo-1*H*-pyrrole-2-carboxylate (**28**) (*67*) which had been first erroneously characterized as the 5-bromo isomer (*68*), and methyl 4,5-dibromopyrrole-2-carboxylate, (**29**, R = CH$_3$) have been isolated along with aldisin (**75a**) and 2-bromoaldisin (**75b**) from specimens of a *Lissodendoryx* sp. sponge of Sri Lanka (*68*) and from *Axinella tenuidigitata* (*69*). They may arise from the more complex phakelins (*cf.* Section 3.2.3.2.) by degradation. Both methyl 5-bromo-1*H*-pyrrole-2-carboxylate and **28** have been obtained by synthesis (*70*). Moreover, 4,5-dibromopyrrole-2-carboxylic acid (**29**, R = H) was isolated from the marine sponges *Agelas oroides* (*71, 72*), *A. flabelliformis* (*73*), and *Axinellidae* sp. (*74*). The corresponding methyl ester (**29**, R = CH$_3$), which was identified with a synthetic sample (*70, 75*), had been reported earlier, together with the corresponding amide (**46**) and nitrile (**30**), as products of methanol extraction from the marine sponge *Agelas oroides* collected in the bay of Naples (*71*), but acetone extraction of the same material demonstrated the ester to be an artefact since only the free acid was isolated by the latter procedure. Later on, **29** (R = CH$_3$) and the corresponding nitrile (**30**) were obtained again from *A. oroides*, this time collected off the coast of Malta (*72*). The structure of **30** was confirmed by single crystal X-ray diffraction analysis.

R = CH$_3$ (**31a**)

R = C$_2$H$_5$ (**31b**)

32

Makaluvic acid A (**33**)   Makaluvic acid B (**34**)

R = CH$_2$–CH=CH$_2$ (**31c**)

On the other hand, 4,5-Dibromo-1-methyl-1*H*-pyrrole-2-carboxylic acid (**31a**) (*76*) and its methyl ester (*77*) have been identified as the major secondary metabolites of the sponge *Agelas mauritiana*, an organism that contains also midpacamide (**61**). The same pyrrole derivative has been isolated, along with its corresponding homolog (**31b**) and 1-allyl-4,5-dibromopyrrole-2-carboxylic acid (**31c**) from an Australian marine sponge *Axinella* sp., which also contains, 4,5-dibromopyrrole-2-carboxylic acid (**29**, R = H), as well as 3,4,5-tribromopyrrole-2-carboxylic acid (**32**, R = H), and the *N*-methyl derivative (**32**, R = CH$_3$) of the latter (*67*). The structures of these bromopyrrole derivatives were secured by spectroscopic analysis and synthesis.

Makaluvic acids A (**33**) and B (**34**) are two non-halogenated pyrrole-2-carboxylic acids, which have been isolated from the sponge *Zyzzya fuliginosus* collected in Chuuk Atoll (Micronesia). Their structures were elucidated by spectroscopic analysis and, in the case of makaluvic acid A, secured by X-ray diffraction analysis (*78*). Probably, both **33** and **34**

References, pp. 140–188

are formed by oxidation of some pyrrolo[4,3,2-*de*]quinoline derivatives, which are also found in the sponge.

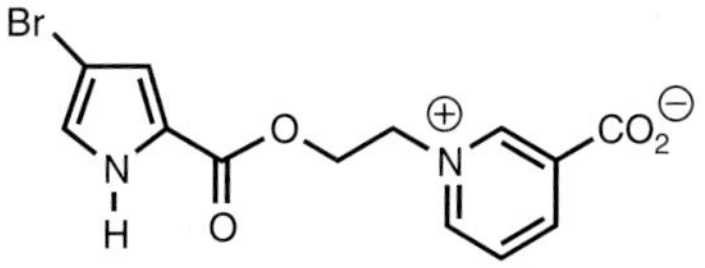

Agelongine (**35**)

Agelongine (**35**) is an ester of 4-bromo-1*H*-pyrrole-2-carboxylic acid possessing antiserotonergic activity *in vivo*, which was isolated, along with the dispacamides (*cf.* Section 3.2.2.1.) and longamide (**68a**) from the sponge *Agelas longissima*, collected at a depth of 15 m along the coasts of Little San Salvador island (*79*). In contrast to oroidin (**53**), which was isolated from the same sponge, agelongine contains a pyridinium ring instead of imidazole commonly found in related bromopyrrole derivatives isolated from *Agelas* sp. Moreover, **35** contains a quite different central segment where an ester linkage replaces the usual amidic bond.

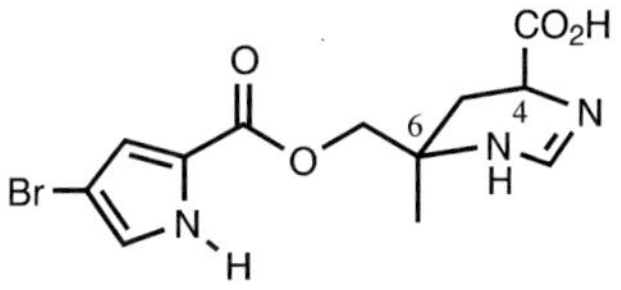

(–)-Manzacidin A (**36**)

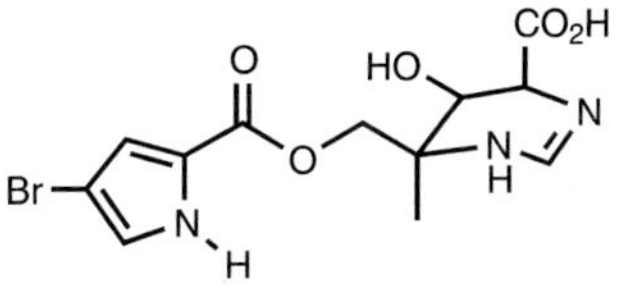

Manzacidin B (**37**)

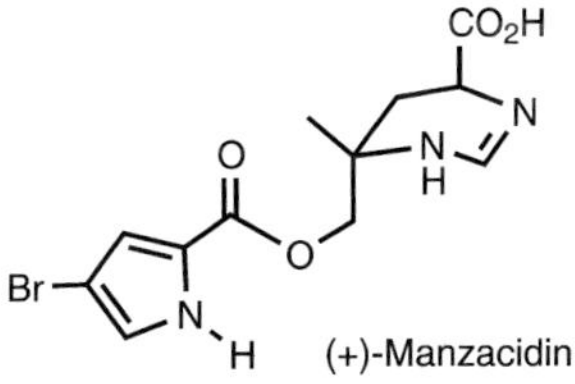

(+)-Manzacidin C (**38**)

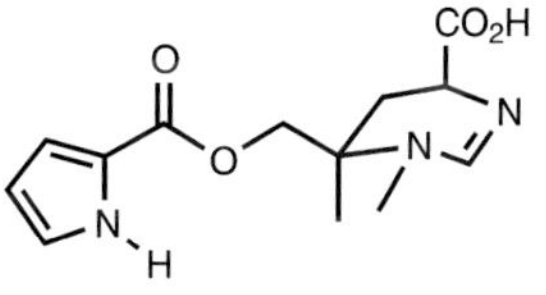

Manzacidin D (**39**)

The manzacidins A–C (**36–38**) are also esters of 4-bromopyrrole-2-carboxylic acid, which were obtained from the Okinawan marine sponge *Hymeniacidon* sp., collected at Manza Beach. They contain a tetrahydropyrimidine ring (*80*). The relative configuration of manzacidins A–C was established by 2D NMR spectroscopy correlations, which left, however, the absolute configuration of the molecules to be determined. For (−)-manzacidin A (**36**) and (+)-manzacidin C (**38**) the absolute configuration has been established by stereoselective total synthesis using (2*S*)-allylglycinol = (2*S*)-2-aminopent-4-enol as starting material (*81*). Accordingly, (+)-manzacidin C (**38**) is the C6-epimer of (−)-manzacidin A (**36**). Manzacidin D (**39**) has been isolated together with the known substances norzooanemonin and trigonelline from the coralline demosponge *Astrosclera willeyana*, collected at a depth of 26 m, on the northern part of the Great Barrier reef in Australia. Its structure was elucidated by spectroscopic methods, mainly ^{1}H- and ^{13}C NMR data, which established also the relative configuration at the stereogenic centers C4 and C6 (*82*).

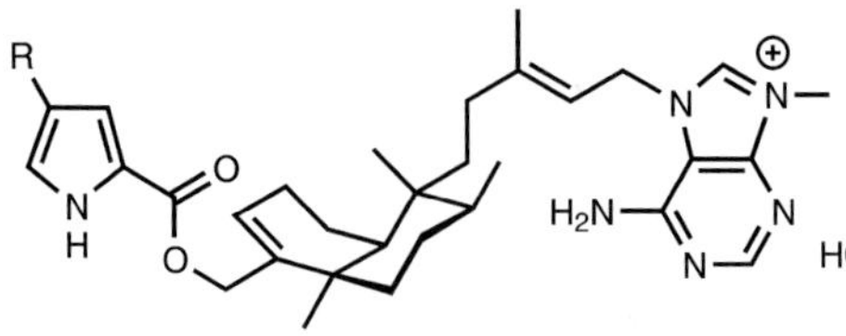

R = Br : (−)-Agelasine G (**40a**)
R = H : (−)-Ageline B (**40b**)

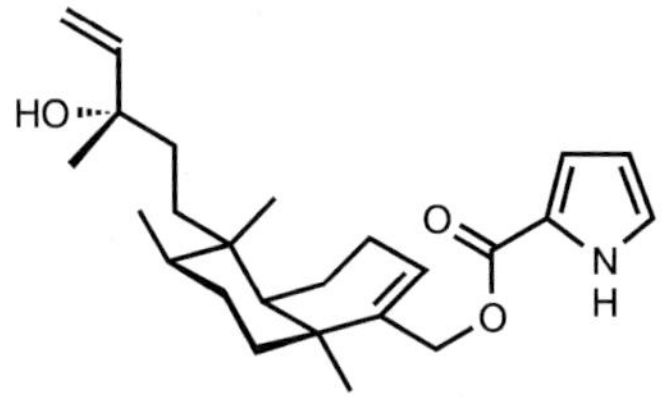

Nakamurol D (**41**)

Agelasine G (**40a**) another ester derivative of 4-bromopyrrole-2-carboxylic acid, was isolated, as the sole member of the agelasine family containing a pyrrole ring, from a marine sponge *Agelas* sp, collected off Konbu, Okinawa (*83*). The structurally related pyrrole alkaloid ageline B (**40b**) was isolated, among other terpenoid 9-methyladenine derivatives, from an unidentified Pacific sponge of the same genus, collected at Palau, Western Caroline islands (*84*). Agelasine G and ageline B are "mixed metabolites" which incorporate pyrrole, terpenoid, and nucleic acid moieties. Ageline B (**40b**) is a mildly toxic to fishes and shows moderate antimicrobial activities. Its structure was elucidated by interpretation of spectral data with particular emphasis of ^{13}C NMR correlations. As both ageline B and agelasine G are laevorotatory, the absolute configuration of the latter was assigned to be the same as that of (−)-ageline B (**40b**), which had been proposed to be enantiomeric to the terrestrial plant metabolite sagittariol. Interestingly, a pyrrol-2-carboxylate of the latter – called nakamurol D (**41**) – has been isolated from the Okinawan marine sponge *Agelas nakamurai* (*85*).

3.2.2. Pyrrole-2-Carboxamides from Sponges

The halogen-free 1*H*-pyrrole-2-carboxamide (**42a**) and its *N'*-formyl derivative (**43**) have been isolated from *Agelas oroides*, collected from the Great Barrier reef in Australia (*86*). The 5-bromo derivative of the

References, pp. 140–188

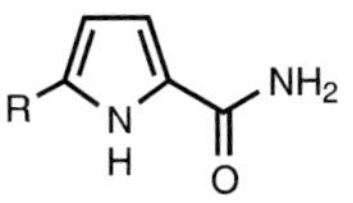

R = H : (**42a**)
R = Br : (**42b**)

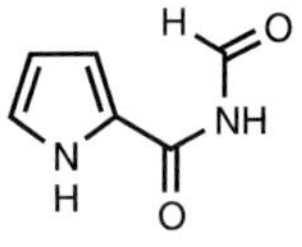

N'-Formylpyrrole-
2-carboxamide (**43**)

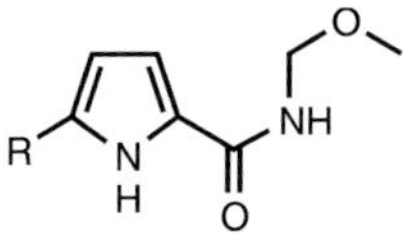

R = H : (**44a**)
R = Br : (**44b**)

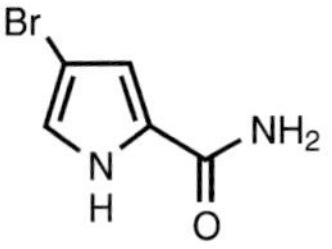

4-Bromopyrrole-
2-carboxamide (**45**)

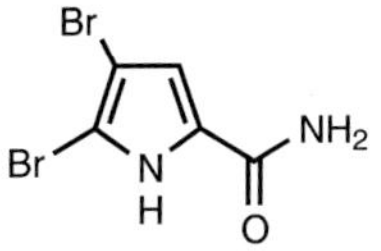

4,5-Dibromopyrrole-
2-carboxamide (**46**)

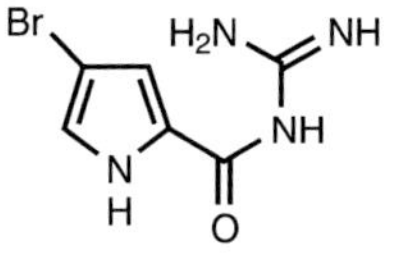

N'-[(4-Bromopyrrol-2-yl)-
carbonyl]guanidine (**47**)

former (**42b**) occurs in the Papua New Guinea sponge *Agelas nakamurai*, together with its N'-methoxymethyl derivative (**44b**) (*87*). The corresponding halogen-free analog of the latter (**44a**) was obtained, together with longamide and the methyl ester of longamide B (see below) from the marine sponge *Homaxinella* sp. collected by netting at a depth of 70 m off the coast of Tokushima (*88*). On the other hand, 4-bromo-1*H*-pyrrole-2-carboxamide (**45**) has been isolated, along with oroidin (**53**) and hanishin (**67b**) from the axinellid sponge *Acanthella carteri* (*89*). The latter also contains 4,5-dibromo-1*H*-pyrrole-2-carboxamide (**46**) which had been earlier isolated, along with pseudoceratidine (**50**) from the marine sponges *Pseudoceratina purpurea* (*90*) and *Agelas* sp. (*65, 91*). Interestingly, 4,5-Dibromo-2-carboxamide promotes larval metamorphosis of the ascidian *Ciona savignyi* and induces larval metamorphosis of the ascidian *Halocynthia roretzi*, so that it may be of interest, like pseudoceratidine (**50**) itself, as an antifouling agent. Structurally related to 4-bromo-1*H*-pyrrole-2-carboxamide is *N*-(4-bromopyrrol-2-yl)carbonylguanidine (**47**) which has been found in unidentified *Agelas* species (*92*).

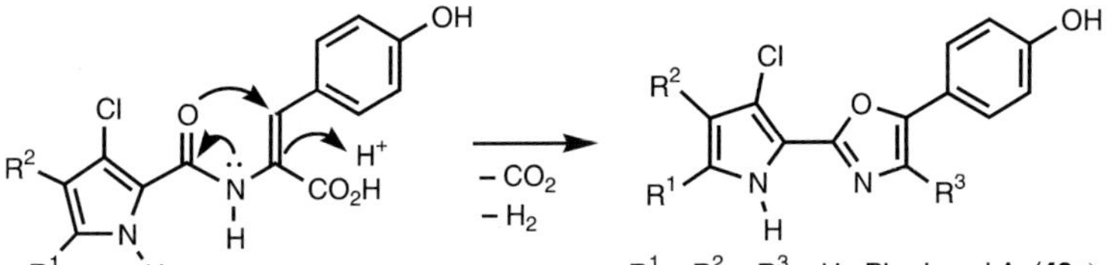

$R^1 = R^2 = R^3 = H$: Phorbazol A (**48a**)
$R^1 = R^2 = Cl; R^3 = H$: Phorbazol C (**48b**)
$R^1 = H; R^2 = R^3 = Cl$: Phorbazol B (**48c**)
$R^1 = R^3 = Cl; R^2 = H$: Phorbazol D (**48d**)

Scheme 3.2. Possible biogenetic origin of the phorbazoles (**48a–d**)

As mentioned before, most halogenated pyrrole metabolites contain bromine atoms. Until now, phorbazoles A–D (**48a–d**) and cyclocinamide A (**49**) are the unique chloropyrroles which have been found in marine organisms. In cyclocinamide A, 4-chloro-1-methylpyrrole-2-carboxylic acid is part of a cytotoxic hexapeptide extracted from the New Guinean marine sponge *Psammocinia* sp. (*93*). The absolute configurations of the four asymmetric C-atoms present in the molecule have been determinated, as given in formula **49**, by total synthesis of a stereoisomer of the natural hexapeptide (*94*).

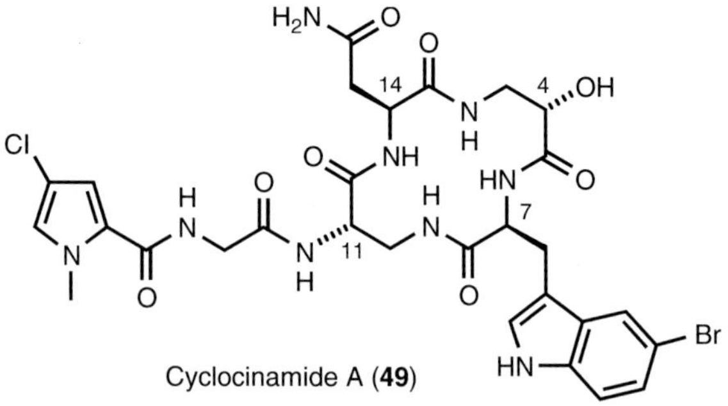

Cyclocinamide A (**49**)

Phorbazoles A–D have been isolated from the marine sponge *Phorba* aff. *clathrata*, collected in Sodwana Bay, South Africa (*95*). The structure of the main component, phorbazole A (**48a**), was unambiguously determined by X-ray diffraction analysis of its *O,N*-dimethyl derivative. The structures of phorbazoles B–D, which all together are minor components (about 10%) of the mixture, were determined by exhaustive use of 1D and 2D NMR techniques. The oxazole ring of the porphazoles may be assumed to proceed biogenetically from a pyrrolecarboxamide derived from tyrosine (*cf.* Scheme 3.2.).

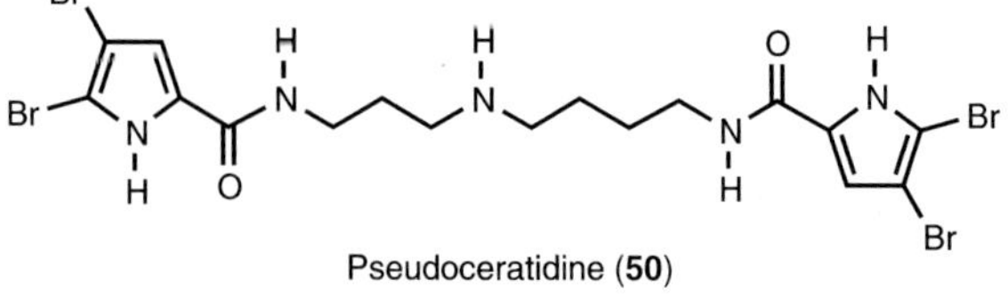

Pseudoceratidine (**50**)

Pseudoceratidine (**50**) is a spermidine derivative isolated from the Japanese marine sponge *Pseudoceratina purpurea* possessing a significant antifouling activity against *Balanus amphitrite* larvae (*96, 97*). Owing to the interest of low-cost antifouling agents preventing barnacles, mollusks, and other organisms from becoming attached to ships' hulls thus blocking corrosion and improving fuel efficiency,

References, pp. 140–188

several chemical syntheses of pseudoceratidine have been achieved and the properties of this compound as an useful alternative for organo tin derivatives (mostly tributyltin methacrylate) which are widely used as antifouling agents today, have been investigated (*97, 98*).

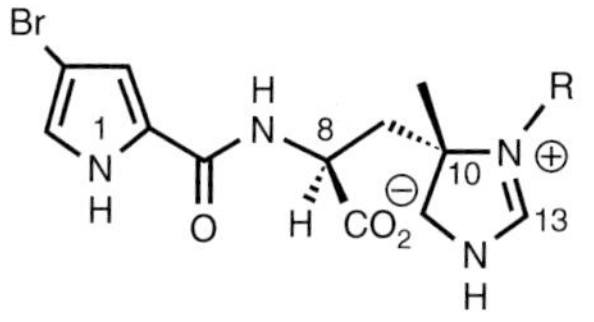

R = CH$_3$: Clathramide A (**51a**)
R = H: Clathramide C (**51b**)

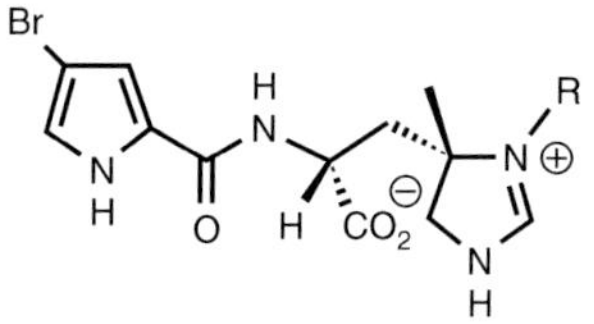

R = CH$_3$: Clathramide B (**52a**)
R = H: Clathramide D (**52b**)

Clathramide A (**51a**) and B (**52a**) are pyrrole alkaloids isolated from the Caribbean sponge *Agelas clathrodes* (*99*). Their structures were elucidated by spectroscopic methods, especially NMR spectroscopy. In order to interpret the NOE enhancements, a molecular dynamics/ molecular mechanics conformational study of the two possible diastereomers was carried out in the CHARMm force field. The absolute configuration of the clathramides was elucidated using Acylase I, a commercially available enzyme that catalyzes the enantioselective hydrolysis of *N*-acyl-L-aminoacids. As this enzyme hydrolyzed the peptide bond of clathramide A but failed to hydrolyze that of clathramide B, it was concluded that the diastereomeric compounds differ in the configuration at C8, in particular indicating the (*S*) configuration for this C-atom in clathramide A, and (*R*) in clathramide B. The corresponding *N*14-dimethyl derivatives, clathramide C (**51b**) and clathramide D (**52b**) were found later in the sponge *Agelas dispar*, collected at 14 m depth in the lagoon of Little San Salvador island (Bahamas) (*100*). In spite of their rather similar structure, it appears very difficult to hypothesize that clathramides originate following the same biogenetic pathway of oroidins (s. below); most likely, clathramides derive from a direct coupling of one proline (or pyrrole carboxylic acid) and one histidine residue, with the subsequent inclusion of one or two methyl group(s).

3.2.2.1. Pyrrole-2-Carboxamides of the Oroidin Type

Initially isolated from the Mediterranean axinellid sponge *Agelas oroides* (*71*) oroidin (**53**) is historically the central example in a series of similar alkaloids isolated from marine sponges. Among other brominated

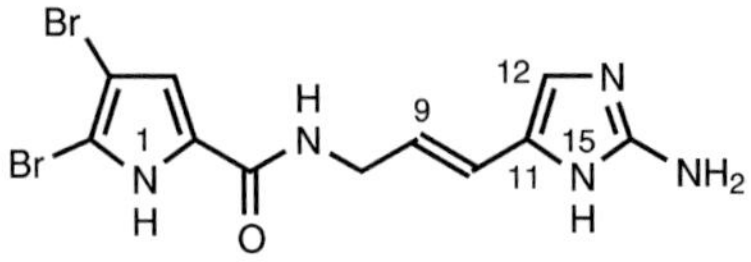

Oroidin (**53**)

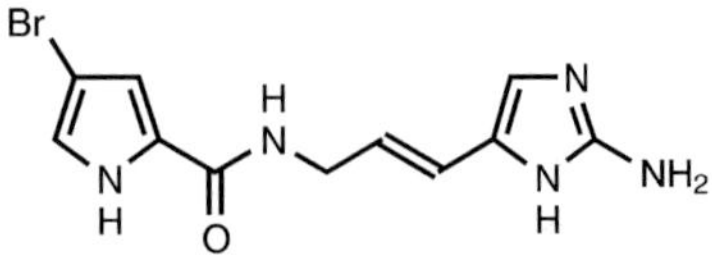

Hymenidin (**54**)

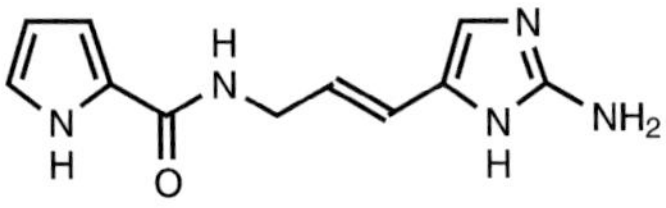

Clathrodin (**55**)

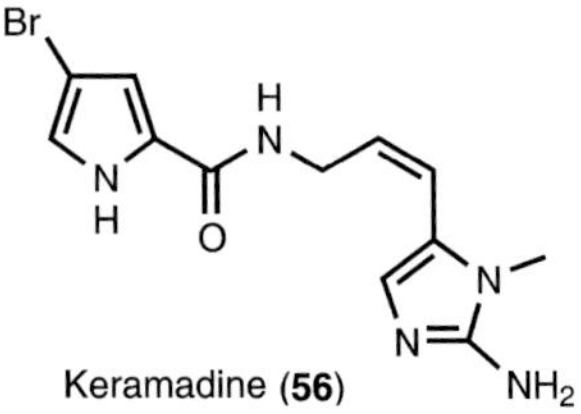

Keramadine (**56**)

pyrrole alkaloids, oroidin, which has been also isolated from four Caribbean *Agelas* sponges (*A. clathrodes*, *A. conifera*, *A. dispar*, and *A. longissima*) (*101*), as well as from *Axinella damicornis* (*102*), *A. carteri* (*103*), *A. verrucosa* (*104*), *Acanthella aurantiaca* (*104*), *Pseudaxinyssa cantharella* (*105*), and from a *Goreauiella* sp. collected at 698 m depth (*106*) is of ecological importance because of its function as a chemical defendant in marine sponges (*107*). After some confusion (*71*), the structure of oroidin, as rectified by Garcia *et al.* (*108*), was confirmed by X-ray diffraction analysis (*109*) and by synthesis (*cf.* Scheme 3.3.) (*110–113*).

Structurally related to oroidin are hymenidin (**54**) (*114*) and clathrodin (**55**), the latter isolated from *Agelas clathrodes* (*115*). On the other hand, keramadine (**56**) (*116, 117*), in which one of the *N*-atoms of the imidazole ring is methylated, differs from other pyrrole alkaloids of the oroidin type on the *cis*-geometry at the C9=C10-bond. The structures of clathrodin (*112, 118*), hymenidin (*111*), and keramadine (*111, 113, 117*) have been confirmed by synthesis. Oroidin and hymenidin are presumably the biogenetic precursors of a large number of sponge metabolites (*e.g.* phakellins, phakellstatins, ageliferins, agelaspongins, agelastatins, axinellamines, hymenialdisines, *etc.*) which may be formed by both intramolecular and intermolecular cycloadditions of hymenidin or oroidin (*119*)

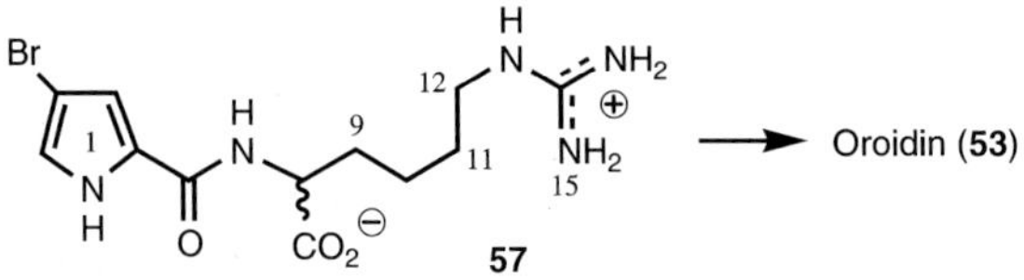

57 ⟶ Oroidin (**53**)

References, pp. 140–188

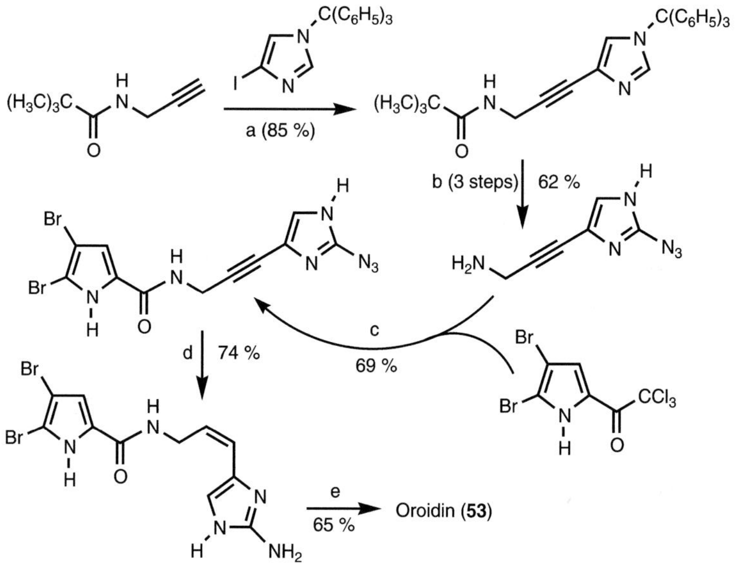

Scheme 3.3. A recent synthesis of oroidin (**53**) through its (Z)-isomer (*119*): **a)** Pd[P(C$_6$H$_5$)$_3$]$_2$Cl$_2$/CuI/di-*iso*-butylamine in tetrahydrofuran. **b)** *i*: n-C$_4$H$_9$Li/p-H$_3$C-C$_6$H$_4$-SO$_2$-N$_3$ in tetrahydrofuran; *ii*: trifluoroacetic acid in CH$_2$Cl$_2$; *iii*: HCl in diethylether/methanol. **c)** Na$_2$CO$_3$ in dimethylformamide. **d)** H$_2$/Pd (Lindlar's catalyst) in methanol. **e)** 6N HCl in methanol

Although biosynthetic experiments have not yet been performed, it is admitted that ornithine is the common precursor of alkaloids of the oroidin type (*120*). One ornithine molecule is supposed to cyclize giving proline (and subsequently pyrrole-2-carboxylic acid), while another molecule should form, by addition of a guanidine molecule, 2-amino-4-(3-aminopropenyl)imidazole. The latter has been isolated from the Axinellidae sponges *Teichaxinella morchella* and *Ptilocaulis walpersi* (*121*), thus strengthening the biogenetic hypothesis that the two units should join together through an amide bond, giving rise to the final structure with two heteroaromatic rings linked by a three-carbon chain (*cf.* Scheme 3.4.). An alternative biosynthetic precursor of the hymenidin/oroidin-related alkaloids may be a condensation product (**57**) of 4-bromopyrrole-2-carboxylic acid and homoarginine which has been recently isolated from the marine sponge *Agelas wiedenmayeri*, collected off the coast of the Florida Keys (*122*). However, a recent study of the biosynthesis of stevensine (**70a**), a presumed metabolite of oroidin, proves the incorporation of ^{14}C-labelled histidine, proline and ornithine into the alkaloid. In the light of these experiments, it is

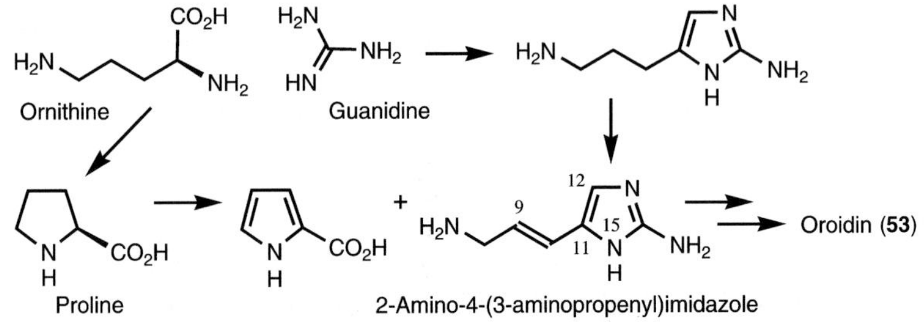

Scheme 3.4. Possible biogenetic precursors of oroidin (53)

proposed that both ornithine and proline can be transformed into the 4,5-dibromopyrrole-2-carboxylate subunit of oroidin, whereas histidine may be the precursor of the 4-(3-aminoprop-1-enyl)imidazole moiety of the same alkaloid.

Analogs to oroidine are the dispacamides, in which the 2-aminoimidazole ring has been replaced by an iminohydantoin subunit. Dispacamides A (**58a**) and B – also called monobromodispacamide (**58b**) – are antihistaminc alkaloids which have been isolated from Caribbean sponges of the genus *Agelas* (*e.g. A. conifera, A. clathrodes, A. dispar* and *A. longissima*), collected along the coasts of Little San Salvador island (*123*). From the same *Agelas* species, dispacamides C (**59a**) and D (**59b**) have been isolated and their structures determined with spectroscopic methods (*101*). Both **58a** (*112, 124, 125*) and **58b** (*112*) have been obtained by chemical synthesis. Dispacamide is assumed to be the direct precursor of marine pyrrolo azepinones such as the hymenialdisines (**72a–c**) through bond formation between C4 and C10.

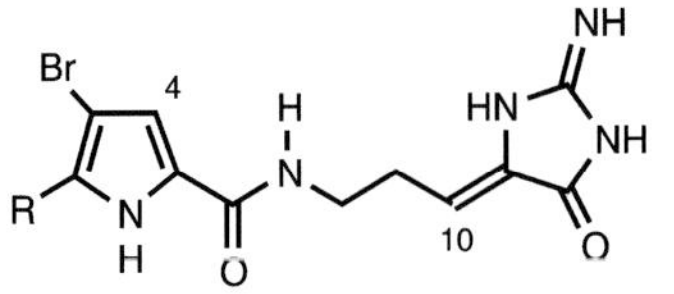

R = Br : Dispacamide A (**58a**)
R = H : Dispacamide B (**58b**)

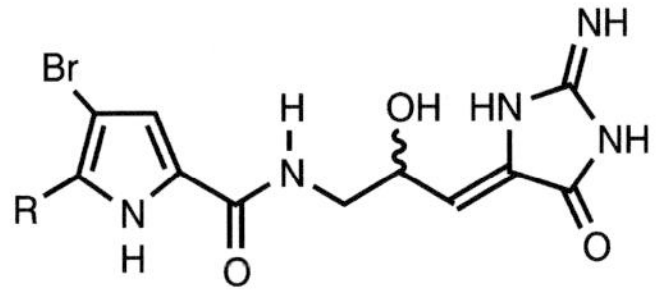

R = Br : Dispacamide C (**59a**)
R = H : Dispacamide D (**59b**)

Closely related to the dispacamides are mukanadin A and mukanadin B, which were isolated from the Okinawan marine sponge *Agelas nakamurai* (*126*). Irrespective of the absolute configuration at C9, which is (*S*) in (+)-mukanadin A but unknown for dispacamide D (**59b**), both compounds proved to be identical. On the other hand, mukanadin B (**60**), which contains a hydantoin ring at the end of the pyrrole side chain, may be a product of partial hydrolysis of dispacamide B (**58b**). Midpacamide (**61**) (*76, 77*) and its 2-debromoderivative (**61**, R=H) (*77*) are derivatives of 10,11-dihydromukanaidin B. They are, together with mauritamide A (**62**), the sole *N*-methylpyrrole derivatives in this series. Midpacamide, which was named in recognition of the Mid-Pacific Marine Laboratory for providing collection facilities, has been isolated along with 4.5-dibromo-1-methylpyrrole-2-carboxylic acid (**31a**), as the major constituent, from an unidentified orange marine sponge collected in the Marshall islands (*76*). Its structure has been confirmed by synthesis of the racemate (*125, 127*).

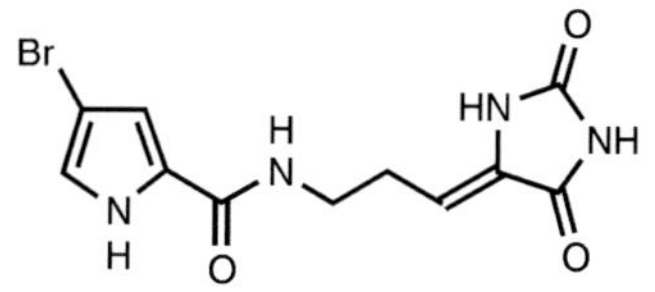

Mukanadin B (**60**)

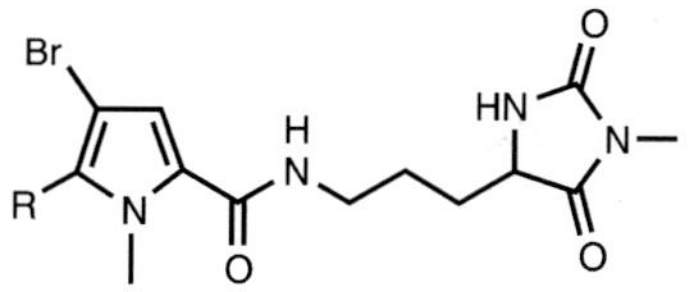

R = Br : Midpacamide (**61**)

Mauritamide A (**62**) is the taurine-containing alkaloid, which occurs together with midpacamide (**61**) and dibromophakellin (**76b**) in the Fijian sponge *Agelas mauritiana* (*128*). Taurodispacamide A (**63a**), on the other hand, contains a diaminoimidazol moiety instead of the iminohydantoin ring, which is characteristic for the dispacamides. It has been isolated recently, along with cyclooroidin (**66**), from *Agelas oroides* collected off the Bay of Naples (*129*). Two more taurine-containing alkaloids, tauroacidin A (**63b**) and tauroacidin B (**63c**) have been isolated from the Okinawan sponge *Hymeniacidon* sp. (*130*). Like taurodispacamide A (**63a**), both **63b** and **63c** possess a diaminoimidazol moiety. Tauroacidins exhibit tyrosine kinase inhibitory activity. The structures of **63b** and **63c** were elucidated on the basis of spectral data

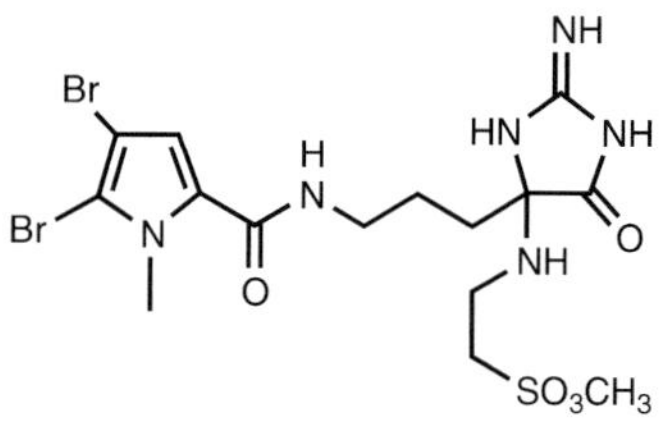

Mauritamide A (**62**)

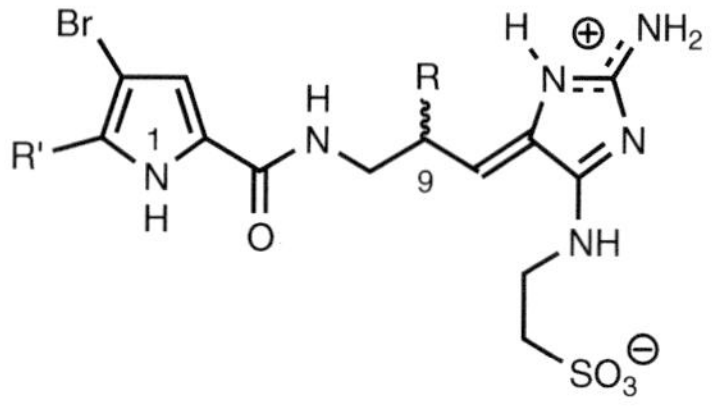

R = H, R' = Br : Taurodispacamide A (**63a**)
R = OH, R' = Br : Tauroacidin A (**63b**)
R = OH, R' = H : Tauroacidin B (**63c**)

and chemical transformations. In order to determine the absolute configuration at the stereogenic center C9 of the alkaloids, both tauroacidin A and B were subjected to ozonolysis yielding 1 mole each of taurine and isoserine. Transformation of the latter into the corresponding isopropyl ester and subsequent reaction with (S)-(+)-α-methoxy-α-trifluoromethylphenylacetic acid chloride (MTPACl) afforded a mixture of diastereomeric N-MTPA derivatives, which was analyzed by HPLC. Both (S)- and (R)-enantiomers of isoserine were found in the hydrolysates of **63b** and **63c** in the ratio of ca. 6:4 and 1:1, respectively.

3.2.2.2. Cyclized Oroidin Metabolites

Dispacamides may be also the biogenetic precursors of the slagenins A–C (**64–65**), which have been isolated from the marine sponge *Agelas nakamurai* (*131*), which is also a source of mukanadins (s. above). Actually, the structural similarity between dispacamides and slagenins suggests that the tetrahydrofuran ring present in the latter could be formed by intramolecular nucleophilic addition of the oxygen atom at C4 of the hydantoin subunit of some dispacamide analog, followed by addition of either water or methanol to yield the corresponding slagenin (Scheme 3.5.). As a matter of fact, slagenins B (**64b**) and C (**65**) have opposite absolute configurations at C11 and C15 but the same at C9, thus indicating that whereas the formation of the tetrahydrofuran ring is enantioselective, the *syn*-addition of water or methanol can take place from both sides of the imidazolone ring. Both the structures of slagenins A–C (*132*) and the absolute configurations of slagenins B and C (*133*) have been proved by chemical synthesis.

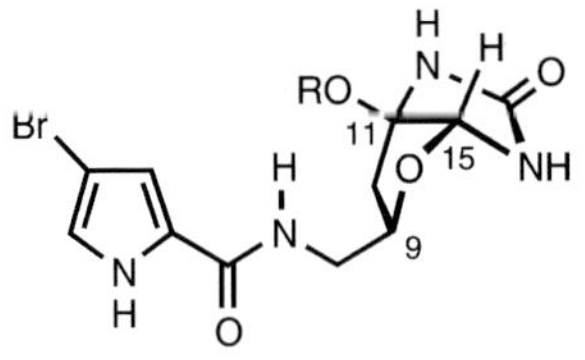

R = H : Slagenin A (**64a**)
R = CH$_3$: Slagenin B (**64b**)

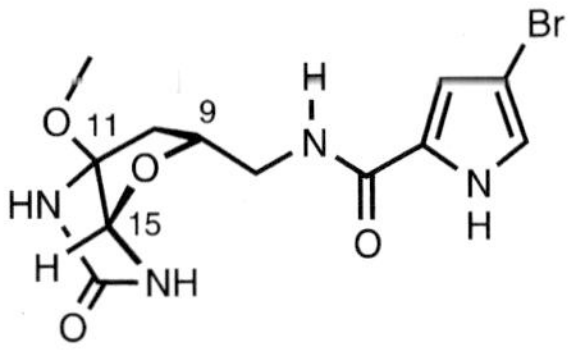

Slagenin C (**65**)

Intramolecular cyclisations of oroidin and its analogs in which the pyrrole moiety acts as a nucleophile towards the exocyclic C=C-bond may lead to two different types of metabolites containing either a pyrrolo[1,2-*a*]ketopiperazine or a pyrrolo[2.3-*c*]azepinone skeleton. To

References, pp. 140–188

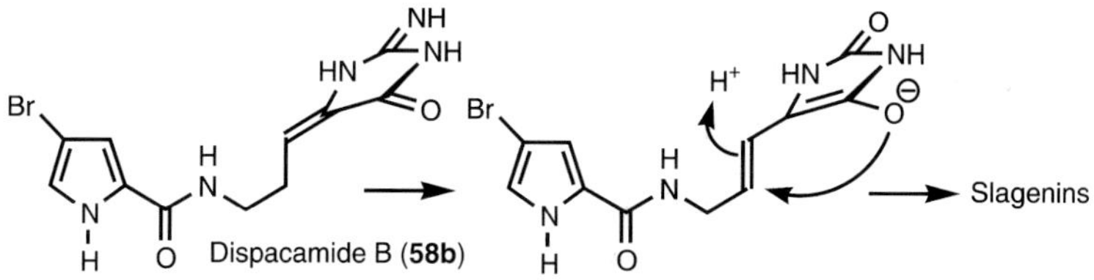

Scheme 3.5. Possible biogenetic pathway from dispacamides to slangenins

the former belongs longamide B (**67a**), which has been isolated, as a racemate, from the Caribbean marine sponge *Agelas dispar* (*100*). The corresponding ethyl ester – hanishin (**67b**) – was obtained from extracts of the highly polymorphic sponge *Acanthella carteri* (*89*) from the northern coast of the Hanish islands (Yemen), and most recently the S-(+)-methyl ester **67c** – rather than the racemate reported previously in *Homaxinella* sp. from Japan (*88*) – has been isolated from *Agelas ceylonica* collected off the Mandapam coast of India (*134*). Moreover, the methylester of 3-debromolongamide B (**67d**) has been isolated from the sponge *Axinella tenuidigitata* (*135*). On the other hand, both enantiomers of longamide B methyl ester were obtained by HPLC on the chiral stationary phase CHIRACEL OJ-R of the product isolated again as a racemate, from a Japanese marine sponge *Homaxinella* sp. (*88*) which also contained racemic longamide (**rac. 68a**), The latter, which is structurally distinct from the longamide previously isolated from the Indian plant *Piper longum* (*cf.* Ref. *136*) had been originally obtained in optical active (+) form, along with oroidin (**53**), agelongine (**35**) and dispacamide (**58**) from the Caribbean sponge *Agelas longissima* (*137*) collected along the coasts of Little San Salvador island.

Longamide (**68a**) is a bicyclic pyrrolecarboxamide containing a pyrroloketopiperazine nucleus, which is quite unusual among *Agelas* bromopyrrole alkaloids. Nevertheless, the Okinawan marine sponge *Agelas nakamurai* contains besides mukanadins A and B (s. above), mukanadin C (**68b**), which proved to be identical with racemic 2-debromolongamide (*126*). The latter had been previously isolated, along with aldisin (**75a**) and 2-bromoaldisin (**75b**), from the marine sponge *Axinella proliferans*, collected from Chuuk Atoll in Micronesia (*138*).

The structure of (+)-longamide (**68a**) was established on the basis of spectroscopic data. The absolute configurations at the sole asymmetric C9 atom of longamide and longamide B methyl ester (*88*) have been established by comparison with the chiroptic data available for dibromophakellin (**76b**), the structure of which had been previously established by X-ray diffraction analysis. As dibromophakellin and longamide possess the same chromophore, the absolute configuration of the latter could be established through a molecular dynamics analysis of both metabolites in the CHARMm force field, which evidenced a preferred half-chair conformation of the six-membered ring of longamide, which is slightly skewed around the C5–C6 bond, giving rise to an inherently chiral chromophore of the same helicity as that of dibromophakellin. As the polycyclic structure of the latter is very rigid, the consistency of CD data for both compounds suggested the same chromophore helicity, and therefore the (*S*) configuration at C9 for (+)-longamide. The structures of longamide (*139*, *140*), longamide B (*140*), its methyl ester (*140*), and hanishin (*140*) have been confirmed by chemical synthesis of the corresponding racemates.

As cyclooroidin (**66**) has been isolated recently, along with taurodispacamide A (**63a**) from *Agelas oroides* from the Bay of Naples (*129*), a plausible biogenetic pathway to longamide B (**67a**) may proceed by cyclization of oroidine (**53**) followed by oxidative breakdown of the 2-aminoimidazole moiety as represented in Scheme 3.6. Alternatively, breakdown of the aliphatic chain may precede cyclization. Further enzymatic oxidative degradation of longamide B might afford longamide. More likely, however, the latter may originate from a

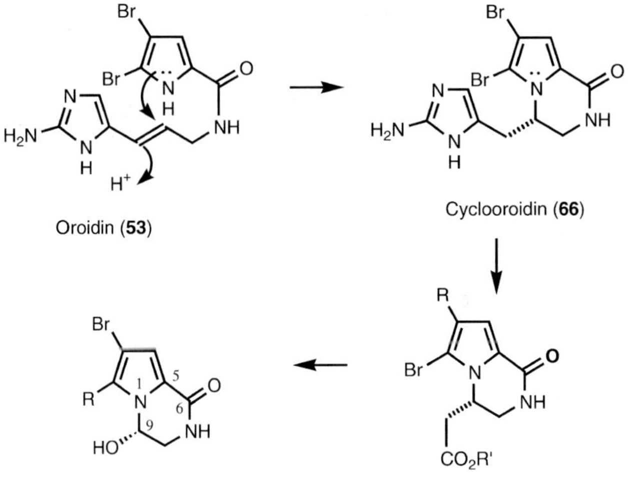

Oroidin (**53**)

Cyclooroidin (**66**)

R = Br: (+)-Longamide (**68a**)
R = H: Mukanadin C (**68b**)

R = Br, R' = H : Longamide B (**67a**)
R = Br, R' = C$_2$H$_5$: Hanishin (**67b**)
R = Br, R' = CH$_3$ (**67c**)
R = H, R' = CH$_3$ (**67d**)

Scheme 3.6. Plausible biogenetic pathway from oroidin (**53**) to the longamides

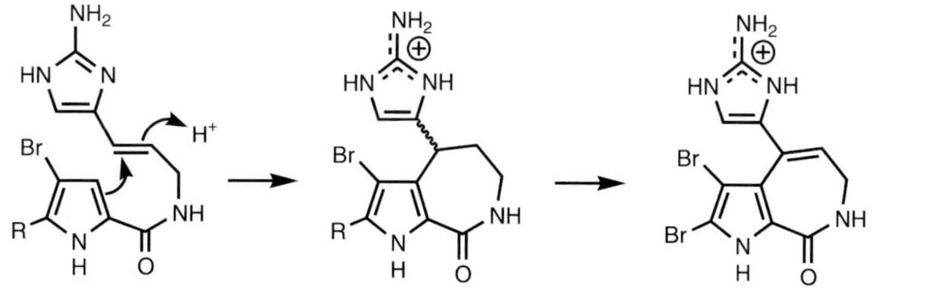

R = Br : Oroidin (**53**)   R = Br : Hymenin (**69a**)   R = Br : Odiline (Stevensine) (**70a**)
                             R = H : 2-Debromohymenin (**69b**)   R = H : 2-Debromostevensine (**70b**)

Scheme 3.7. Presumed biogenetic origin of hymenin (**69a**) and odiline (**70**)

condensation between 4,5-dibromopyrrole-2-carboxylic acid, an abundant metabolite of *Agelas longissima*, and a glycine unit.

A different kind of intramolecular cyclisation (*cf.* Scheme 3.7.), in which a β-position of the pyrrole ring is involved, may transform oroidin (**53**) and hymenidin (**54**) into hymenin (**69a**) and 2-debromohymenin (**69b**), respectively. (−)-Hymenin has been isolated from a sponge *Hymeniacidon* sp. collected at the Okinawan Ishigaki island (*141*), whereas 2-debromohymenin was found, among oroidin (**53**) and eight other already known bromopyrrole alkaloids, in the sponge *Stylissa (Axinella) carteri*, collected in Indonesia (*142*).

Although the absolute configuration of (−)-hymenin remains to be determined, both the racemic alkaloid and stevensine (**70a**) have been obtained by chemical synthesis through regioselective addition of 2-aminoimidazole to the exocyclic C=C-bond of 2,3.dibromo-6,7-dihydro-1*H*-pyrrole[2,3-*c*]azepin-8-one, as the key step (*cf.* Scheme 3.8.) (*143, 144*). Two features of this reaction are worth of comment. Thus, the regioselectivity of the nucleophilic addition of 2-aminoimidazole to the C4=C5-bond is explained by the fact that this double bond is conjugated to the pyrrole ring system and, therefore, the formation of the corresponding azafulvenium ion (s. insert in Scheme 3.8.) on regioselective protonation of the substrate conditions the attack of the nucleophile on C4. The second aspect concerns the regioselectivity of the reaction with respect to the 2-aminoimidazole reagent itself. Actually, substitution of the latter for the methoxy group of the 1*H*-pyrrole[2,3-*c*]azepin-8-one derivative which serves as intermediate for the synthesis of stevensine (*cf.* Scheme 3.8.) takes place only at the exocyclic NH$_2$ group of 2-aminoimidazole, when trifluoroacetic acid is used as the proton source. On the other hand, when methanesulfonic acid was used, the thermodynamically more stable

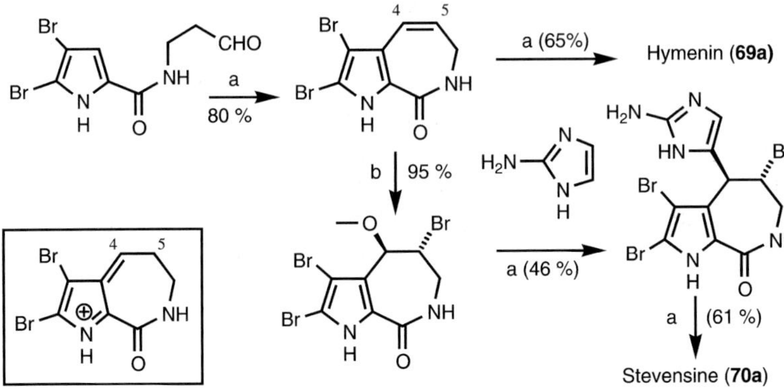

Scheme 3.8. A synthesis of hymenin (**69a**) and stevensine (**70a**) (*145*): a) H₃CSO₃H. b) Br₂ in methanol

C–C coupling product was obtained in 46% yield. Moreover, although hymenin could be synthesized from the acid-catalyzed coupling of 2,3-dibromo-6,7-dihydro-1*H*-pyrrolo[2,3-*c*]azepin-8-one with 2-aminoimidazole (see above), no reaction ensued, under analogous conditions, upon combining 2-aminoimidazole with the corresponding 2,3,5-tribromopyrrolo[2,3-*c*]azepin-8-one derivative, thus indicating that the formation of the corresponding azafulvenium ion does not take place merely by elimination of the methoxy group. Finally, the methanesulfonic acid-catalyzed elimination of HBr to yield stevensine (**70a**) was only successful working in an unstoppered reaction flask, whereas the same reaction carried out in a sealed tube lead to the loss of bromine atom on C3, and formation of 3-debromostevensine and 5-bromo-3-debromostevensine in 14% and 47% yield, respectively.

Enzymatic dehydrogenation of (−)-hymenin leads presumably to odiline = stevensine (**70a**), a structurally related 6,7-dihydropyrrolo[2,3-*c*]azepin-8-one derivative which has been isolated from the New Caledonian sponge *Pseudaxinyssa cantharella* (*105*). Odiline was also obtained (as stevensine) from an unidentified Micronesian sponge (*145*) as well as from a *Goreauiella* sp. sponge collected at 698 m depth (*106*). More recently, stevensine has been obtained together with 2-debromostevensine (**70b**) and oroidin (**53**), as well as eight other related bromopyrrole alkaloids from the Indopacific sponge *Stylissa* (*Axinella*) *carteri* (*142*).

A recent study carried out with cell cultures of the sponge *Teichaxinella morchella* proves the incorporation of ^{14}C-labelled histidine, proline and ornithine into stevensine. In the light of these

References, pp. 140–188

experiments, it is proposed that both ornithine and proline can be transformed into the 4,5-dibromopyrrole-2-carboxylate subunit of oroidin (*cf.* Scheme 3.4.), whereas histidine may be the precursor of the 4-(3-aminoprop-1-enyl)imidazole moiety of the same alkaloid, which generates stevensine by cyclization (*146*). The structure of stevensine has been confirmed by synthesis (*cf.* Scheme 3.8.). Moreover, stevensine has been obtained by dehydrogenation of hymenin *in vitro*, by means of a regioselective protodebromination and bromine migration process (*147*).

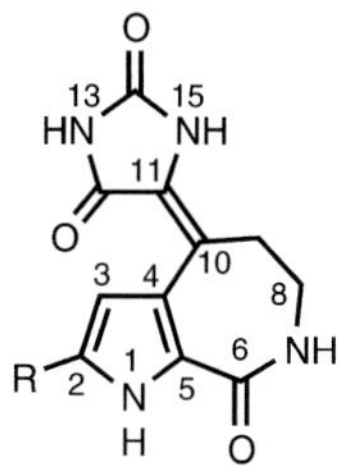

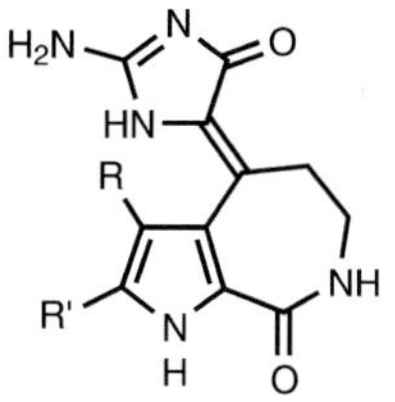

R = Br : Axinohydantoin = Fuscin (**71a**)
R = H : Debromoaxinohydantoin (**71b**)

R = H; R' = Br : Hymenialdisine (**72a**)
R = R' = H : Debromohymenialdisine (**72b**)
R = R' = Br : 3-Bromohymenialdisine (**72c**)

Another series of pyrrole alkaloids containing a fused pyrrolo[2.3-*c*] azepin-8-one skeleton, but biogenetically rather related to the dispacamides has been isolated also from marine sponges. Thus, the yellow metabolite axinohydantoin (**71a**) was isolated from the sponge *Axinella* sp. collected in Palau, on the Western Caroline islands, and its structure was assigned by X-ray crystallographic methods (*148*). Axinohydantoin is identical with fuscin, which was isolated from *Phacellia fusca* collected in the South China sea (*149*). The structure deduced for axinohydantoin was found to be closely related to that of hymenialdisine (**72a**), with reversal of the geometry of the C10=C11 bond being the most characteristic difference. This suggests that **71a** is not simply a product of partial hydrolysis of hymenialdisine (**72a**). The latter has been isolated, together with debromohymenialdisine (**72b**), which had been previously found in the Great Barrier Reef sponge *Phakellia flabellata* (*150*) and in an unidentified Korolevu (Fiji) sponge (*68*), from the marine sponges *Acanthella carteri* (*151*), *A. aurantica* (*104*), and *Pseudaxinyssa cantharella* (*105*), as well as from sponges of the genera *Axinella* (*74, 104, 148*) and *Hymeniacidon* (*148, 152*). Both axinohydantoin (**71a**) and hymenialdisine (**72a**) occur together with their

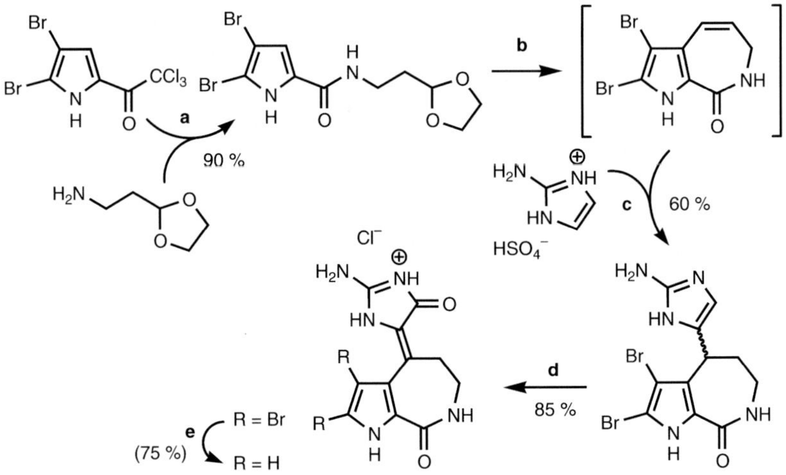

Scheme 3.9. An improved synthesis of debromohymenialdisine (**72b**): **a**) in acetonitrile 23°C, 16 h. **b**) H$_3$C–SO$_3$H, 45°C, 4 days. **c**) 2-aminoimidazole · HSO$_4^-$, 45°C, 4 days. **d**) 2 eq. Br$_2$ in acetic acid containing sodium acetate, 23°C, 1 h. **e**) Pd/C (10%) in methanol containing sodium acetate, 10 h

corresponding debromo derivatives, debromoaxinohydantoin (**71b**) and debromohymenialdisin (**72b**), respectively, in another *Hymeniacidon* sponge, *Monanchora*, from Papua New Guinea (*153*). The structures of hymenialdisine (**72a**) and debromohymenialdisine (**72b**) have been confirmed by chemical synthesis (*cf.* Scheme 3.9.) (*144, 154*). Moreover, the structure of hymenialdisine (**72a**) has been unequivocally established by X-ray diffraction analysis (*104, 152, 155*). The relatively unstable (10*E*)-isomers of both hymenialdisine and debromohymenialdisine, on the other hand, have been isolated from the common shallow-water marine sponge *Stylotella aurantium* from Palau (*156*).

A further derivative in this series of related pyrrolo[2.3-*c*]azepin-8-one derivatives, 3-bromohymenialdisine (**72c**), has been found in an *Axinellidae* sponge collected in Tanzania (*74*) as well as in the Indonesian marine sponge *Axinella carteri* (*103*). Its structure has been confirmed by synthesis (*144b*). Later on, spongiacidin A (**73a**) and spongiacidin B (**73b**) have been identified in an Okinawan marine sponge *Hymeniacidon* sp. (*157*). From the carbon chemical shift of C9 was deduced that spongiacidin A (**73a**) is the (*E*)-isomer at the exocyclic C10=C11-bond of 3-bromohymenialdisine (**72c**). Likewise, spongiacidin C (**74a**) and spongiacidin D (**74b**), which were isolated from the same *Hymeniacidon* sponge, were characterized as the (*Z*)-isomers of debromoaxinohydantoin (**71b**) and axinohydantoin (**71a**), respectively.

References, pp. 140–188

Most recently, both stereoisomers of each of the compounds **72a–c** have been isolated from the sponge *Stylissa* (*Axinella*) *carteri*. Upon standing in dimethylsulfoxide all (*E*)-isomers at the exocyclic C10=C11 bond converted into the respective, thermodinamically more stable, (*Z*)-isomers (*142*).

R = Br : Spongiacidin A (**73a**)   R =H : Spongiacidin C (**74a**)   R= H : Aldisin (**75a**)
R = H : Spongiacidin B (**73b**)   R = Br : Spongiacidin D (**74b**)   R= Br : 2-Bromoaldisin (**75b**)

Pyrrole alkaloids containing a fused pyrrolo[2.3-*c*]azepin-8-one skeleton show α-adrenoceptor blocking activity (*158*). On the contrary, alkaloids like oroidin, keramadine, and clathrodine, in which two heterocyclic nuclei are linked by a linear chain, show serotonergic and/or cholinergic antagonist activities (*159*). Interestingly, the same structural modifications leading from dispacamide A (**58a**) to oroidin (**53**) cause a strong increase in the α-blocking activity in the cyclic bromopyrrole alkaloids. In fact, hymenin (**69a**), with the nucleus of a cyclized oroidin is much more active than hymenialdisine (**72a**), possessing the nucleus of a cyclized dispacamide A.

Oxidative degradation of debromohymenialdisine (**72b**) affords the more simple pyrrolo[2.3-*c*]azepin-8-one derivative aldisin (**75a**) (*150*). The latter has been isolated, together with the corresponding degradation product of hymenialdisine, 2-bromoaldisin (**75b**), from the marine sponges *Pseudaxinyssa cantharella* (*105*) and *Hymeniacidon aldis* (*68*), as well as from an unidentified sponge from Korolevu (Fiji) (*68*). Aldisin and 2-bromoaldisin have been also obtained from an *Axinellidae* sponge collected in Tanzania (*74*) and from a sponge of the genus *Lissodendoryx* collected in Sri Lanka (*68*), respectively. It must be pointed out, however, that both aldisin and 2-bromoaldisin may be formed, as artefacts, by addition of water to the internuclear double bond followed by *retro*-aldol loss of the guanidine function (*68*). Noteworthy both aldisin and 2-bromoaldisin have been characterized in one report (*105*) as optical active compounds ($[\alpha]_D = -6$ and +5, respectively, in methanol), despite the fact that their molecules are not chiral. Although the synthesis of **75a** has been reported (*160*), its condensation with glycocymidine (2-amino-2-imidazolin-4-one) to produce debromohymenialdisine (**72b**) failed (*161*).

Intramolecular cyclisations of oroidin and its analogs in which the pyrrole moiety acts as a nucleophile towards the C=C-bond of the 2-aminoimidazole ring lead to the formation of group of marine alkaloids known as phakellins (*162, 163*). Such cyclisations are probably preceded by a shift of the C=C-bond of the linking chain between the pyrrole and imidazole, thus explaining the formation of the tetracyclic system containing a pyrroloketopiperazine moiety, which is characteristic for the phakellins (*163b*) (*cf.* Ref. *152*). The same biosynthetic pathway may lead to the isophakellins, which are probably formed by cyclization of the same biogenetic precursor in the *syn*-conformation of the pyrrole–CO-bond (*cf.* Scheme 3.10.).

Bromophakellin (**76a**) and dibromophakellin (**76b**) have been isolated from the marine sponges *Phakellia flabellata*, found on the Great Barrier reef (*162*). Both alkaloids were characterized by spectroscopic methods (*163*). The structure of dibromophakellin, which has been also isolated from an *Axinellide* sponge (*74*), was confirmed by X-ray diffraction analysis of a single crystal of the monoacetyl derivative (*163a*), as well as through a biomimetic total synthesis (*164*). Thus, dihydrooroidin hydrochloride was treated with bromine in acetic acid yielding an insoluble highly unstable salt which combined rapidly with

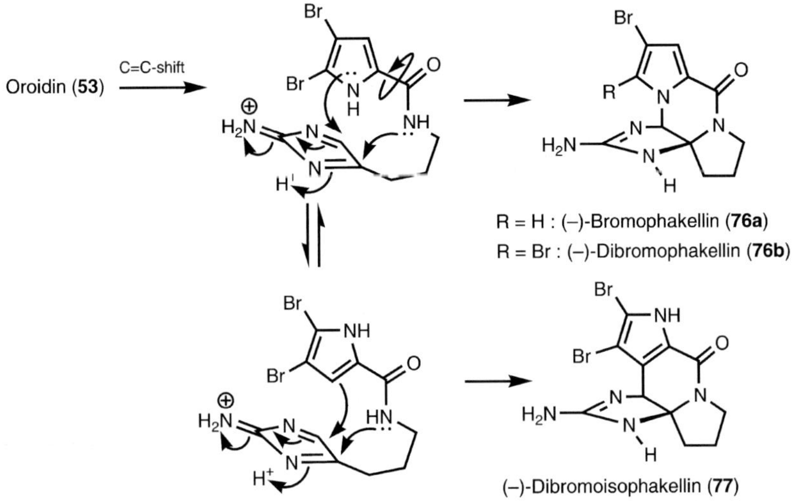

Scheme 3.10. Presumed biogenetic origin of phakellins (**76**) and isophakellins (**77**)

References, pp. 140–188

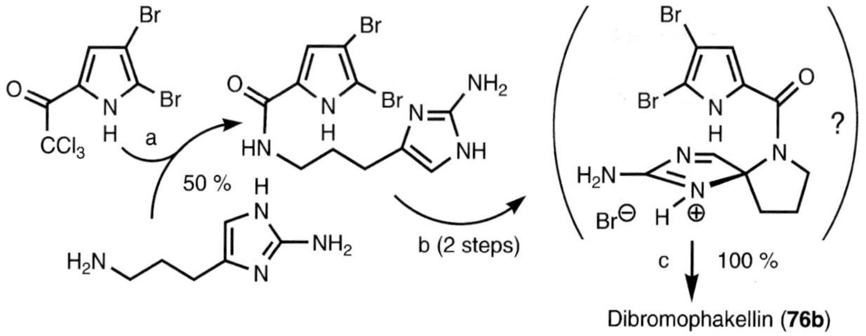

Scheme 3.11. A biomimetic synthesis of racemic dibromophakellin (**76b**) (*164*): a) Na$_2$CO$_3$ in dimethylformamide; b) *i*: Br$_2$ in acetic acid; *ii*: methanol. c) (CH$_3$)$_3$COK in 2-butanol

methanol to afford an equally unstable product of unknown structure. When treated with potassium *tert*-butoxide this product was quantitatively converted to racemic dibromophakellin (*cf.* Scheme 3.11.).

(−)-Dibromoisophakellin (**77**) has been isolated from the marine sponge *Acanthella carteri*, collected off the Madagascar coast (*165*). Its enantiomer, (+)-dibromocantharellin (**78**), as well as the (+)-enantiomer of dibromophakellin have been isolated, along with odiline (**70a**), from the New Caledonian sponge *Pseudaxinyssa cantharella* (*105*). The structure and absolute configuration of both (+)-dibromocantharellin (*105*) and (−)-dibromoisophakellin (*165*) have been confirmed by X-ray diffraction analysis, as well as by a total synthesis of the latter, as a racemate (*166*). Although all phakellins contain a guanidine group, their basicity (pK$_s$ ≈ 7.9) is considerably lower than that of guanidines (pK$_s$ > 13.4). As the high basicity of guanidines, in general, is attributed to resonance stabilization of the corresponding protonated species, this anomaly may be explained by inhibition of the resonance in the phakellinium cation. As a matter of fact, planarity of the guanidine group, which is a requirement for charge delocalization in the guanidinium cation, is not possible in the phakellinium cation, since it would introduce severe conformational strains in the ketopiperazine ring of the latter (*163b*).

To the phakellin group also belongs dibromophakellstatin (**79**), a highly cytotoxic alkaloid isolated from the Republic of Seychelles sponge *Phakellia mauritiana* (*167*), which also contains dibromophakellin (**76b**) and debromohymenialdisine (**72b**). Dibromophakellstatin, the structure and absolute configuration of which have been secured by X-ray diffraction analysis (*167*) and by chemical synthesis (*166*), is an analog of (−)-dibromophakellin (**76b**) in which the guanidino group has been hydrolyzed to a 3-imidazolone moiety.

(+)-Dibromocantharellin (**78**)   Dibromophakellstatin (**79**)

Oroidin (**53**)

Dibromoagelaspongin (**80**)

Scheme 3.12. Suggested biogenetic pathway to dibromoagelaspongin (**80**)

An analogous intramolecular cyclization reaction as suggested for the formation of the tetracyclic skeleton of the phakellins (*cf.* Scheme 3.10.) may explain the transformation of oroidin (**53**) into dibromoagelaspongin (**80**), a structurally related metabolite which has been isolated from a sponge *Agelas* sp. found off the coast of Tanzania (Scheme 3.12.). The structure of dibromoagelaspongin has been determined by spectroscopic methods and confirmed by X-ray diffraction analysis (*168*).

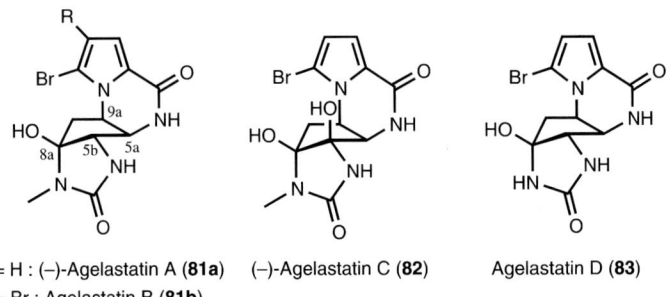

R = H : (−)-Agelastatin A (**81a**)
R = Br : Agelastatin B (**81b**)

(−)-Agelastatin C (**82**)

Agelastatin D (**83**)

References, pp. 140–188

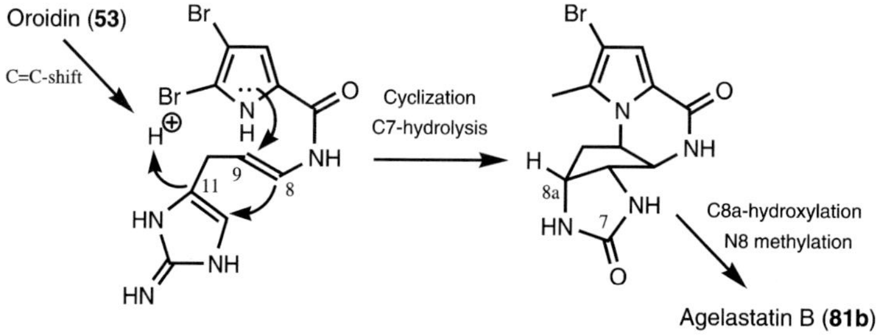

Scheme 3.13. Possible biogenetic pathway from oroidin (53) to agelastatin B (81b)

A different kind of intramolecular cyclisation of members of the oroidin family may be initiated by protonation of C11 and enzyme-driven C8 attack at the 4 position of the imidazole ring in a oroidin isomer as substrate, in which the exocyclic double bond is shifted from C9=C10 to C8=C9, followed by pyrrole nitrogen attack at the developing C9-carbocation (*cf.* Scheme 3.13.). This reaction sequence seems to be a plausible biogenetic pathway to the agelastatins, the most frequently encountered member of which is agelastatin A (**81a**), a markedly cytotoxic alkaloid with antitumor activity, which has been isolated from the deep water marine sponge *Agelas dendromorpha* collected in the Coral Sea near New Caledonia (*169*). In the sponge, agelastatin A is accompanied by a small amount of agelastatin B (**81b**), the corresponding dibromopyrrole derivative. Both agelastatins A and B are inseparable, but the N^5,N^6,O-trimethyl derivative of the latter could be purified for characterization purposes. The structures and relative configurations of agelastatin A (**81a**) and its N^5, N^6, O-trimethyl derivative were established by a series of spectroscopic experiments. The absolute configuration at the four stereogenic centers around the carbocyclic ring was determined to be (5a*S*, 5b*S*, 8a*S*, 9a*R*) by a combination of molecular modeling of the preferred conformation of the compound and application of the CD exciton-coupling method (*170*). Later on, the isolation of two new closely related minor metabolites, agelastatin C (**82**) and agelastatin D (**83**) along with agelastatin A (**81a**) from the West Australian sponge *Cymbastela* sp. was reported (*171*).

Agelastatin C is identical to agelastatin A, except that it is hydroxylated at C5b. Interestingly, this compound does not seem prone to spontaneous loss of *N*-methylurea. The total synthesis of agelastatin A (**81a**) has been achieved in about 14 steps performed in 12 operations in approximately 7% over-all yield starting from cyclopentadine as the source of the carbocyclic ring of the natural compound (*172*). Agelastatin A (**81a**) has been reported to have significant *in vitro* activity against some kinds of tumor cells (*173*). Structure-activity studies indicate that the C8a hydroxyl group and both NH groups are needed for optimal activity.

3.2.2.3. Dimeric Oroidin Metabolites

The structurally most simple oroidin dimer is mauritiamine (**84**), which was isolated along with oroidin (**53**) and 4,5-dibromopyrrole-2-carboxamide (**46**) from the marine sponge *Agelas mauritiana* collected off Hachijo-jima Island, Japan (*91*).

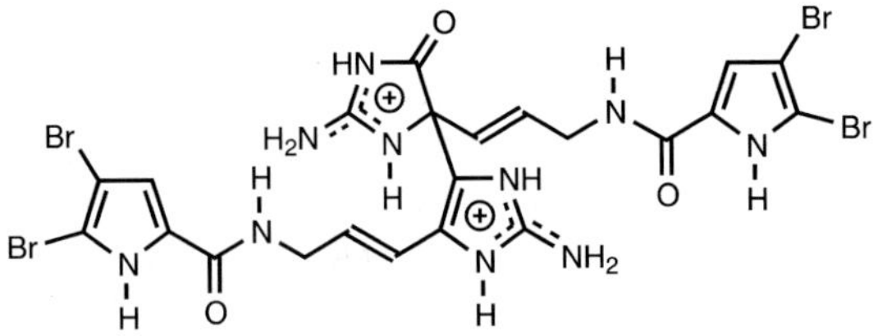

Mauritiamine (**84**)

Both **84** and oroidin (**53**) were shown to inhibit metamorphosis of the barnacle *Balanus amphitrite*. As pseudoceratidine (**50**), mauritiamine, which is probably formed from two molecules of oroidin (**53**), deserves interest as a potential antifouling agent. The synthesis of mauritiamine in which the key step centers on the oxidative dimerization of a 2-aminoimidazole derivative in a manner that may prove relevant to its biosynthesis, has been reported (*174*).

A different kind of mechanism probably operates in the biosynthesis of sceptrin (**85**), which formally could be formed by enzyme-catalyzed [2+2] cycloaddition of two hymenidin (**54**) molecules as depicted in Scheme 3.14. However, as *in vitro* the suprafacial-suprafacial process,

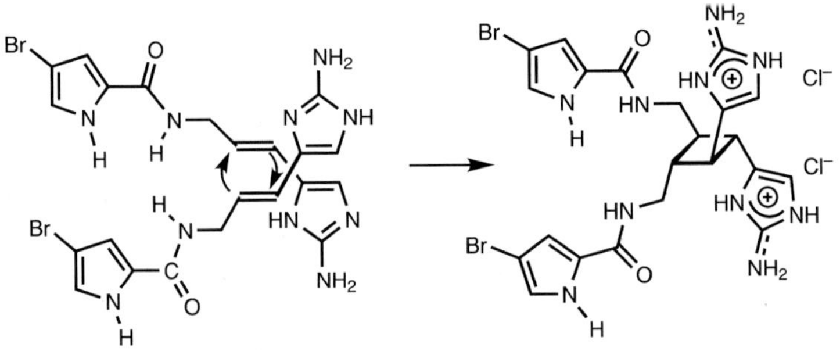

Hymenidin (**54**) → Sceptrin (**85**)

Scheme 3.14. Formal dimerization of hymenidin yielding sceptrin (**85**)

References, pp. 140–188

which accounts for the all-*trans* arrangement of the substituents on the cyclobutane ring, is thermally forbidden, a multi-step mechanism in which carbenium ions are involved as intermediates, seems to be more likely (*175*).

Sceptrin (**85**) was first isolated from the Caribbean sponge *Agelas sceptrum*, collected at Glover Reef, Belize and its structure was elucidated by MS, IR, ^{1}H- and ^{13}C-NMR-spectroscopy, as well as X-ray diffraction analysis (*109*). The same compound was found later among other oroidin-related alkaloids in an unidentified Micronesian sponge (*145*). Sceptrin (**85**) itself and two further analogs bearing either only one or four bromine atom(s) have been isolated from the caribbean sponge *Agelas conifera* (*175*, *176*). This sponge contains also two sceptrin analogs – called oxysceptrins – derived from dispacamide A (**58a**), as the monomer, in which one of the 2-aminoimidazole rings is replaced by a 2-imidazolone moiety. Oxysceptrin, which is a potent actomyosin ATPase activator, has been isolated also from the Okinawan marine sponge *Agelas cf. nemoechinata* (*177*). Moreover, nakamuric acid (**86**), presumably a product of partial degradation of sceptrin, has been isolated from the Indopacific sponge *Agelas nakamurai* along with sceptrin itself, debromosceptrin, and ageliferin (**87**: R=R′=H) (*178*).

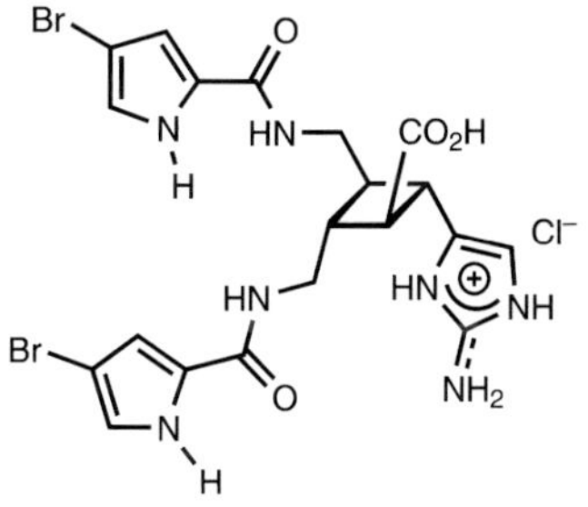

Nakamuric acid (**86**) 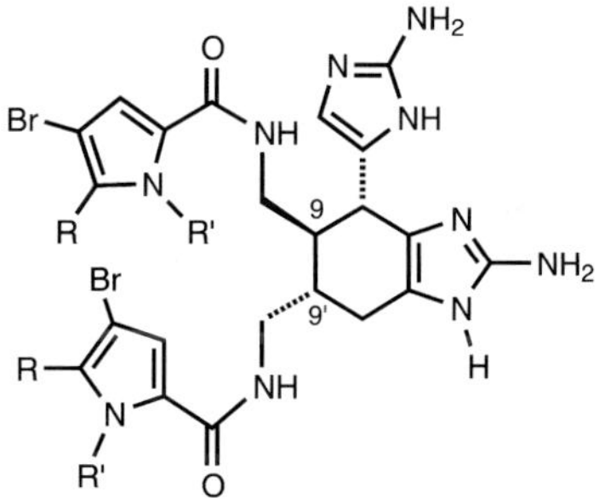   R = H or Br; R' = H or CH$_3$: Ageliferins (**87**)

On the other hand, enzyme-catalyzed (and hence enantioselective) intermolecular [4+2] cycloaddition involving either two oroidin (**53**) or two hymenidin (**54**) molecules, as well as one molecule of each, is a possible pathway for the biosynthesis of ageliferins (*cf.* Scheme 3.15.), which has been simulated in the laboratory with the synthesis of the 12,12′-dimethyl derivative of ageliferin (*179*). However, a formal multi-step mechanism in which carbenium ions act as intermediates (*175*) cannot be ruled out *a priori*.

Three members of this class of marine alkaloids, ageliferin (**87**: R=R′=H), bromoageliferin and dibromoageliferin (**87**: R=Br; R′=H),

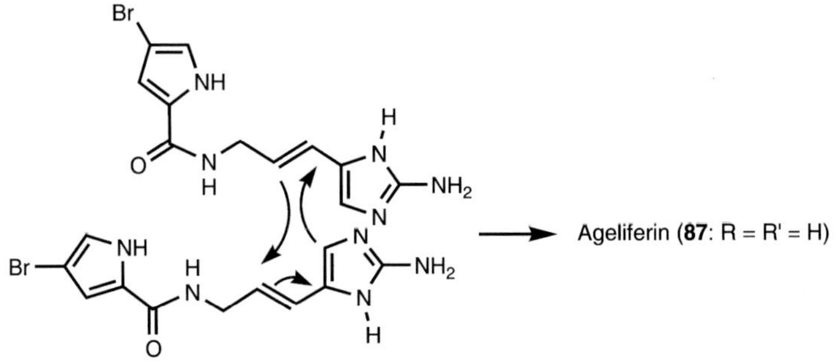

Scheme 3.15. Possible biogenetic pathway from hymenidin (**54**) to ageliferin (**87**: R=R′=H)

have been isolated as potent actomyosin ATPase activators from the Okinawan marine sponge *Agelas* sp. (*180*) and later on from the Caribbean sponge *A. conifera* (*175*). Bromo and dibromoageliferin have been also obtained, along with up to seven ageliferins (**87**: R′=CH$_3$) which are methylated on one or both of the pyrrole *N*-atoms from the sponge *Astrosclera willeyana* (*181*). They structures were elucidated on the basis of their spectral data, especially two-dimensional NMR spectroscopy. Ageliferins had been isolated previously from the sponges *Agelas conifera* and *A. mauritiana* (*106*), but the detailed structure elucidation including the relative configuration at C9' and the position of the bromine atoms in bromoageliferin had not been reported.

More complex dimeric structures are obtained through a synergism of intramolecular and intermolecular cyclization processes. Thus, in the molecules of konbu'acidin A (**88**) a bromophakellin moiety is part of a fused hexacyclic skeleton containing two guanidine subunits. (−)-Konbu'acidin A, which was isolated from the marine sponge *Hymeniacidon* sp., collected off Konbu on the Okinawa island, exhibits inhibitory activity against cyclin dependent kinase (cdk4) (*182*). Most likely, the biosynthesis of the konbu'acidin A skeleton takes place by addition of an oroidin molecule to a molecule of the 9-*cis* isomer of hymenidin (*cf.* Scheme 3.16.). The absolute configuration represented by structure **88** is, however, arbitrary, since the absolute configuration of (−)-konbu'acidin A is unknown.

An analogous biosynthetic pathway may lead to konbu'acidin A analogs containing less brominated phakellin or even isophakellin substructures, among which only the products of hydrolysis of the pyrrole-carboxamide group are known until now. Thus, (−)-palau'amine (**89a**) and two brominated derivatives therefrom (**89b** and **89c**) have been isolated from the sponge *Stylotella aurantium* (originally incorrectly

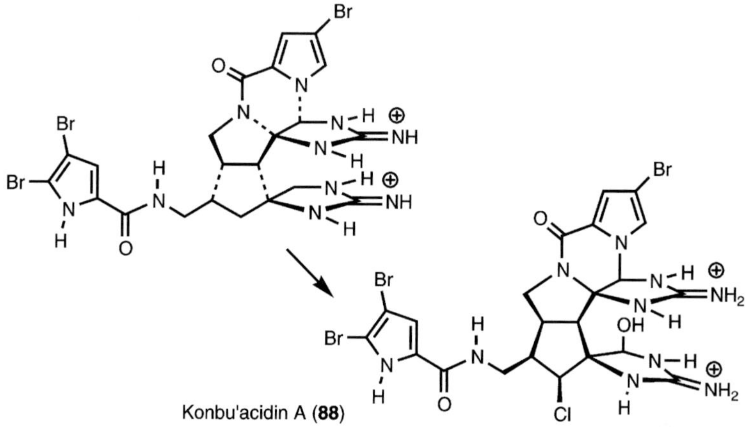

Scheme 3.16. Presumed origin of the two molecule moieties of kombu'acidin A (**88**)

identified as *Stylotella agminata*), collected at Palau, Western Caroline islands (*183*). Their structures were elucidated mainly by NMR spectroscopic methods, which provided also the relative stereochemistry of the molecules. The absolute configuration of (−)-palau'amine is, however, unknown although the fact that its CD spectrum displays the same chacteristics as that of (−)-bromophakellin (**76a**) hydrochloride suggests that the two compounds have the same absolute configuration (*183b*). Accompanying palau'amine, three isophakellin derivatives called styloguanidines (**90a–c**) were isolated from the sponge *Stylotella aurantium* collected on two different sites (*183b, 184*). They are chitinase inhibitors that arrest the moulting process of the cyprid larvae of barnacles. Incomplete synthetic approaches to palau'amine and the styloguanidines have been reported (*185*).

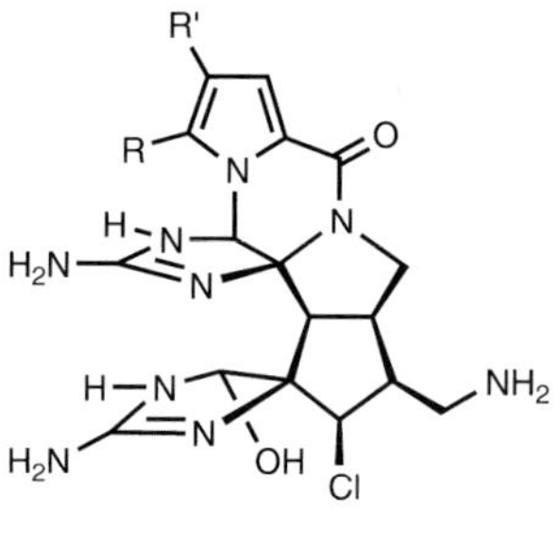

R = R' = H : (−)-Palau'amine (**89a**)
R = H, R' = Br : 4-Bromopalau'amine (**89b**)
R = R' = Br : Dibromopalau'amine (**89c**)

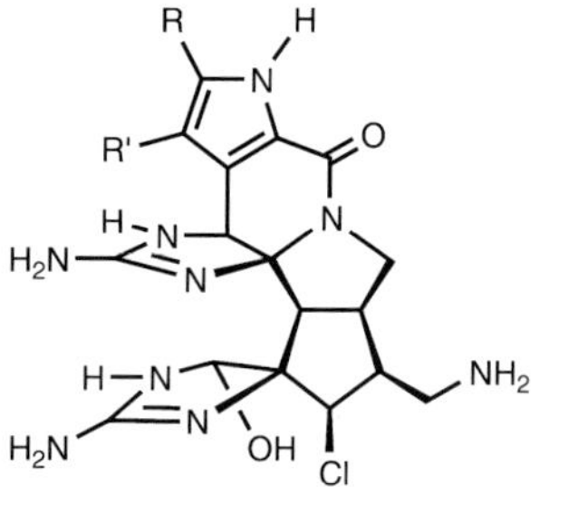

R = R' = H : (+)-Stiloguanidine (**90a**)
R = H; R' = Br : Bromostiloguanidine (**90b**)
R = R' = Br : Dibromostiloguanidine (**90c**)

Another family of oroidine dimers is represented by four bromopyrole alkaloids known as axinellamines. They contain a fused tetracyclic core formed by condensation of the guanidine moieties of two oroidine molecules. The structures of axinellamines A (**91a**) and axinellamine B (**91b**), which have been isolated together with their *syn*-isomers axinellamine C (**92a**) and axinellamine D (**92b**), respectively, from the Australian marine sponge *Axinella* sp. (*186*) are however different from those of axinellamines A and B isolated from the Caribbean marine sponge *Axinella* sp. (see below). Axinellamines B–D exhibit bactericidal effects against *Heliobacter pylori*.

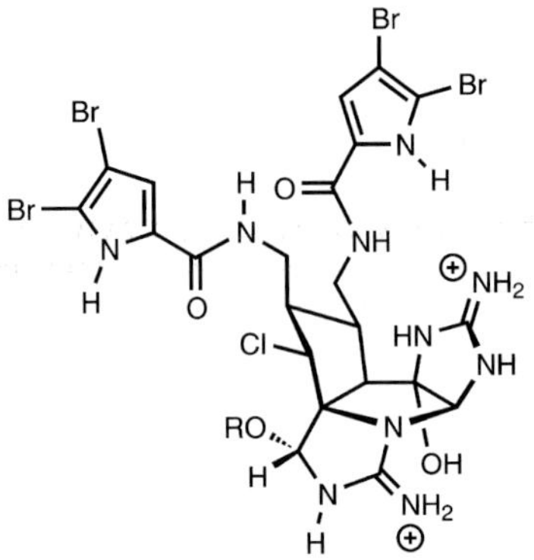

R = H : Axinellamine A (**91a**)
R = CH$_3$: Axinellamine B (**91b**)

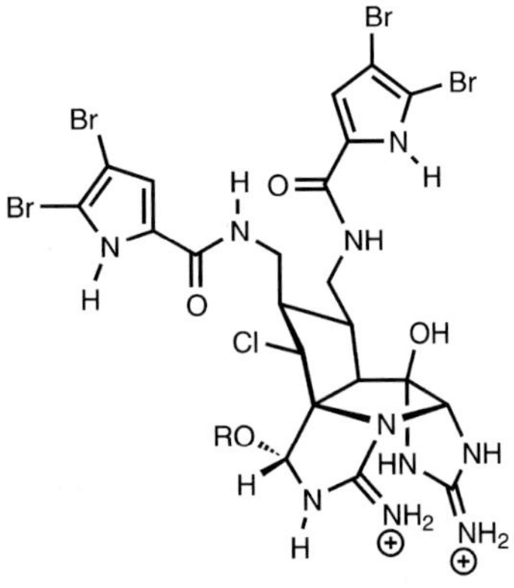

R = H : Axinellamine C (**92a**)
R = CH$_3$: Axinellamine D (**92b**)

3.2.3. Alkyl Pyrroles from Sponges

As mentioned before, the name axinellamine has been given to two different class of alkaloids, both isolated from marine sponges of the genus *Axinella*, which were characterized in the same year (1998) by two independent research groups. Thus, the structures represented in formulae **93** and **95** for axinellamine A and B, respectively, which were isolated from the Caribbean marine sponge *Axinella* sp. (*187*) are different from those of axinellamines A–D from the Australian marine sponge *Axinella* sp. (*vide supra*), Both axinellamine A and B, the structures of which were established on the basis of spectroscopic data, are optical active compounds ([α]$_D$ = −45.4 and [α]$_D$ = −8.9, respectively, in CHCl$_3$). As settled by chemical synthesis of its enantiomer, the absolute configuration of (−)-axinellamine A at C11 is *R* (*188*). Owing to their similar structures and their common origin, the same absolute configuration can be anticipated

References, pp. 140–188

for (−)-axinellamine B as well as for (−)-axinellamide (**94**), a structurally related metabolite, which had been isolated earlier from the same sponge collected off the coast of Trinidad (*189*).

(−)-Axinellamine A (**93**)

Axinellamide (**94**)

(−)-Axinellamine B (**95**)

For the first time in 1975 Cimino *et al.* (*190*) reported the isolation of a series of 3-alkylpyrrole-2-carboxaldehydes obtained as a mixture of C_{19} (3.3%), C_{20} (12.5%), C_{21} (49%), C_{22} (25%) and C_{23} (10.2%) homologs, from the marine sponge *Oscarella lobularis*, which also contains some unsaturated analogs and the corresponding pyrrole-2-carboxylic acids and their methyl esters. A chemical synthesis of 3-octadecyl-2-pyrrolecarboxaldehyde (in mixture with its 4-octadecyl isomer) has been reported, but no identification with the purported natural compound was attempted (*191*). However, the substitution pattern of these metabolites was later questioned, since comparison of their spectral data with those of some 5-alkylpyrrole-2-carboxaldehydes (**96b**) obtained later from the sponge *Laxosuberites* sp. by Stierle and Faulkner (*192*) revealed surprisingly few differences. From the latter marine sponge an inseparable mixture of four 5-alkylpyrrole-2-carboxaldehydes (**96b**), consisting of the homologues C_{15} (46%), C_{16} (12%), C_{17} (23%), C_{19} (19%) was isolated along with 5-(12'-cyano-6'-dodecenyl)pyrrole-2-carboxaldehyde (**96c**) and 5-(23'-cyano-23'-hydroxy-6'-tricosenyl)pyrrole-2-carboxaldehyde (**96e**). Their structures were determined by interpretation of spectral data and by chemical transformations. Some of the above 5-alkyl-2-pyrrolecarboxaldehydes have been also isolated from the sponges *Mycalecarmia monanchrorata* and *Mycale mytilorum* collected off the South Indian coast (*193*), and from the caribbean sponges *Mycale microsigmatosa* and *Desmapsamma anchorata*, as well (*194*). The latter sponges contain also a series of 5-alkenyl-pyrrole-2-carboxaldehydes (**96g–n**) analogs to the above alkyl derivatives. Furthermore, *Mycale tenuispiculata* from the Thiruvananthapuram coast of India contains, along with mycaleoxime (**98**), two additional 5-alkenylpyrrole-2-carbaldehydes, namely **96d**, **96f** (*195*).

R = : – (CH$_2$)$_8$–CH$_3$ (**96a**)

R = : – (CH$_2$)$_n$–CH$_3$ (n = 14, 15, 16, 18, 19, 22) (**96b**)

R = : – (CH$_2$)$_5$–CH=CH–(CH$_2$)$_5$–CN (Z) (**96c**)

R = : – (CH$_2$)$_5$–CH=CH–(CH$_2$)$_{16}$–CN (Z) (**96d**)

R = : – (CH$_2$)$_5$–CH=CH–(CH$_2$)$_{15}$–CH(OH)CN (Z) (**96e**)

R = : – (CH$_2$)$_5$–CH=CH–(CH$_2$)$_7$–CH$_3$ (E) (**96f**)

R = : – (CH$_2$)$_8$–CH=CH–(CH$_2$)$_6$–CH$_3$ (Z) (**96g**)

R = : – (CH$_2$)$_8$–CH=CH–(CH$_2$)$_8$–CH$_3$ (Z) (**96h**)

R = : – (CH$_2$)$_8$–CH=CH–(CH$_2$)$_{10}$–CH$_3$ (Z) (**96i**)

R = : – (CH$_2$)$_{11}$–CH=CH–(CH$_2$)$_8$–CH$_3$ (Z) (**96j**)

R = : – (CH$_2$)$_8$–CH=CH–(CH$_2$)$_{12}$–CH$_3$ (Z) (**96k**)

R = : – (CH$_2$)$_{13}$–CH=CH–(CH$_2$)$_7$–CH$_3$ (Z) (**96l**)

R = : – (CH$_2$)$_{10}$–CH=CH–(CH$_2$)$_{12}$–CH$_3$ (Z) (**96m**)

R = : – (CH$_2$)$_{14}$–CH=CH–(CH$_2$)$_8$–CH$_3$ (Z) (**96n**)

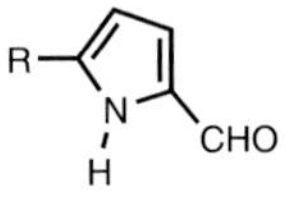

96

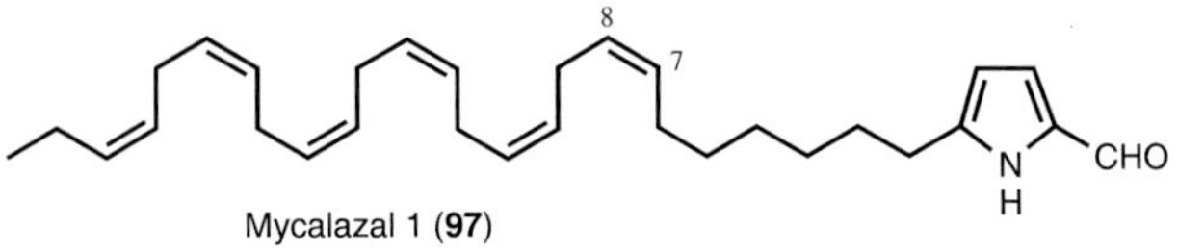

Mycalazal 1 (**97**)

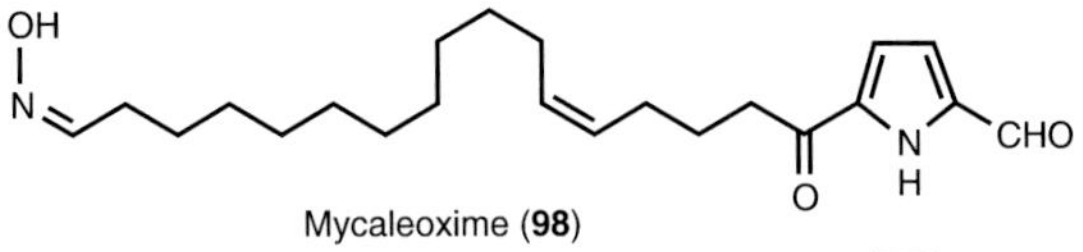

Mycaleoxime (**98**)

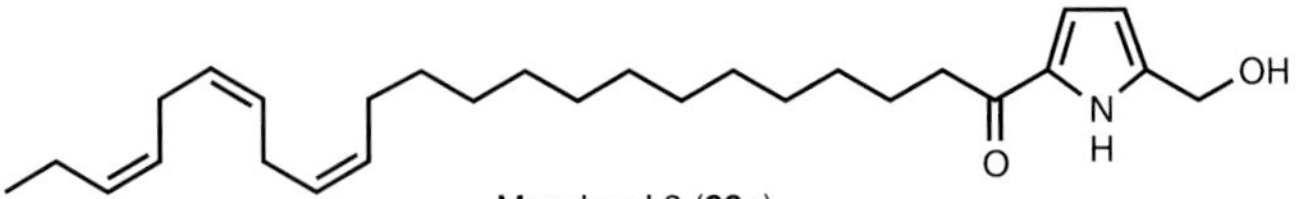

Mycalazol 8 (**99a**)

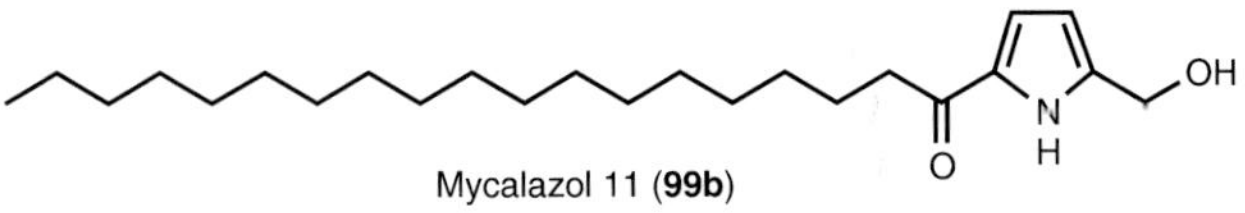

Mycalazol 11 (**99b**)

On the other hand, 5-nonyl-2-pyrrolecarboxaldehyde (**96a**) has been isolated from a soft coral (*Telesto* sp.) and its epibiotic, but unidentified, demosponge (*196*). However, since the pyrrole perfused both organisms it is not possible to identify the exact source of the natural product. The structure of 5-nonyl-2-pyrrolecarboxaldehyde was confirmed by synthesis, thus supporting Faulkner's assignment, who proposed 2,5- rather than 2,3-disubstitution at the pyrrole ring for this kind of sponge metabolites.

Structurally, the above 5-alkylpyrrole-2-carboxaldehydes are closely related to the mycalazals 1 (**97**) and mycalazal 2 (7,8-dihydromycalazal 1),

References, pp. 140–188

which have been isolated along with a series of twelve 5-acyl-2-hydroxymethylpyrroles – named mycalazols 1–12 (*e.g.* **99a** and **99b**) – from the north-eastern Atlantic sponge *Mycale micracanthoxea* (*197*). They have been reported to show significant *in vitro* cytotoxicity against different cell lines. Mycalazol 11 (**99b**) has been obtained by chemical synthesis (*198*). Presumably, the biosynthetic precursors of the aliphatic side chains of both the mycalazals and the above mentioned alkylpyrrolecarboxaldehydes are fatty acids.

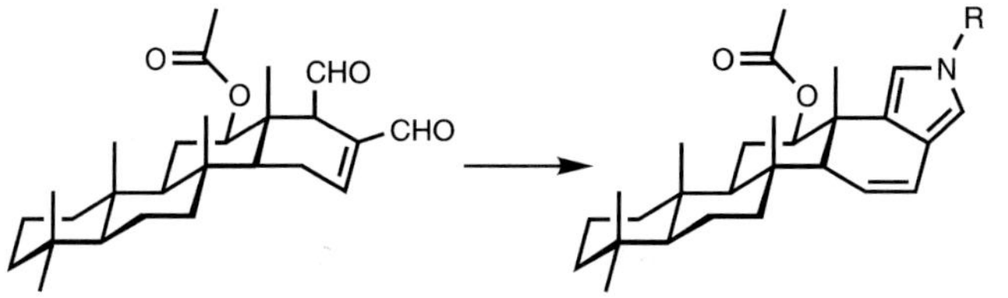

Scalaradial (**100**)

R = CH$_2$-CH(CH$_3$)-CH$_2$-CH$_3$: (+)-Molliorin A (**101a**)

R = CH$_2$-CH(CH$_3$)$_2$: (–)-Molliorin C (**101b**)

R = CH$_2$-CH$_2$-C$_6$H$_5$: (+)-Molliorin D (**101c**)

R = CH$_3$: (+)-Molliorin E (**101d**)

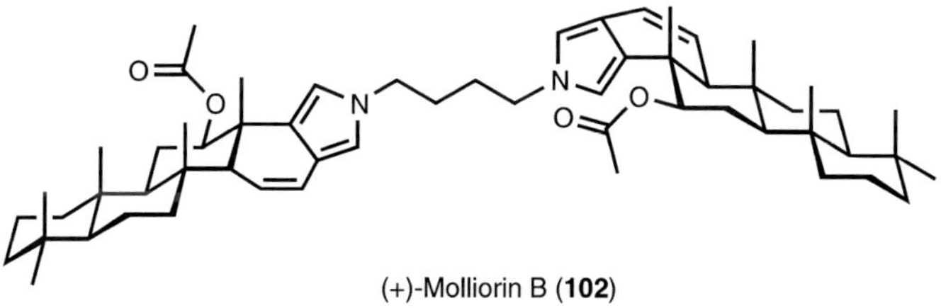

(+)-Molliorin B (**102**)

On the other hand, terpene derivatives containing a pyrrole ring are in sponges quite rare (*cf.* **40** and **41**). The unusual pyrrole metabolites molliorins A (**101a**), C (**101b**), D (**101c**) and E (**101d**) as well as the dimeric molliorin B (**102**) have been isolated from the marine sponge *Cacospongia mollior*, collected in the bay of Taranto (Italy) (*199*). Their structures, which were elucidated by spectroscopic methods, were confirmed by partial chemical syntheses from scalaradial (**100**) and the corresponding primary amines under the experimental conditions of the Paal-Knorr synthesis of pyrrole derivatives. As **100** is found among the tetracyclic sesterpenes present in marine sponges, it is not unreasonable to suppose that, also *in vivo*, the molliorins may derive from scalaradial through reaction with primary amines arising from the decarboxylation of the corresponding amino acids.

3.2.4. Aryl Pyrroles from Sponges

A series of polyarylpyrrole derivatives, which all appear to be derived from the assemble of three tyrosine subunits, has been isolated in recent years from marine sources. Intriguingly, some of these structurally closely related metabolites have been found in taxonomically quite distinct organisms such as sponges, ascidians, and prosobranch mollusks, collected at remote places in the Indian of Pacific ocean, thus raising the question, whether such compounds originate from microorganisms symbiotically associated with these animals (*cf.* Ref. (*200*)).

The most characteristic group of this kind of arylpyrrole derivatives is formed by the lamellarins which are structurally – and probably biogenetically – closely related to the storniamides (**106**), ningalins (**125–128**) and lukianols (**123**). A comprehensive review on the structures, origins, biogenesis, synthesis, and bioactivity of the lamellarins and their analogs has been published recently by S. Urban *et al.* (*201*).

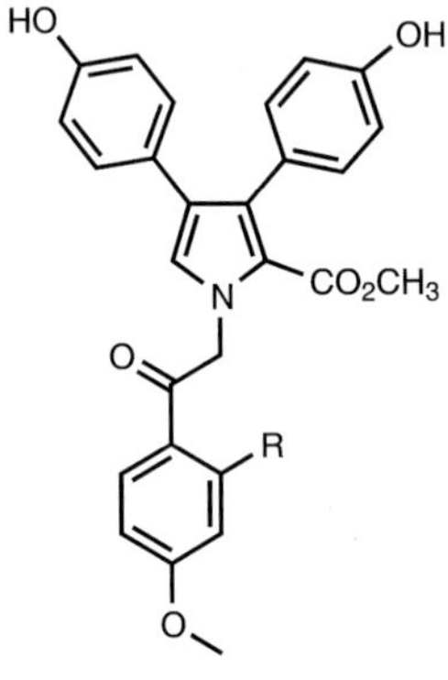

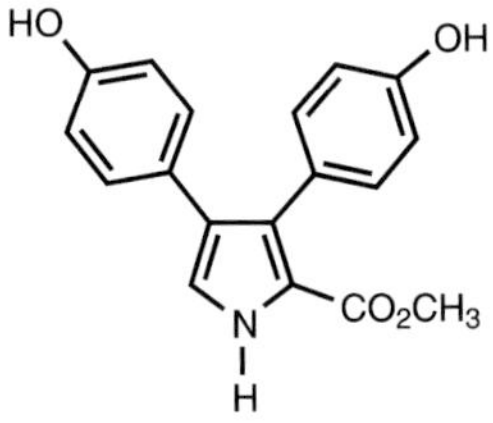

Lamellarin Q (**104**)

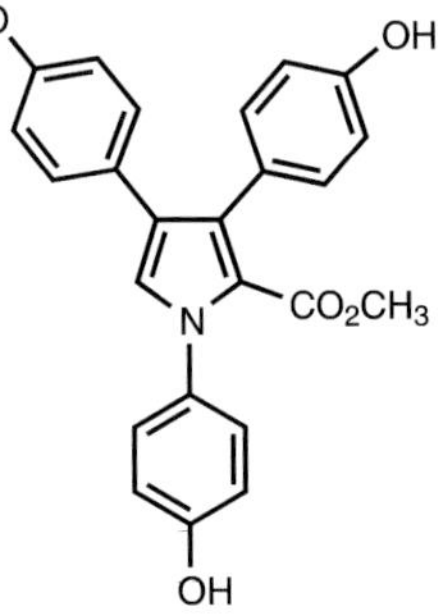

Lamellarin R (**105**)

R = H : Lamellarin O (**103a**)
R – OH : Lamellarin P (**103b**)

According to their structure, lamellarins may be classified into two different groups: *i*) those in which the pyrrole ring is part of a benz[*g*]indolizine system (*cf.* Section 3.3.1.), and *ii*) those with more simple structures, in which the pyrrole ring is not fused to adjacent aromatic rings. As a matter of fact, only members of the second group have been found until now in sponges. Thus, lamellarin O (**103a**) and lamellarin P (**103b**) have been isolated from the marine sponge *Dendrilla cactos*, living in the south Australian Bass Strait (*202*). From the same sponge, but collected on a geographically distinct region (off the coast of New South Wales), lamellarin Q (**104**) and lamellarin R

(**105**) were isolated later (*203*). In general, lamellarins from sponges contain less phenolic HO-groups than lamellarins from other sources. An intriguing structural relationship exists between the lamellarins and the lukianols (**123**). Thus, lamellarin O (**103a**) might be structurally the product of methanolysis of a lukianol A (**123a**) analog.

Both lamellarin O (**103a**) and lamellarin P (**103b**) belong to the lamellarin structure class, previously reported from tunicates and a prosobranch mollusk *Lamellaria* sp. (*cf.* Section 3.3.1.). In **103a** and **103b**, however, the pyrrole ring is not fused to adjacent aromatic rings. The structures of lamellarins O and P have been secured by spectroscopic analysis and partial synthesis (*202*). Several total syntheses of lamellarin O (**103a**) (*204*) and lamellarin Q (**104**) (*204b*) have been reported (*cf.* Scheme 3.17.). Neither lamellarin Q (**104**), the simplest of these aryl pyrroles isolated from marine sponges, nor lamellarin R (**105**) display

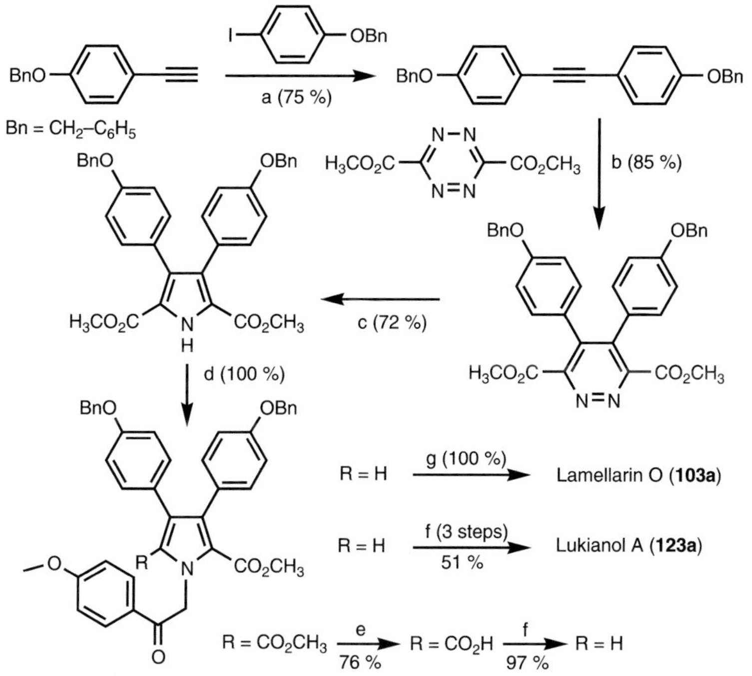

Scheme 3.17. A versatile synthesis of lamellarin O (**103a**), lukianol A (**123a**) and related aryl pyrroles (*204c*): **a**) PdCl$_2$[P(C$_6$H$_5$)$_3$]/CuI/triethylamine]; **b**) dimethyl 1,2,4,5-tetrazine-3,6-dicarboxylate in toluene. **c**) Zn in acetic acid; **d**) 2-bromo-4'-methoxyacetophenone/ K$_2$CO$_3$ in dimethylformamide; **e**) LiOH in tetrahydrofuran/methanol/water; **f**) trifluoro-acetic acid in CH$_2$Cl$_2$; **g**) H$_2$/Pd(C); **f**) *i*: as **e**; *ii*: sodium acetate in acetanhydride; *iii*: BBr$_3$ in CH$_2$Cl$_2$

antibiotic properties. Lamellarin R represents a curious departure from the common structural features of the lamellarins, since its *N*-substituent appears not to be derived from tyrosine, an observation that, if correct, leaves open alternative possibilities for the biosynthetic origin of the *p*-hydroxyphenyl substituent common to so many lamellarins.

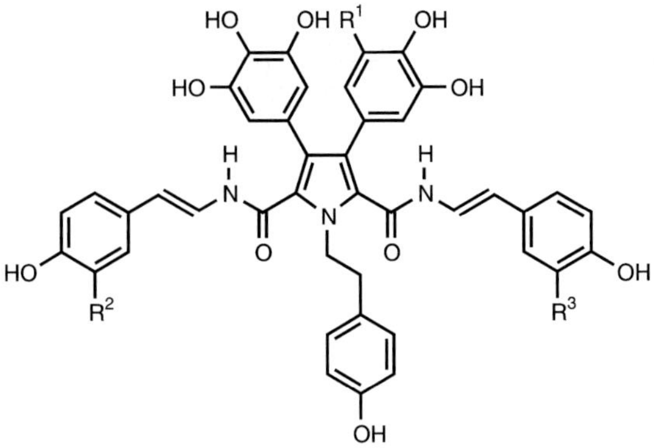

R^1 = OH; R^2 = R^3 = H : Storniamide A (**106a**)
R^1 = R^3 = OH; R^2 = H : Storniamide B (**106b**)
R^1 = H; R^2 = R^3 = OH : Storniamide C (**106c**)
R^1 = R^2 = R^3 = OH : Storniamide D (**106d**)

Structurally related to the lamellarins are the storniamides A–D (**106a–d**), a group of alkaloids isolated from a Patagonian soft, burrowing yellow sponge *Cliona* sp. collected off the coasts of Rio Negro (Argentina). Their name was coined from the name of the biological station of the Instituto de Biologia Marina y Pesquera (Institute of Marine and Fishing Biology) "Alte Storni" which is close to the site of sponge collection. The structures of storniamides A–D (**106a–d**) were established by spectroscopic methods (*205*). Although nona-*O*-methyl storniamide A, has been obtained by synthesis (*204c, 206*), its transformation into the natural product was as yet unsuccessful. Storniamides showed antibiotic activity against several Gram-positive bacteria.

Reminiscent of the structure of lamellarins is that of halitulin (**107**), a bisquinolinylpyrrole alkaloid which was isolated from the marine sponge *Haliclona tulearensis* collected in Sodwana Bay, in Durban, South Africa (*207*). Its structure was elucidated mainly on the basis of spectroscopic data as well as chemical modifications. A rather chemical inspired biogenetic pathway to the halitulin has been suggested (*cf*. Scheme 3.18.). Halitulin was found to be cytotoxic against several types of tumor cells.

References, pp. 140–188

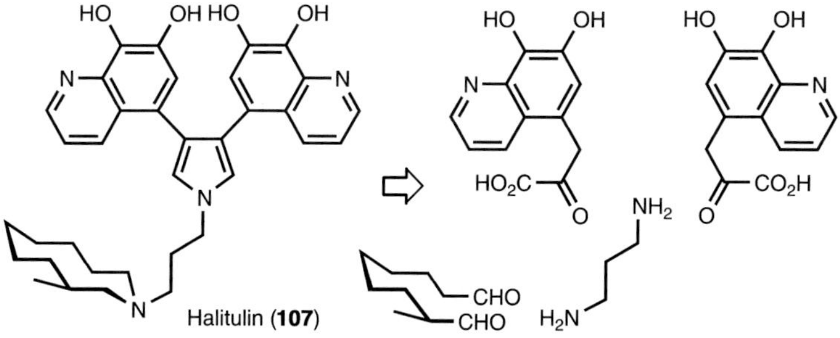

Scheme 3.18. Suggested biosynthetic precursors of halitulin (**107**)

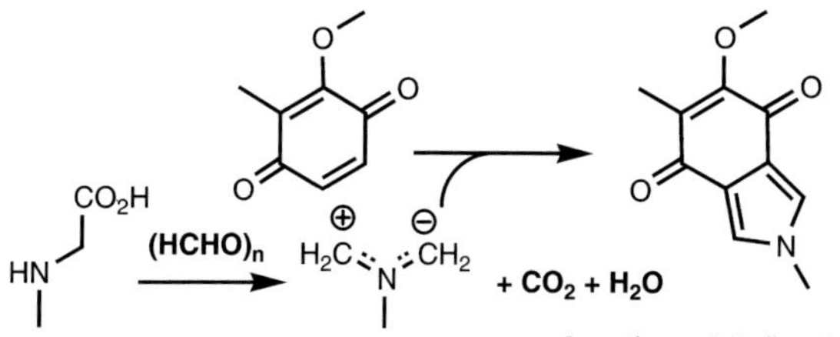

Scheme 3.19. A straightforward synthesis of alkaloid **108** from the marine sponge *Reniera*

A pyrroloquinone derivative – 6-methoxy-2,5-dimethyl-2*H*-4,7-dihydroisoindole-4,7-dione (**108**) – has been isolated, along with other quinoid metabolites; from a bright-blue marine sponge of the genus *Reniera*, found near Isla Grande in Mexico (*208*). The unusual structure of this alkaloid, actually the first naturally occurring isoindole derivative so far known, was confirmed by three independent chemical syntheses (*209*), the most recent one yielding the alkaloid in one-pot reaction with an over-all yield of 81% (*cf.* Scheme 3.19.). The alkaloid is active against some microorganisms.

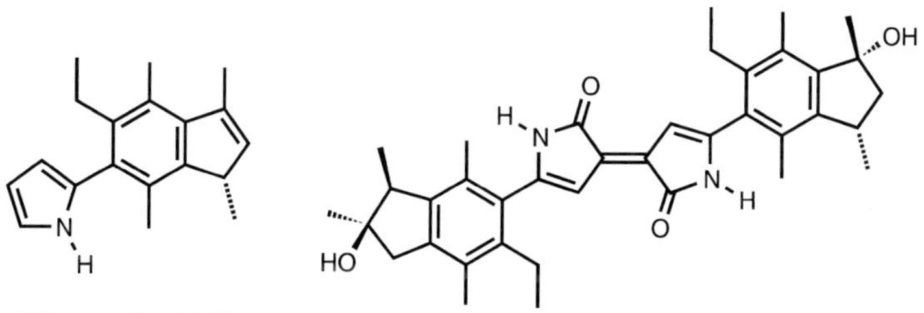

Trikentramine (**109**)   Trikendiol (**110**)

The unusual pyrrole alkaloid, trikentramine (**109**), is an arylpyrrole derivative which was isolated from the sponge *Trikentrion loeve* found at Thiouriba, Dakar in Senegal. Its structure was elucidated by spectroscopic and X-ray diffraction analysis (*210*). From the same species, the deep red isoindigoid pigment (+)-trikendiol (**110**) was obtained later (*211*). Trikendiol may be the dimerization product of a pyrrolin-2-one analog of trikentramine (**109**) (*cf.* Ref. *212*). Until now, the absolute configuration of the indanol moiety of trikendiol is unknown, so the enantiomer shown represents an arbitrary choice. On standing in the presence of traces of acid, **110** undergoes easy dehydration at both indanol moieties thus yielding another red pigment with two additional C=C bonds. The skeleton of trikentramine is believed to be completely new for either marine or terrestrial organisms; no biosynthetic pathways have been postulated.

3.2.5. Hydroxy Pyrroles from Sponges

Until now, only a few pyrrolinone derivatives have been isolated from sponges. Thus, sarcotragin A (**111a**) and sarcotragin B (**111b**) are two trisnorsesterterpene alkaloids of an unusual structure class which have been isolated only recently from a *Sarcotragus* sp. sponge collected from the korean Jaeju Island (*213*). Sarcotragin A (**111a**) contains a phenethylamine lactam residue, whereas in sarcotragin B (**111b**) the latter is replaced by a glycine lactam moiety. Both structures were elucidated by combined chemical and spectroscopic methods. The absolute configuration at C11 of sarcotragin A (**111a**) was determined by oxidative degradation, which yielded (*S*)-2-methylglutaric acid. Due to the unstable nature of compound **111a**, however, the absolute configuration at the C17 asymmetric center still remains to be determined.

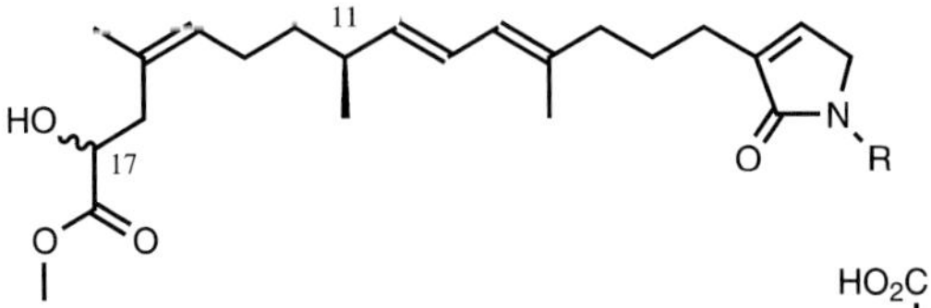

R = CH$_2$-CH$_2$-C$_6$H$_5$: Sarcotragin A (**111a**)

R = CH$_2$-CO$_2^{\ominus}$ Na$^{\oplus}$: Sarcotragin B (**111b**)

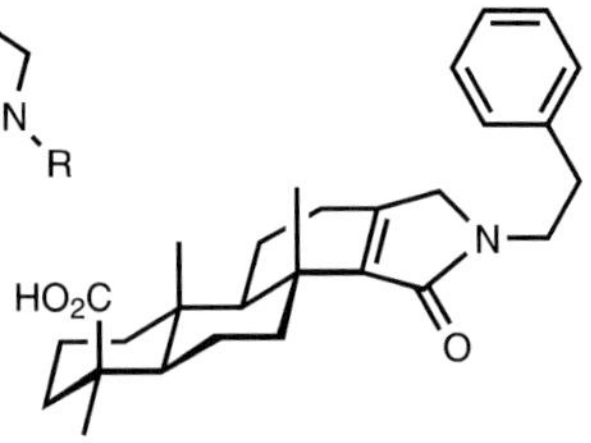

(−)-Haumanamide (**112**)

As the molliorins (*cf.* **101a–d** and **102**), sarcotragins A and B are undoubtedly derived from a linear sesterpene precursor frequently found

References, pp. 140–188

in Dictyoceratid sponges and the corresponding amino acids. To the same group of terpene alkaloids belongs haumanamide (**112**), a diterpene alkaloid containing a fused pyrrolinone ring, which has been isolated from another Dictyoceratid sponge of the genus *Spongia* (*214*).

3.2.6. Tetramic Acid Derivatives from Sponges

Derivatives of tetramic acid (2,4-dioxopyrrolidine) are included in the present review because at least one of their tautomeric forms (*cf.* formula **g** in Scheme 3.20.) has the structure of a dihydroxy pyrrole (*215*). It is well known, however, that the lactam tautomer of 2-hydroxypyrroles is preferred, in general, to the hydroxypyrrole tautomer. Contrarily, 3-hydroxypryrroles occur often as such and not in the tautomeric keto form. Derivatives of 3-acetyltetramic acid may exist in nine different tautomeric forms resulting from the shift of the ring-bound H-atoms (*cf.* Scheme 3.20.). According to NMR spectroscopic data, however, the energetically preferred tautomers are pyrrolindiones with enolized acetyl group (**b** and **e** in Scheme 3.20.) *vs* the corresponding

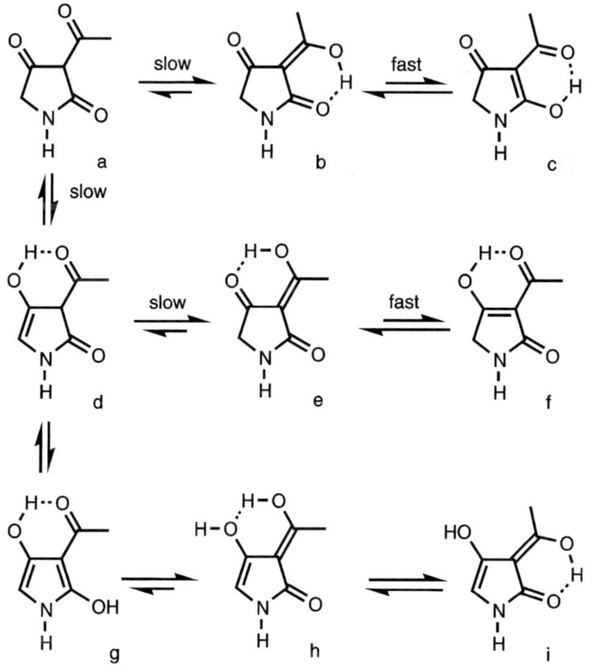

Scheme 3.20. Possible tautomeric forms of 3-acetyltetramic acid

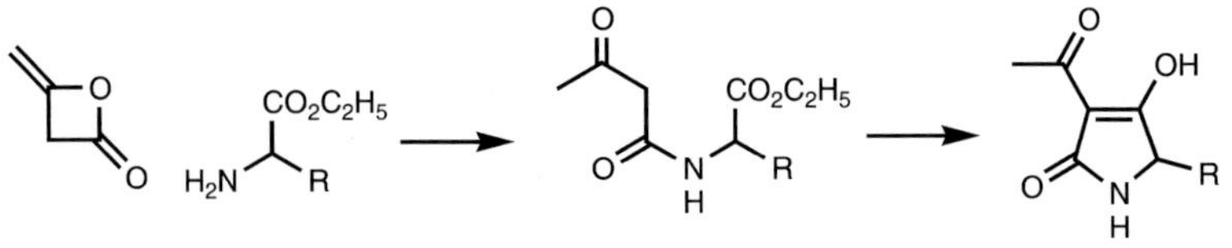

Scheme 3.21. A general synthesis of α-acetyltetramic acids

endo-cyclic enols **c** and **f**, respectively. The approximate ratios of the individual tautomers **b:c:e:f** for simple 3-acetyltetramic acids have been determined to be 80:0:15:5 (*216*). The predominance of the higher frequency C2 hydrogen bonded carbonyl signal led to the conclusion that the *exo*-enol **b** is the main tautomeric form. Nevertheless, structures **b** and **f** are undiscriminatingly used in the literature on representing the structures of 3-acetyltetramic acid derivatives.

Typically, the UV spectrum of acyltetramic acids in acidic medium shows an absorption maximum at $\lambda_{max} = 275 \pm 10$ nm, which does not shift significantly in going from acidic to alkaline pH. Contrarily to enols, however, a new maximun of approximately equal intensity at $\lambda_{max} = 230$–260 nm appears in the basic or near neutral pH-range (*217*).

General syntheses of 3-acetyltetramic acids consist either in the reaction of the suitable α-amino acid ester with diketene and subsequent Dieckmann cyclization of the obtained *N*-acetoacetyl-α-amino acid derivative (*217, 218*) (*cf.* Scheme 3.21.) or in the acylation of pyrrolidine-2,4-diones (*219*). Some derivatives of 3-acyltetramic acid have been reported to be DNA gyrase inhibitors, although this enzyme has not been established as the target of these agent in whole cells (*220*).

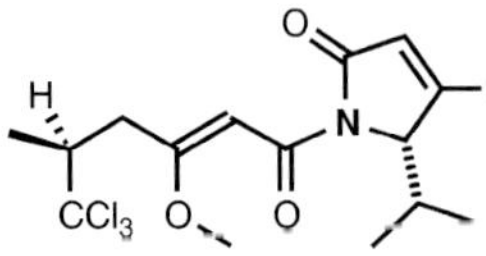

Dysidin (**113**)

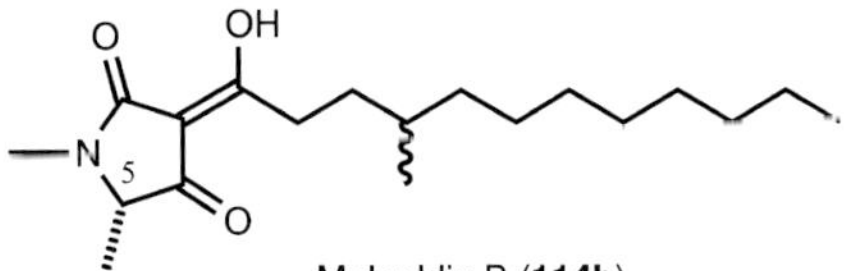

Melophlin B (**114b**)

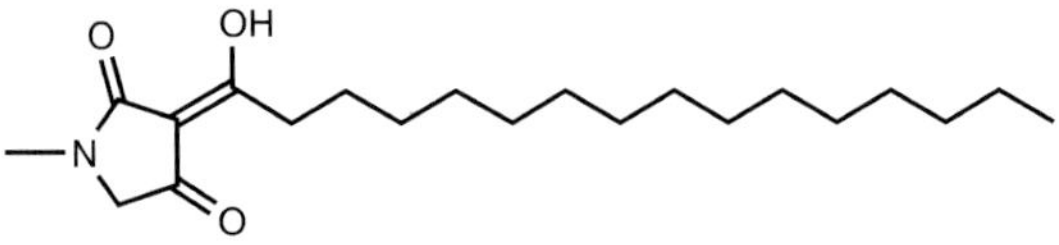

Melophlin A (**114a**)

Tetramic acid derivatives are more common in fungi (*cf.* Section 5.5.) and bacteria (*cf.* Section 6.9.) than in taxonomically higher organisms.

References, pp. 140–188

Dysidin (**113**) has been isolated from the sponge *Dysidea herbacea* (*221*). Its structure was elucidated by NMR-, IR- UV- and mass spectroscopy, as well as degradation experiments. The absolute configuration has been established by X-ray diffraction analysis.

Melophlins A (**114a**) and B (**114b**) are two tetramic acid derivatives of potential interest in cancer chemotherapy. They have been recently isolated from the marine sponge *Melophlus sarassinorum* collected at Spermonde Islands, Ujung Pandang (Indonesia), and they structures were elucidated by spectroscopic methods (IR, UV, HR-MS, as well as ^{1}H- and ^{13}C-NMR spectroscopy). Moreover, the absolute configuration of **114b** at C5 was determined by oxidative degradation to *S*-alanine. As other tetramic acid derivatives, **114a** and **114b** occur as a mixture of two tautomers (corresponding to **b** and **e** in Scheme 3.20.) in a ratio of ca. 9:1 (*222*).

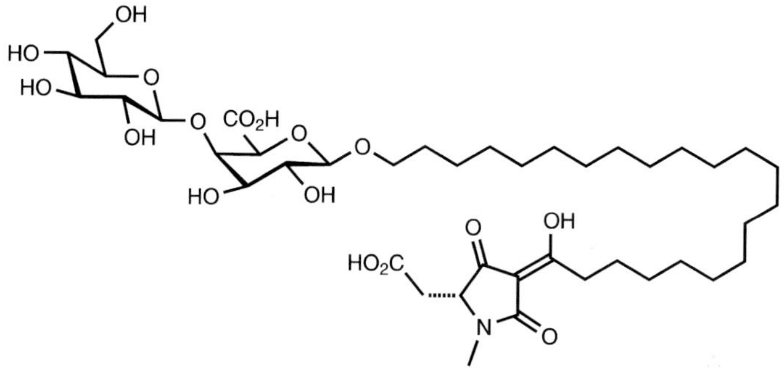

Ancorinoside A (**115**)

Ancorinoside A (**115**) is a acyltetramic acid glycoside which has been isolated from the marine sponge *Ancorina* sp. collected off the coast of Tokushima, Japan. Its structure was determined by NMR spectroscopic methods (*223*).

Aurantoside A (**116a**) and aurantoside B (**116b**) have been isolated, both as orange (Latin *auranticus* = orange) amorphous powders, from the marine sponge *Theonella swinhoei*. Their structures were established by FAB mass spectrometric and NMR spectroscopic data (*224*). Thus, the presence of a conjugated hexaene system was inferred from the COSY, HMQC, and HMBC spectra. The two chlorine atoms were placed on C17 and C19 on the basis of their ^{13}C chemical shifts of δ 129.8 and 137.4 ppm, respectively. According to the ^{1}H–^{1}H coupling constants and NOESY data, the double bonds have all-*trans* geometry. However, the geometry of the terminal C18=C19-bond has been later established to

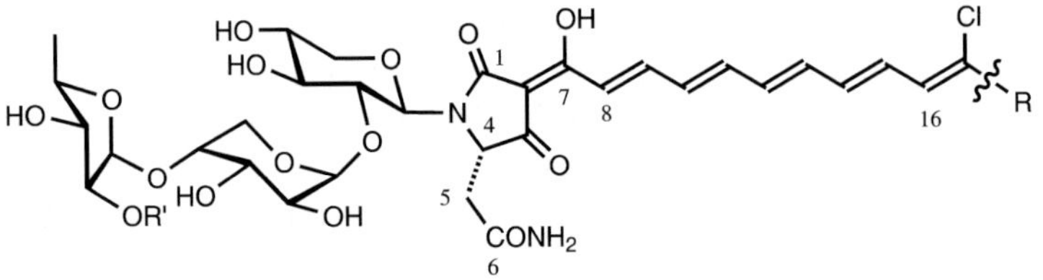

R = R₁; R' = CH₃ : (−)-Aurantoside A (**116a**)
R = R₁; R' = H : (−)-Aurantoside B (**116b**)
R = R₂; R' = H : (−)-Aurantoside D (**116c**)
R = R₂; R' = CH₃ : (−)-Aurantoside E (**116d**)
R = R₃; R' = CH₃ : (−)-Aurantoside F (**116e**)

R = R₄; R' = CH₃ : (−)-Rubroside A (**116f**)
R = R₅; R' = CH₃ : (−)-Rubroside B (**116g**)
R = R₅; R' = H : (−)-Rubroside C (**116h**)
R = R₆; R' = CH₃ : (−)-Rubroside F (**116i**)
R = R₆; R' = H : (−)-Rubroside G (**116j**)

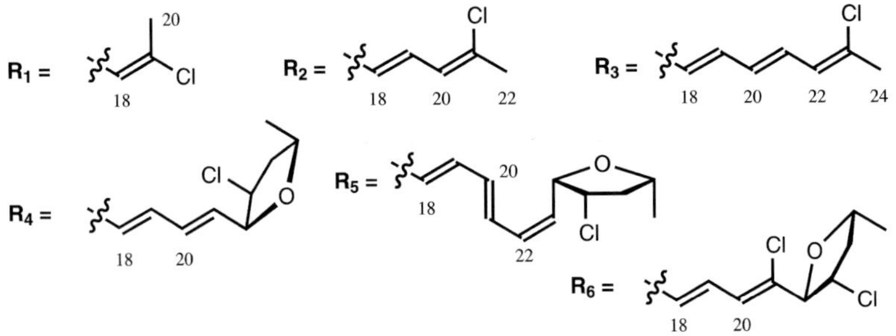

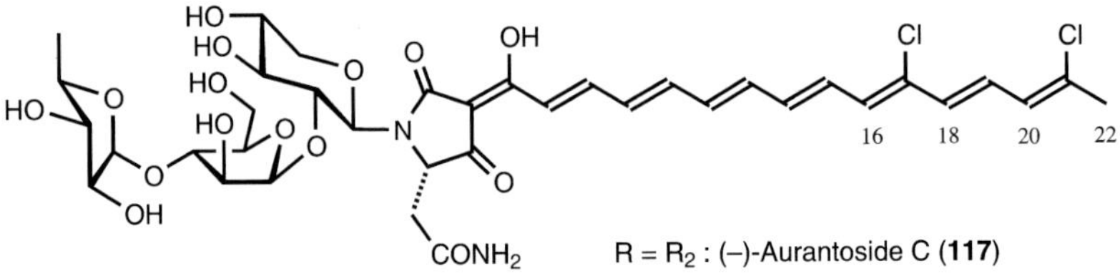

R = R₂ : (−)-Aurantoside C (**117**)

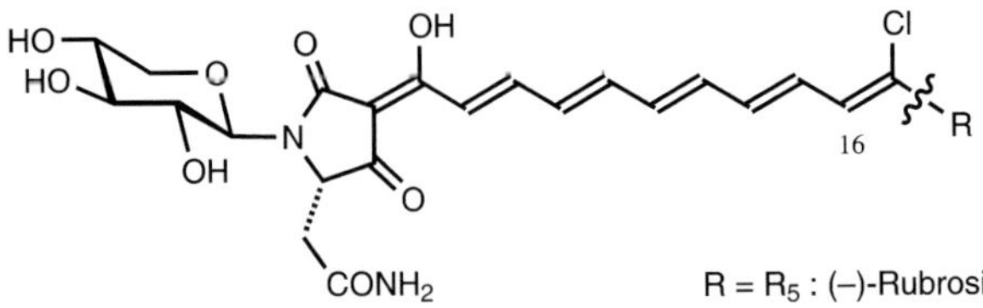

R = R₅ : (−)-Rubroside D (**118a**)
R = R₆ : (−)-Rubroside H (**118b**)

be (*E*) (*cf.* Ref. *225*). Aurantoside B (**116b**) is a tetramic acid glycoside, which contains D-xylopyranose, D-arabinopyranose and 5-deoxyarabinofuranose residues. The C2-hydroxy group of the latter is in aurantoside A (**116a**) methylated.

References, pp. 140–188

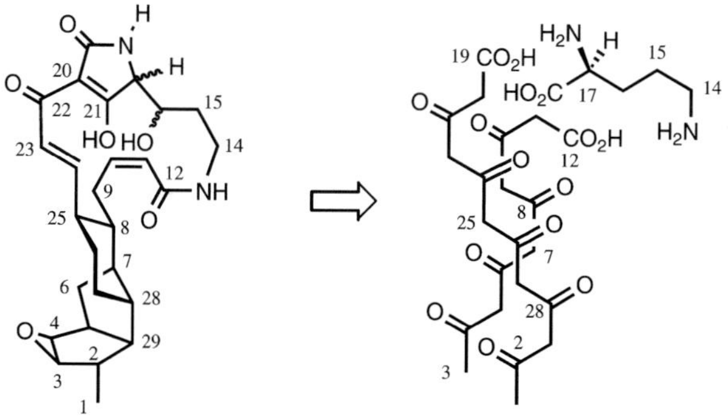

Discodermide (**119**)

Scheme 3.22. Possible biogenetic orign of discodermide (**119**)

Later on, aurantoside C (**117**) and aurantosides D, E, and F (**116c–e**) have been isolated from the marine sponges *Homophymia conferta* (*226*) and *Siliquariaspongia japonica* (*225*), respectively. The latter sponge contains also rubrosides A–C (**116f–h**) as well as rubroside D (**118a**), rubroside H (**118b**), rubroside F (**116i**), and rubroside G (**116j**), all of them structurally related to the aurantosides. The absolute configuration of the tetramic acid and sugar moieties in all rubrosides was deduced by chiral GC analysis of the respective products of chemical degradation (*227*).

As other tetramic acid derivatives found in sponges (see below), aurantosides and rubrosides are probably secondary metabolites from microbial symbionts. Aurantosides and most rubrosides exhibit potent antifungal activity against *Aspergillus fumigatus* and *Candida albicans*. Aurantosides A and B show high activity against certain leukemia cells.

Discodermide (**119**) is an enoyltetramic acid derivative which was isolated from the Caribbean marine sponge *Discoderma dissoluta*. Its structure was elucidated through a combination of spectroscopic techniques, in particular NMR spectroscopy (*228*). It appears likely that discodermide, as ikarugamycin (**292**), is biosynthetically derived from two pentaacetate chains and ornithine, which contributes to the N-C5-C4-fragment of the tetramic acid subunit. The C2-C3-fragment of the pyrrolidindione ring of the tetramic acid subunit originates probably from one of the pentacetate chains (*cf.* Scheme 3.22.). Discodermide (**119**) is antifungal, particularly against *Candida albicans*, and cytotoxic.

Cylindramide (**120**) is a cytotoxic tetramic acid derivative, which has been isolated from the marine sponge *Halichondria cylindrata*. Its structure was determined by interpretation of 2D NMR spectra as a

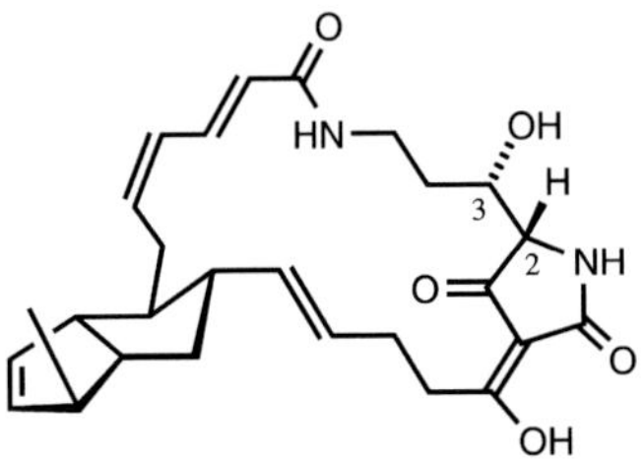

Cylindramide (**120**)

macrocyclic lactam including an acyltetramic acid and a trisubstituted bicyclo[3.3.0]octene moiety (*229*). The absolute configuration is undetermined except at C2 and C3 which were both assigned as (*S*) because, as in the case of alteramide A (**289**), L-β-hydroxyornithine was obtained after oxidative ozonolysis of the antibiotic. Like alteramide A, **120** is most likely a metabolite of symbiotic associated bacteria and not of the sponge.

3.3. Pyrroles from Other Invertebrates

3.3.1. Aryl Pyrroles

As mentioned before (*cf.* Section 3.2.4.), lamellarins belong to a group of pyrrole alkaloids which have been isolated from widely varying locations and marine organisms. Lamellarins which have been isolated from invertebrates other than sponges are characterized either by a 5,6-dihydrobenz[*g*]indolizine (**121**) or by a benz[*g*]indolizine (**122**) substructure, in which the pyrrole ring is connected to an adjacent aromatic ring by a lactone bridge.

Lamellarins A–D were first obtained in 1985 from the pacific mollusc of the genus *Lamellaria* from Palau (Micronesian Caroline Islands) (*230*). Later on, in 1988, four additional members, lamellarins E–H were found in the sea squirt (ascidian) *Didemnun chartaceum*, which was collected near Aldabra Atoll, Republic of the Seychelles in the Indian Ocean (*231*). Further six new alkaloids, lamellarins I–N were isolated, along with lamellarins A–D, from a Great Barrier Reef colonial ascidian, *Didemnun* sp. by Carroll *et al.* (*232*) who speculated that the Palauan mollusk *Lamellaria* sp. may have sequestered lamellarins from a didemnid ascidian food source. Also from an Australian tunicate of the genus *Didemnun* lamellarin S (*233*) and lamellarin Z (*234*) were isolated later. Moreover, lamellarin N together with lamellarins T–Y were obtained

References, pp. 140–188

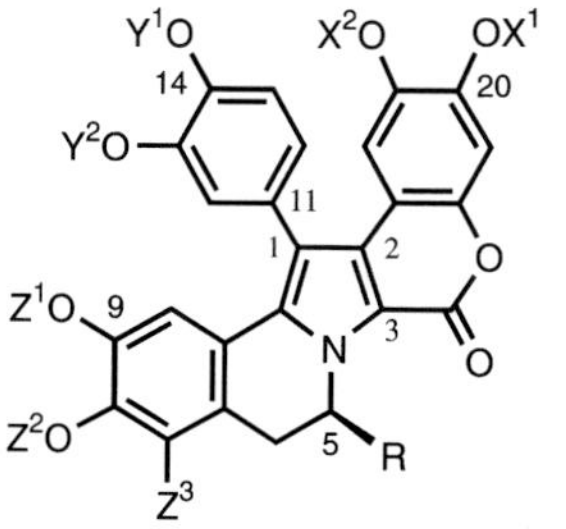

Lamellarins (Type I) (121)

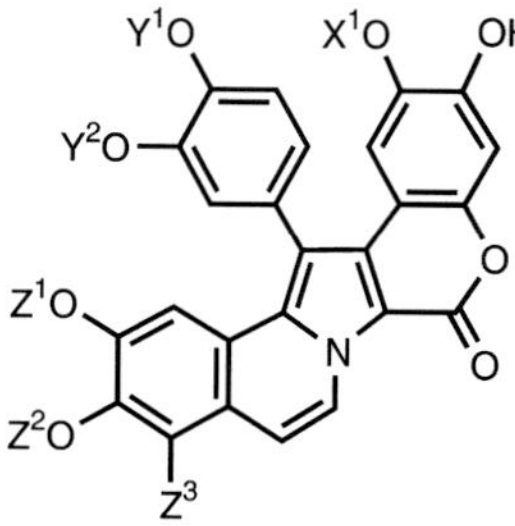

Lamellarins (Type II) (122)

	X^1	X^2	Y^1	Y^2	Z^1	Z^2	Z^3	R
A	H	CH_3	H	CH_3	CH_3	CH_3	OCH_3	OH
C	H	CH_3	H	CH_3	CH_3	CH_3	OCH_3	H
E	H	CH_3	CH_3	H	CH_3	CH_3	OH	H
F	H	CH_3	CH_3	CH_3	CH_3	CH_3	H	H
G	CH_3	H	CH_3	H	CH_3	H	H	H
I	H	CH_3	CH_3	CH_3	CH_3	CH_3	OCH_3	H
J	H	CH_3	CH_3	CH_3	CH_3	H	H	H
K	H	CH_3	H	CH_3	CH_3	CH_3	OH	H
L	H	CH_3	CH_3	H	CH_3	H	H	H
S	H	H	H	H	CH_3	H	H	H
T	H	CH_3	CH_3	H	CH_3	CH_3	OCH_3	H
U	H	CH_3	CH_3	H	CH_3	CH_3	H	H
V	H	CH_3	CH_3	H	CH_3	CH_3	OCH_3	OH
Y	H	CH_3	CH_3	H	H	CH_3	H	H
Z	CH_3	H	H	H	CH_3	H	H	H

	X^1	Y^1	Y^1	Z^1	Z^2	Z^3
B	CH_3	H	CH_3	CH_3	CH_3	OCH_3
D	CH_3	H	CH_3	CH_3	H	H
H	H	H	H	H	H	H
M	CH_3	H	CH_3	CH_3	CH_3	OH
N	CH_3	CH_3	H	CH_3	H	H
W	CH_3	CH_3	H	CH_3	CH_3	OCH_3
X	CH_3	CH_3	H	CH_3	CH_3	OH
α	CH_3	H	CH_3	CH_3	CH_3	H

(some of them as the 20-sulfate derivatives) from an unidentified ascidian collected near Trivandrum at the Arabian Sea coast of India (*235*). From the same source, the 20-sulfate of lamellarin α has been isolated, as the latest in this series of related benz[*g*]indolizine derivatives. In cell cultures, lamellarin α 20-sulfate inhibits HIV-1 replication, through binding to the integrase protein of the virus (*236*). A comprehensive review on the structures, origins, biogenesis, synthesis

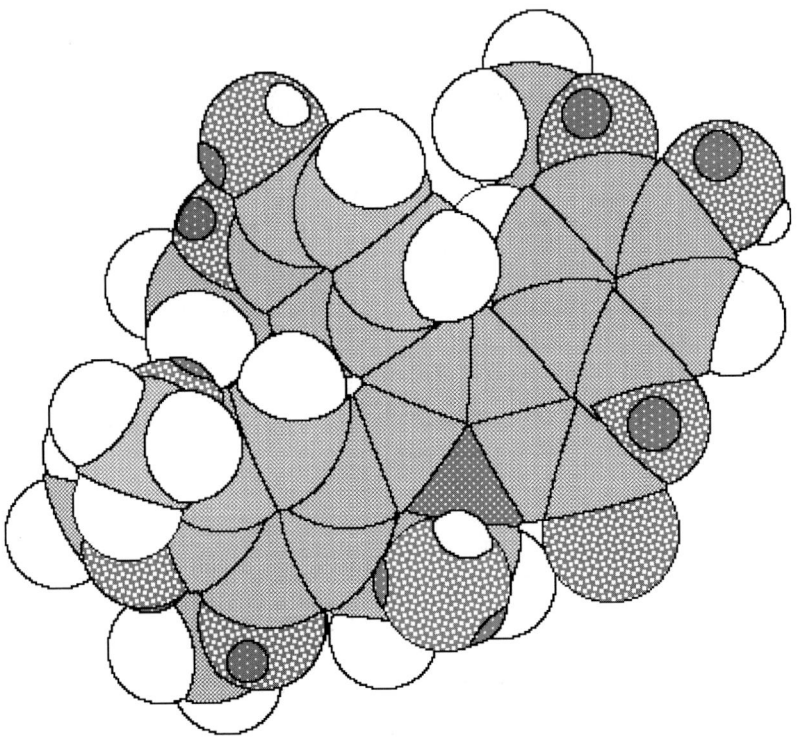

Fig. 1. Computer-generated perspective drawing of the X-ray structure of lamellarin A

and bioactivity of the lemellarins has been published recently by Urban et al. (*201*).

The structures of lamellarin A (*230*) and lamellarin E (*231*) have been determined by X-ray diffraction analysis (Fig. 1). Characterization of the other lamellarins was accomplished by NMR spectral studies (*cf.* Ref. 237). Lamellarins D and H (*238*), as well as lamellarin G (as trimethyl ether) (*239*) lamellarin K (*240*) and lamellarin L (*241*) have been obtained by total synthesis. Lamellarins exist in solution as racemic mixtures of two enantiomeric atropisomers due to restriction of the rotation about the C1–C11 bond. Because molecular mechanics calculations furnish the unrealistic value of $>2512\,\mathrm{kJ\,mol^{-1}}$ for the rotation barrier, it has been suggested that interconversion is only possible after opening of the carbinolamine ring (*230*). However, lamellarin S displays a positive optical rotation, indicating it to be at least enantiomerically enriched. Repeated optical rotation measurements over several months suggested slow racemisation of lamellarin S, with a half-live calculated to be about 90 days. Molecular modeling calculations indicated that the lowest energy rotation barrier about the C1–C11 bond in lamellarin S is 84 kJ mol^{-1}, with both atropisomers being of comparable energy. Nevertheless, the absolute configuration of lamellarin S remains unassigned (*233*).

Before the discovery of lamellarins nothing was known of the diet of prosobranch mollusks, which are vulnerable because they have reduced shells covered by a mantle of flesh. They are believed to feed on chemically rich ascidians and to use these diet-derived compounds as chemical defense against predators (*231*). However, the lamellarins do not resemble any known metabolites or marine natural products, and their biosynthetic pathway, which is presumably derived from tyrosine *via* 3,4-dihydroxyphenylalanine (DOPA), is unknown. Some lamellarins have been shown to possess interesting cytostatic properties.

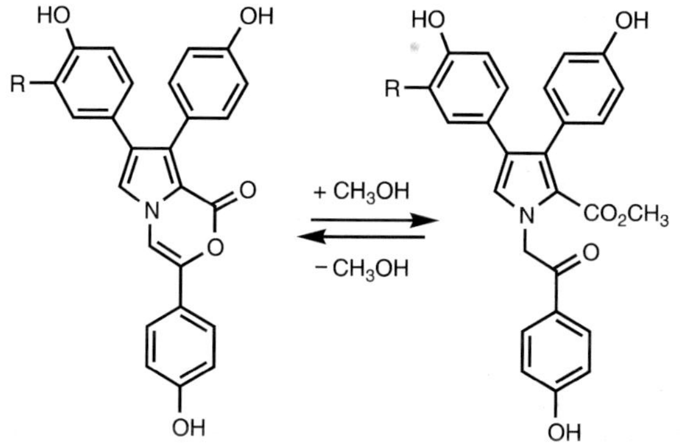

R = H : Lukianol A (**123a**)
R = I : Lukianol B (**123b**)

R = H : *O*-Demethyl-lamellarin O (**124**)

An intriguing structural relationship exists between the lamellarins and the lukianols, which have been isolated from an unidentified encrusting tunicate collected in the lagoon of the Palmyra atoll (*242*). Thus, lukianol A (**123a**) may be considered as the intramolecular enol ester of a lamellarin O analog (**124**). The structures of lukianol A (**123a**) and lukianol B (**123b**), which was isolated as a minor constituent, have been suggested mainly on the basis of high resolution mass spectrometric as well as ^{1}H- and ^{13}C-NMR spectroscopic data. Different chemical syntheses have confirmed the structure of lukianol A (*cf.* Scheme 3.17.) (*204, 243*).

A close structural relationship exists also between lamellarins and ningalins. Thus, ningalin B (**126**) could afford *O*-C9 demethyl lamellarin

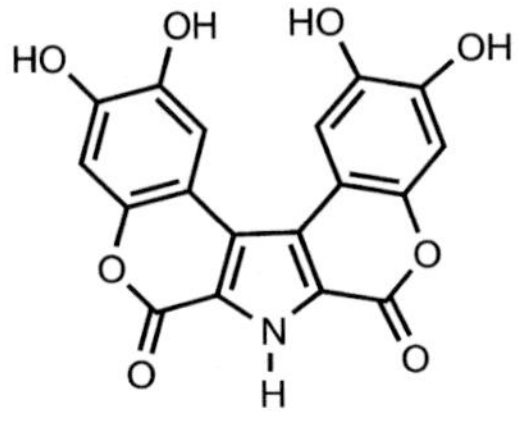

Ningalin A (**125**)

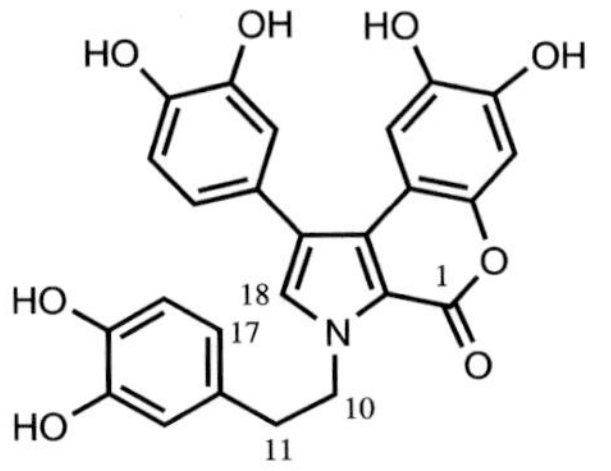

Ningalin B (**126**)

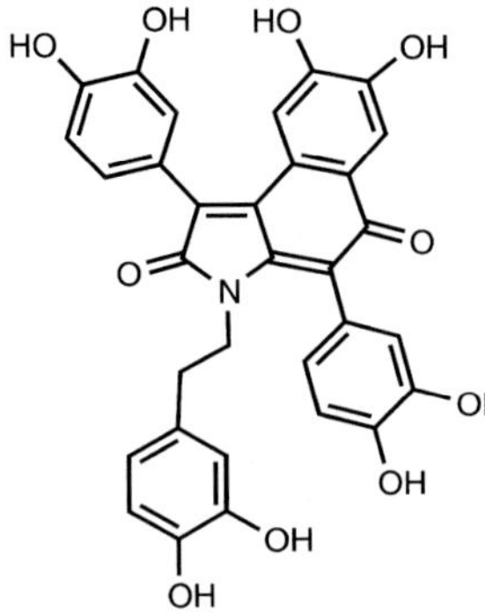

Ningalin C (**127**)

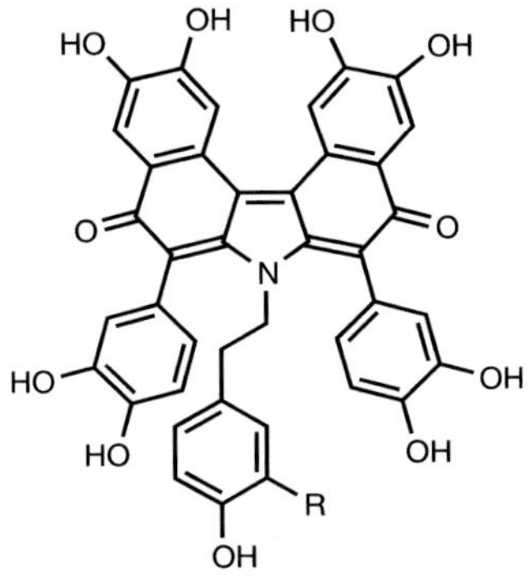

R = OH: Ningalin D (**128**)
R = H: Purpurone (**128b**)

S by ring closure between C17 and C18. Ningalin B has been isolated together with ningalins A (**125**), C (**127**), and D (**128a**) from an ascidian of the genus *Didemnum* collected in western Australia near Ningaloo Reef (*244*). Structurally, ningalin D is closely related to purpurone (**128b**), a purple compound reported from the marine sponge *Iotrochota* sp. collected in Palau (*245*). The structures of the ningalins were elucidated by interpretation of overall spectral data and 2D NMR correlation methods (*244*). The structures of ningalin A (**125**) (*204c*), ningalin B (**126**) (*246*), ningalin C (**127**) (*247*), and purpurone (**128b**) (*247a*) have been confirmed by total syntheses.

As the lamellarins, ningalins appear to be derived from condensation of 3,4-dihydroxyphenylalanine (DOPA). Because the ningalins are related to other metal binding *ortho*-catechols, it is conceivable that they too participate in the chelation of metal ions (*e.g.* V^{5+} or Fe^{3+}) which is characteristic of this class of marine invertebrates.

Another group of arylpyrrole metabolites related to the lamellarins is composed by polycitone A (**129a**) and polycitrins A (**130a**) and B (**130b**) which have been isolated from the marine ascidian *Polycitor* sp. found in

References, pp. 140–188

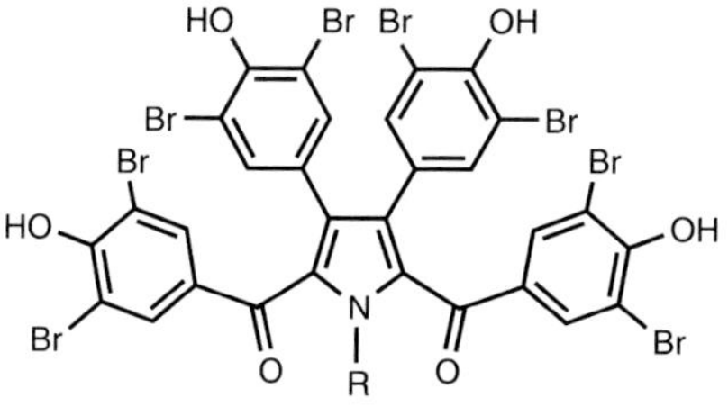

R = –[CH$_2$]$_2$–⟨C$_6$H$_4$⟩–OH : Polycitone A (**129a**)

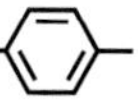

R = H : Polycitone B (**129b**)

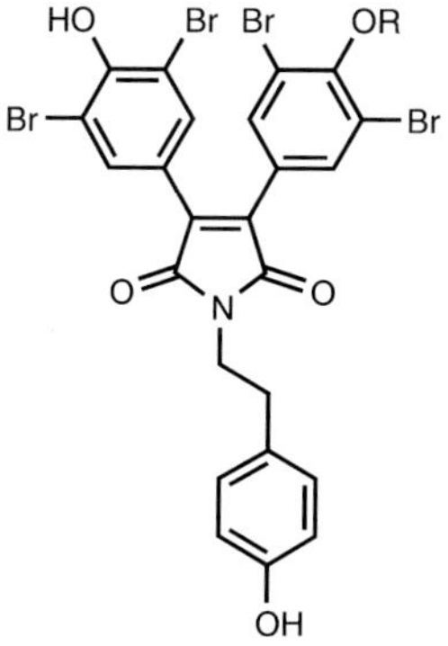

R = H : Polycitrin A (**130a**)

R = CH$_3$: Polycitrin B (**130b**)

Sodwana Bay, South Africa (*248a*). Later, polycitone B (**129b**), which lacks the phenylethyl substituent at the N-atom of the pyrrole ring, has been isolated as a minor metabolite of *Polycitor africanus* from Madagascar (*248b*). Their structures were established mainly on the basis of NMR spectroscopic data and, in the case of polycitone A, also by single-crystal X-ray diffraction analysis, which was carried out on its penta-*O*-methyl derivative. The structures of polycitrin A (*249*) and polycitrin B (*250*) have been confirmed by total synthesis. Neither the biogenesis nor the biological roles of polycitones and polycitrins, the content of which in the ascidian is about 10^{-2} times smaller than that of polycitone A, are so far known. A chemical synthesis of polycitrin A from 3-arylpyruvic acids (*249*) suggests, however, a possible biogenetic pathway.

The alkaloid rigidin (**131**), the structure of which is reminiscent of those of lamellarins (s. above), has been isolated from the Okinawan tunicate *Eudistoma* cf. *rigida* (*251*). Rigidin seems to inhibit calmodulin-activated brain phosphodiesterase. However, it is not clear how

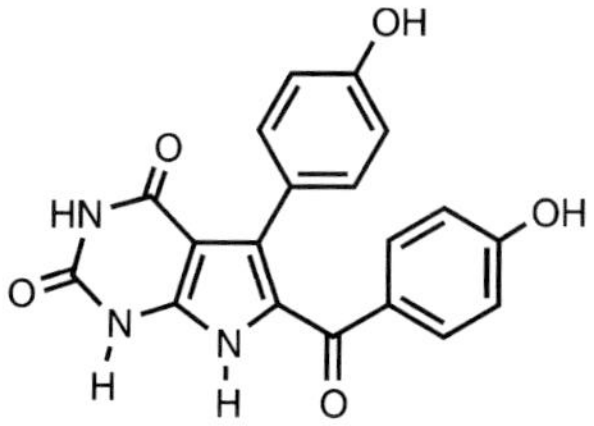

Rigidin (**131**)

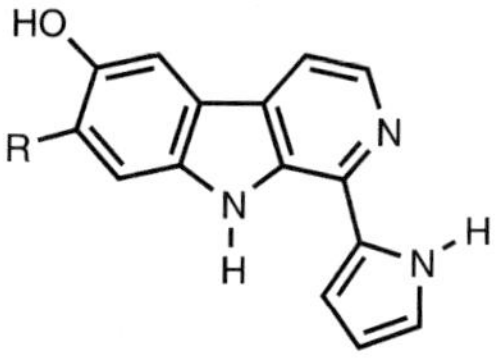

R = Br : Eudistomin A (**132a**)

R = H : Eudistomin M (**132b**)

rigidin functions biologically. Rigidin has been synthesized both from 6-chlorouracil (*252*) and from 2,4-dimethoxypyrrolo[2,3-*d*] pyrimidine (*253*), as starting materials, with overall yields of 26% (9 steps) and 11% (6 steps) respectively.

3.3.2. Other Pyrrole Derivatives

Eudistomin A (**132a**) and eudistomin M (**132b**) are the sole members of a series of metabolites containing the β-carboline ring system, which contain a pyrrole ring as a substituent. They have been isolated from the Caribbean tunicate *Eudistoma olivaceum*. According with high-resolution FAB-mass spectrometric data and the UV, ^{1}H-, and ^{13}C-NMR spectra of the corresponding acetates both structures contain the β-carboline ring system substituted by a 2-pyrrolyl group at C1 (*254*). The structure of eudistomin M (**132b**) has been confirmed by synthesis (*254*).

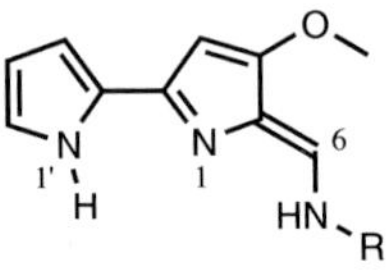

R = H : Tambjamine A (**133a**)
R = CH$_2$–CH(CH$_3$)$_2$: Tambjamine C (**133b**)
R = CH$_2$–CH$_2$–C$_6$H$_5$: Tambjamine F (**133c**)

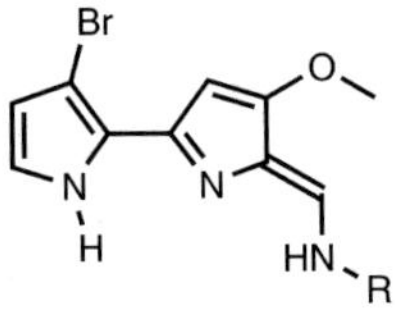

R = CH$_2$–CH(CH$_3$)$_2$: Tambjamine D (**134a**)
R = CH$_2$–CH$_3$: Tambjamine E (**134b**)

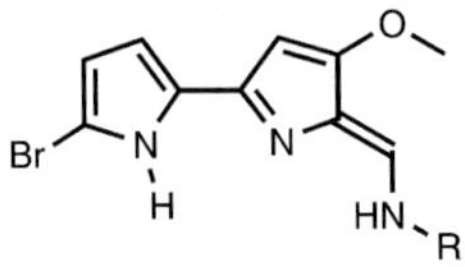

R = H : Tambjamine B (**135a**)
R = CH$_2$–CH$_3$: Tambjamine G (**135b**)
R = CH$_2$–CH$_2$–CH$_3$: Tambjamine H (**135c**)
R = CH$_2$–CH(CH$_3$)$_2$: Tambjamine I (**135d**)
R = CH$_2$–CH(CH$_3$)–CH$_2$–CH$_3$: Tambjamine J (**135e**)

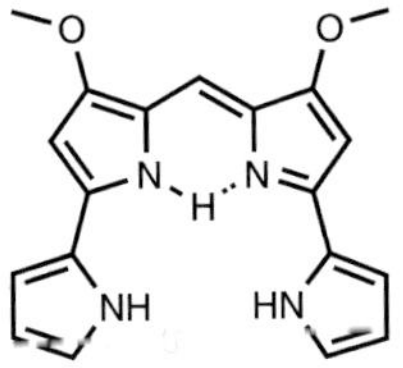

4-Methoxy-5-[(3-methoxy-5-pyrrol-2-yl-2*H*-pyrrol-2-ylidene)methyl]-2,2'-bipyrrole (**136**)

A series of bipyrroles, tambjamines A (**133a**), B (**135a**), C (**133b**), and D (**134a**) have been isolated from the nembrothid nudibranchs *Tambje eliora* and *T. abdere* as well as from the green cheilostome bryozoan *Sessibugula translucens* (*255*). Tambjamine E (**134b**) and tambjamine F (**133c**) occur, together with tambjamines A and C, in the ascidian *Atapozoa* sp. and its nudibranch predators, and were used to

References, pp. 140–188

establish the first chemical link in the predator-prey relationship between ascidians and nudibranchs (*256*). More recently, four new tambjamines – tambjamines G-J (**135b-e**) – have been obtained, along with tambjamines C and E, from the Tasmanian bryozoan *Bugula dentata* (*257*).

The same organism also contain the blue tetrapyrrole **136** (so-called "Bugula pyrrole"), which has been isolated also from *B. dentata* collected in the Gulf of Sagami, Japan (*258*), as well as from an Australian ascidian (*259*) and from *Nembrotha kubaryana* (a nudibranch grazing on Atapozoa), collected in diverse areas of the western Pacific shallow water habitats at several islands in the central Philippines (*256*). This alkaloid is antimicrobial against Gram-positive and -negative bacteria. The structure was solved spectroscopically and it is thought to be identical with a tetrapyrrole produced from a mutant strain of *Serratia marcencens* (*cf.* Section 6.7.1.). Therefore, its is possible that either the bryozoan or microorganisms associated with the bryozoan may biosynthesize **136**. It could also be derived from prodigiosin-producing bacteria in the diet.

The *N*-C6-dodecyl derivative of tambjamine A (**263**) has been obtained from a terrestrial *Streptomyces* sp. The variety of organisms in which tambjamines are found raises questions of possible biosynthesis by symbiotic microorganisms, but Lindquist and Fenical (*256*) believe that at least in the ascidian *Atapozoa* this animal produces them itself.

Tambjamines seem to be involved in complex ecological relationsships since they undoubtedly originate from the dietary source of marine snails. On the other hand, tambjamines serve to the small nudibranchs as defensive compounds against predators, specially against the large carnivorous nudibranch *Roboastra tigris*, which in turn contains the tambjamines originating from its prey. *Tambje abdere* when attacked, produces a yellow mucus that often caused *R. tigris* to break off its attack, while *T. eliora* attempts to escape without apparent use of defensive secretions. Accordingly, *R. tigris* seems to prefer to eat *T. eliora* although it is capable of following the tambjamine containing slime trail let down by the *Tambje* species (*260*).

Tambjamines are yellow to yellowish-brown pigments located in the granular amebocyte blood cells of tambjamine-producing ascidians, and they become concentrated in the nudibranches which are the ascidians' predators. The structures of tamjamines, which has been elucidated by chemical and spectroscopic comparisons with bipyrrolealdehydes of known structure, are closely related to those of prodigiosins (*cf.* Section 6.7.). Since the bipyrroles turn green on standing, it is suspected that the green color of the above mentioned bryozoan originates from dimers of these compounds. Tambjamines A and B inhibit cell division at 1 µg/cm^3 in the fertilized sea urchin egg assay and showed moderate antimicrobial activity at 50 µg/disc in the disc assay method against *Eschericia coli*, *Staphylococcus aureus*, *Bacillus subtilis*, and *Vibrio anguillarum*. As prodigiosin (**267**), tamjamine E (**134b**) binds calf thymus DNA, *in vitro*, by intercalation with a preference for AT sites (*261*).

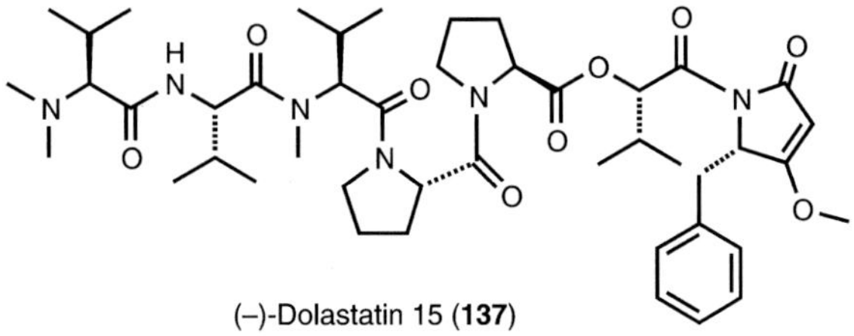

(−)-Dolastatin 15 (**137**)

Dolastatin 15 (**137**) is a tetramic acid derivative containing a depsipeptide chain which was obtained from the Western Indian Ocean shell-less mollusk *Dolabella auriculaira* (*262*). The structure was elucidated by employing primarily high-field 2D-NMR spectroscopy and high-resolution mass spectrometry, which gave the sequence of amino acid residues and identified the pyrrolin-2-one substructure derived from phenylalanine. Total synthesis of **137** using components possessing only the (*S*) configuration gave material that was identical with the natural compound, thus the *all-S* absolute configuration can be inferred (*263, 264*). Dolastatin 15 has been found to strongly and selectively inhibit the growth of 30 human cancer cells lines *in vitro* (*265*).

3.4. Pyrroles from Protozoa

Keronopsins A and B are chemical defense substances of the marine ciliate *Pseudokeronopsis rubra*, a protozoon which is colored deep red and exerts a powerful combination of antibiotic, deterrent, and antifeedant effects (*266*). The organism stores keronopsin A_1 (**138a**) and keronopsin A_2 (**138b**) in the form of sulfate esters, which only on destruction of the cells are converted enzymatically into the more toxic corresponding keronopsins B_1 (**138c**) and B_2 (**138d**). Keronopsins A_1 and A_2 are, as crude compounds, unstable, and are very prone to polymerisation under acidic conditions or in the solid state.

References, pp. 140–188

The structure of the keropsins is analog to that of rumbrin (**194**) but with (*E*) geometry at the C7=C8 bond,

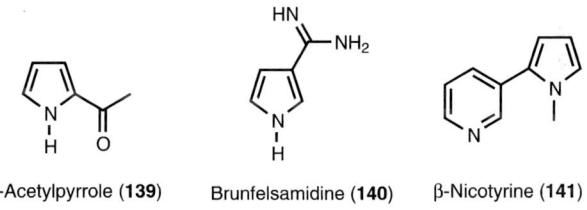

R^1 = H; R^2 = –$SO_3^{\ominus}$ $Na^{\oplus}$: Keronopsin A_1 (**138a**)

R^1 = Br; R^2 = –$SO_3^{\ominus}$ $Na^{\oplus}$: Keronopsin A_2 (**138b**)

R^1 = R^2 = H : Keronopsin B_1 (**138c**)

R^1 = Br; R^2 = H : Keronopsin B_2 (**138d**)

4. Pyrroles from Plants

The most frequently encountered monopyrroles in plants are 2-acetylpyrrole and pyrrole-2-carboxylic acid, the latter often as a minor part of more complex carbocyclic structures of terpenenoid or polyketide origin.

2-Acetylpyrrole (**139**)   Brunfelsamidine (**140**)   β-Nicotyrine (**141**)

2-Acetylpyrrole (**139**) has been identified among the volatile components of different plants, like valerian (*Valeriana officinalis*) (*267, 268*) (*cf.* Ref. *269*), black tea (*Camellia thea*) (*270*), Japanese hops (*271*), and in tobacco leaves (*272, 273*) among others (*e.g.* in the Chinese drugs "Kukoshi" (*Lycium chinense*) (*274*) and "Botanpi" (*Paeonia moutan*) (*275*). It contributes to many aromas including roasted cocoa (*276, 277*) coffee beans (*278, 279*), and tobacco smoke. Moreover, **139** which has been isolated from a culture broth of *Streptomyces* sp. A-5071, was found to protect primary cultured rat hepatocytes against D-galactosamine-induced cytotoxicity (*280*).

Brunfelsamidine = pyrrole-3-carbamidine (**140**) has been isolated from the water extract of root bark of *Brunfelsia grandiflora*, a Solanaceae, which is used by the natives of the upper Amazon region as a medicine narcotic and, in higher doses, as a fish poison (*281*). Pyrrole-3-carbamidine has been also isolated and identified as the lethal constituent of *Nierenbergia hippomanica*, an Argentinian plant toxic to livestock (*282*).

β-Nicotyrine = 3,2′-nicotyrine (**141**) is the first pyrrole derivative encountered in nature. However, although **141**, the structure of which was elucidated in 1894 by Blau (*283*), had been obtained as a product of dehydrogenation of (−)-nicotine as early as in 1880 (*284*) its isolation from cured tobacco (*Nicotiana tabacum*) was only reported in 1937 (*285*). As the pyrrolidine ring of nicotine proceeds from ornithine (*286*) the latter is also the biogenetic precursor of the pyrrole ring in nicotyrine. Both nicotyrine and nicotine possess insecticidal properties (*287, 288*).

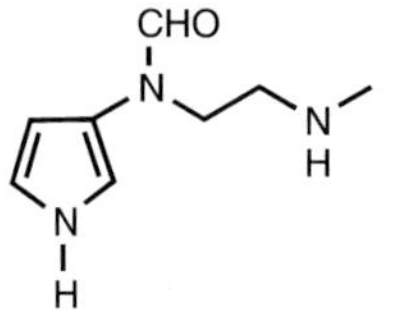

Solsodomine A (**142**) 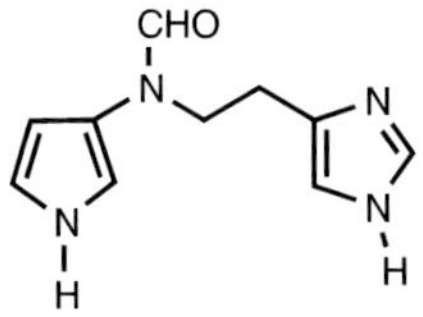   Solsodomine B (**143**)

Most recently, two derivatives of 3-aminopyrrole, solsodomine A (**142**) and solsodomine B (**143**), have been isolated from fresh berries of *Solanum sodomaeum*, collected from the Libyan desert (*289*) Thoghether with β-Nicotyrine they are the sole reported pyrrole antibiotics from *Solanum* sp.

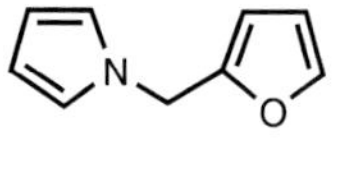

1-Furfurylpyrrole (**144**) 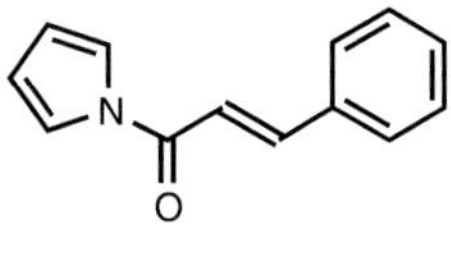   *N*-Cinnamoylpyrrole (**145**) 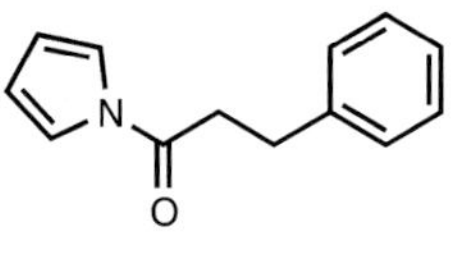   *N*-Dihydrocinnamoylpyrrole (**146**)

1-Furfurylpyrrole (**144**) has been found among the volatile constituents of East Indian sandal-wood oil (*Santalum album*) (*290*).

N-Cinnamoylpyrrole (**145**) has been isolated along with 22 known compounds from the stems, leaves and fruits of *Piper argyrophyllum* (*291*) The corresponding dihydro derivative **146** had been isolated previously along with other six minor piper alkaloids from the dried fruit of *Piper sarmentosum*, a plant which has found much medicinal use in

References, pp. 140–188

Malaya, the Indonesian archipelago, and Thailand (*292*). Its structure has been confirmed by chemical synthesis.

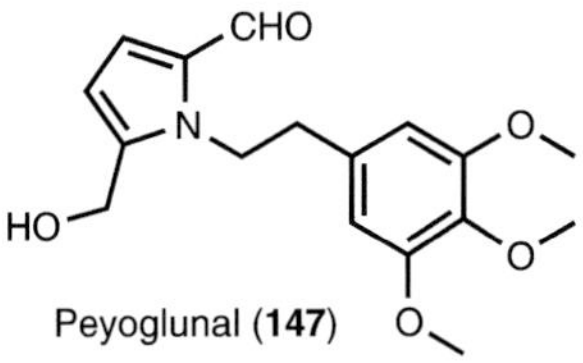

Peyoglunal (**147**)

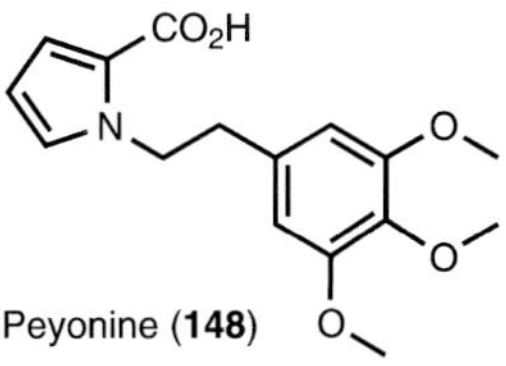

Peyonine (**148**)

Peyoglunal (**147**) and peyonine (**148**) are alkaloids which have been isolated as minor nonbasic constituents of peyote (*Lophophora williamsii*), a member of the family Cactaceae, which grows wild on the Mexican plateau and in the southwestern United States in dry places, on cliffs, or on rocky slopes. The structure assigned to peyonine was confirmed by synthesis through reaction of mescaline with methyl 2,5-dimethoxytetrahydro-2-furoate and subsequent hydrolysis of the obtained methyl ester (*293*).

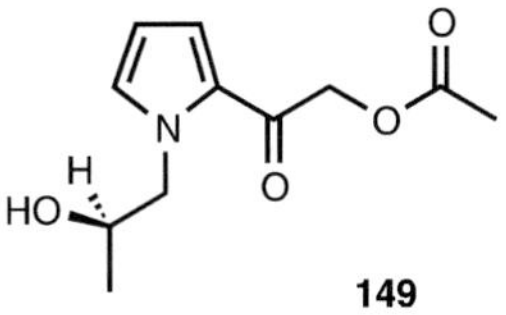

149

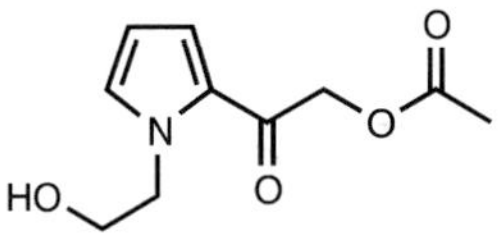

150

Pyrrole derivatives **149** (2-(2-acetoxyacetyl)-1-(2-hydroxypropyl)-pyrrole) and **150** (2-(2-acetoxyacetyl)-1-(2-hydroxyethyl)pyrrole) have been isolated from the marine red algae *Gracilariopsis lemaneiformis* collected in intertidial pools at Cape Perpetua, Oregon (*294*). Their structures have been confirmed by chemical synthesis (*295*) which, in the case of **149**, allowed to assign the (*R*) configuration to the laevrotatory ($[\alpha]_D^{20} = -103.1$ in $CHCl_3$) natural compound.

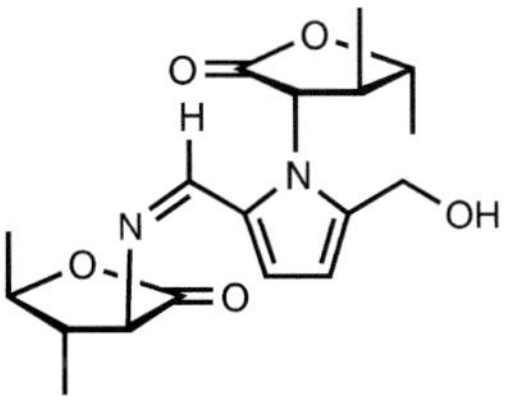

(–)-Funebrine (**151**)

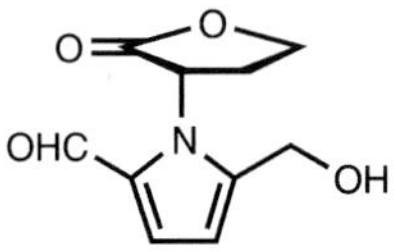

152

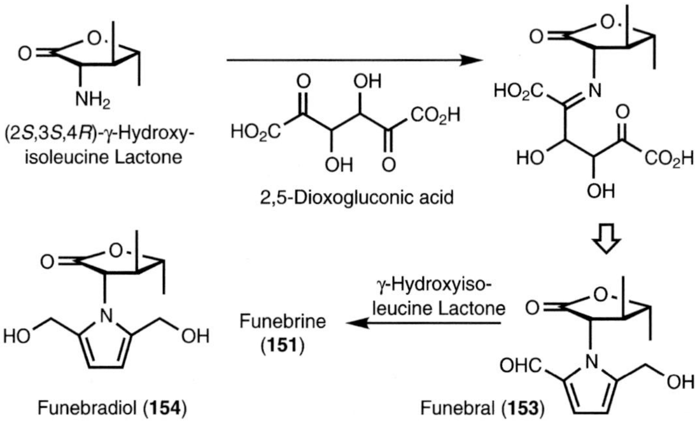

Scheme 4.1. Proposed biogenesis of funebrine (**151**)

Funebrine (**151**) has been isolated from the strongly odorous flowers of the Mexican tree *Quararibea funebris*, which played an important role in the funerary rites of Zapotec Indians from pre-Columbian times, The structure determination of funebrine by single-crystal X-ray analysis is based on the configuration of the congeneric (2*S*,3*S*,4*R*)-γ-hydroxyisoleucinelactone, which is presumed to be a biogenetic precursor (*296*).

The biosynthesis of funebrine has been proposed to proceed by reaction of an α-aminolactone with a six carbon sugar derivative such as 2,5-diketogluconic according to the mechanism of the Paal's synthesis of pyrroles from 1,4-diketones. As a matter of fact, however, the hydroxy groups of the dicarboxylic acid intermediate must be eliminated reductively before formation of the pyrrole ring may occur. Both funebral (**153**) (*297*) which reacts with a second molecule of its α-aminolactone precursor to yield funebrine (*cf.* Scheme 4.1.), and funebradiol (**154**) (*298*) have been isolated and characterized. The total synthesis of both racemic funebrine and funebral has been accomplished by Le Quesne *et al.* (*299*), using the lactone of racemic γ-hydroxyisoleucine, which had be prepared previously by the same group (*300*). All metabolites isolated from Quararibea are structurally unusual. Indeed the only similar natural pyrrole isolated thus far is the bisnorfunebral derivative **152**, which was obtained by Lynn from pea seedlings (*Pisum sativum*), where it was identified as a regulator of cellular cycles (*301*).

References, pp. 140–188

Zeatin Riboside (155)   6-(3-Methylpyrrol-1-yl)adenosine (156)

A metabolite of the plant cytokinin zeatin riboside, 6-(3-methylpyrrol-1-yl)adenosine (**156**), was isolated within the scope of the investigation of the bisynthesis of cytokinins in tobacco crown gall cells. The hypothesis that the pyrrole derivative is formed by enzymatic oxidation of zeatin riboside (**155**) was proved by chemical synthesis of **156** using a biomimetic oxidation reaction (*302*).

Rhazinilam (**157a**) is a *seco*-alkaloid which has been isolated from different members of the Apocynaceae family. It was obtained for the second time in 1970 from *Rhazya stricta* (*303*) and identified later (*304*) with one component (Ld 82) of the alkaloid mixture which had been obtained in 1965 by Linde (*305*) from *Melodinus australis*. More recently, rhazinilam was isolated again from the Malasyan plant *Kopsia singapurensis* (*306*). From a related Malayan *Kopsia* species 5-formylrhazinilam = rhazinal (**157b**) was obtained by Kam *et al.* (*307*). Moreover, two rhazinilam derivatives, namely 5,21-dihydrorhazinilam and 3-oxo-14,15-dehydrorhazinilam (**159**), have been isolated from *Leuconotis eugenifolia* (*308*) and from cell suspension cultures of the South American medicinal plant *Aspidosperma quebracho blanco*, respectively (*309*). As 5, 21-Dihydrorhazinilam on prolonged exposure to air gave rhazinilam, it is likely that the latter is an artefact of the isolation process, and that the natural precursor is the 5, 21 dihydro compound (*304*). The structure of **157a** strongly suggests that it is formed in the metabolic degradation of some *Aspidosperma* alkaloid (*cf.* Scheme 4.2.). Actually, oxidation of (+)-1,2-dehydroaspidospermidine (**158**) with *m*-chloroperbenzoic yields *in vitro* (−)-rhazinilam, thus establishing the absolute configuration of the asymmetric C-atom of the latter (*310a*). Racemic **157a** has been obtained by chemical synthesis (*310b*). Rhazinilam shows only an end absorption and no absorption maximum in the conventional UV region. As confirmed by single-crystal X-ray diffraction studies (*311*) the chromophores of rhazinilam, *i.e.* benzene nucleus, pyrrole ring, and amide moiety, are perpendicular to

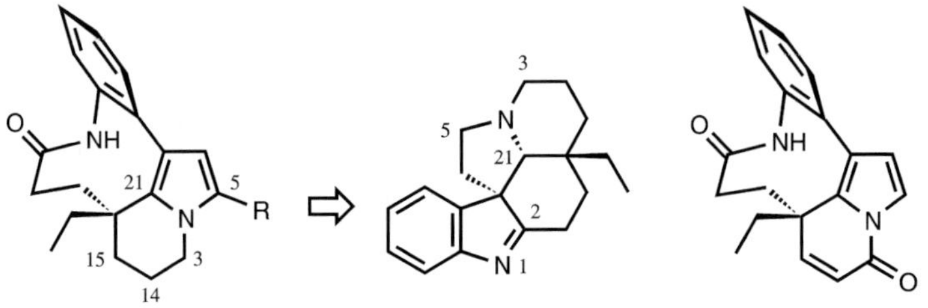

R = H : (−)-Rhazinilam (**157a**)
R = CHO : (−)-Rhazinal (**157b**)

(+)-1,2-Dehydroaspidospermidine (**158**)

3-Oxo-14,15-dehydrorhazinilam (**159**)

Scheme 4.2. Structural relationship between rhazinilam (**157a**) and the aspidosperimidine skeleton

each other in the molecule so that resonances between them are prohibited thus leading to an unusual UV spectrum.

Rhazinilam (**157a**) interacts with the system tubuline-microtubuli through inhibition of the disconnection of the microtubuli (like taxol) and provoking an helical aggregation of tubulin (like vinblastine) (*312*).

4.1. Dihydropyrrolizine Derivatives

Pyrrolizidine alkaloids occur essentially in plants of the families Asteraceae (Compositae) and Boraginaceae. They have been also characterized, although more sporadically, in about ten other families, including the Apocynaceae, Euphorbiaceae, Orchidaceae, Poaceae, and Santalaceae as well as in some tropical and sub-tropical Leguminosae of the genus *Crotalaria*. Boraginaceae, which are particularly abundant around the Mediterranean, have never found in the cold northern areas of the world and seldom in the tropics. A good number of them are common or fairly common in western Europe, *e.g.*: lung-wort (*Pulmonaria officinalis*), comfrey (*Symphytum officinale*), mouse-ear (*Hieracium pilosella*), viper's bugloss (*Echium vulgare*), hound's tongue (*Cynoglossum officinale*), gromwell (*Lithospermum*), and heliotrope (*Heliotropium europaeum*). Some Boraginaceae are ornamental, for example the numerous cultivars of *Heliotropium arborescens* with flowers of delicate fragrance, as well as the various species and varieties of *Myosotis* (*e.g. M. alpestris*, M. *scorpioides*), *Mertensia*, and Siberian bugloss (*Brunnera macrophylla*).

Many Asteraceae species have a reputation as medicines. The best known examples are chamomile and matricaria, mostly used to make herbal teas. They are, however, a fair number of toxic – or at least dangerous – Asteraceae among the species of the genus *Senecio* and *Eupatorium*. The Boraginaceae, on the contrary, are not widely used in pharmacy. Poisonings currently known are due to the abuse of plants (*e.g.* comfrey) to which folklore attributes medicinal virtues. In both families, the danger lies in the occurrence, in the incriminated species, of hepatotoxic pyrrolizidine alkaloids (see below).

Heliotropium europaeum alkaloid (**160**)   Loroquin (**161**)

Most pyrrolizidine alkaloids are derivatives of 1-hydroxymethylpyrrolizidin, which is usually called *necine* in this class of natural compounds. However, evidence that dihydropyrrolizine analogs (which contain a pyrrole ring) may be also formed in plant species along with the known more saturated alkaloids was provided by the isolation of a dimeric alkaloid (**160**) from the seeds of heliotrope (*Heliotropium europaeum*), a Boraginaceae. This dimeric alkaloid presumably results from alkylation of heliotridine, a pyrrolizidin alkaloid, by dehydroheliotridine in the plant and does not appear to be an artefact (*313*) A more simple dihydropyrrolizine alkaloid, named loroquin (**161**), was obtained from the roots of loroco (*Urechites kaerwinsky*), a plant belonging to the family Apocynaceae, the flowers of which are widely used as condiments in Salvadorian food (*314*). Loroquin is related to danaidone (**23**). As necines, in general, are biosynthesized from L-ornithine *via* putrescine and homospermidine (*N*-(4-aminobutyl)-1,4-diaminobutane) (*315*) the same metabolic precursors act most likely as bulding blocks of the pyrrole ring in alkaloids containing the dihydropyrrolizine nucleus.

Pyrrolizidine alkaloids themselves are not toxic until after they undergo oxidation by enzymes containing cytochrome P450, in the liver (Scheme 4.3.). For instance, dehydroheliotridine (X=Y=OH in Scheme 4.3.) has been identified as the major metabolite of heliotridine and lasiocarpine in the rat (*316*). The thus formed pyrrole derivatives act as alkylating agents toward biological nucleophiles (enzyme proteins and nucleic acids) which cause endothelial cell necrosis of the centrilobular veins, which leads to the infiltration and edema of their walls. The toxicity is highly dependent on the structure: the elements required for the toxicity are the esterification at C9 and 1,2-unsaturation

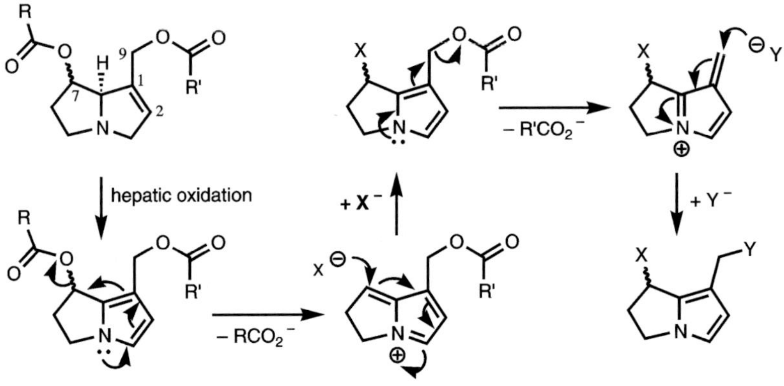

Scheme 4.3. Presumed metabolic transformation of pyrrolizine alkaloids to their dihydropyrrolizine analogs

of the pyrrolizine moiety. Esterification at the C7 position enhances the toxicity: the pyrroles that arise from the diesters react as bifunctional alkylating agents capable of bonding DNA strands and therefore capable of preventing cell division. The transport of a small quantity of metabolites in the bloodstream explains the potential for lung toxicity.

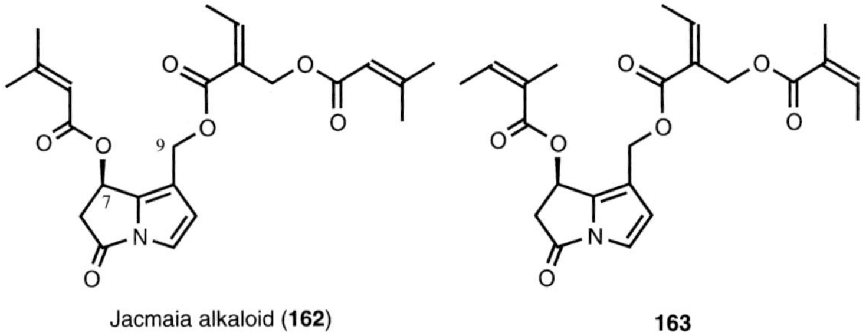

Jacmaia alkaloid (**162**)   **163**

The dehydroheliotridinone derivative **162**, in which the two hydroxyl groups at C7 and C9 are esterified with senecioate and senecioyloxy-angelate, respectively, was isolated from the aerial parts of *Jacmaia incana* (Compositae) (*317*). The roots of South African *Senecio stapeliaeformis*, on the other hand, gave the isomeric *N*-acyl pyrrole **163**, in which the two hydroxyl groups at C7 and C9 are esterified with angelic acid and angeloyloxytiglinic acid, respectively (*318*).

Both above mentioned alkaloids are closely related to the senampelins which belong to a family of isomeric alkaloids isolated from *Senecio cissampelinus*, *S. mikanoides* among other *Senecio* species. Thus, each pair of compounds senampelines A (**164**) and B

References, pp. 140–188

(**165**), senampelines C and D (*319*) and senampelines E and F occur as a mixture of isomers that could not be separated. Senampeline G (**166**) is also contained in the mixture of senampelines E and F (*320*).

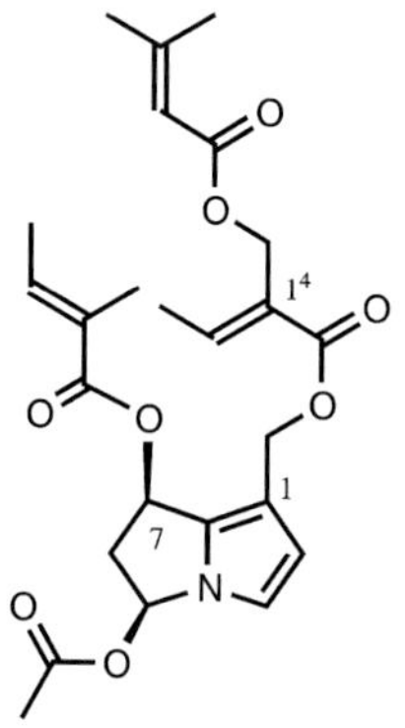

Senampeline A (**164**) 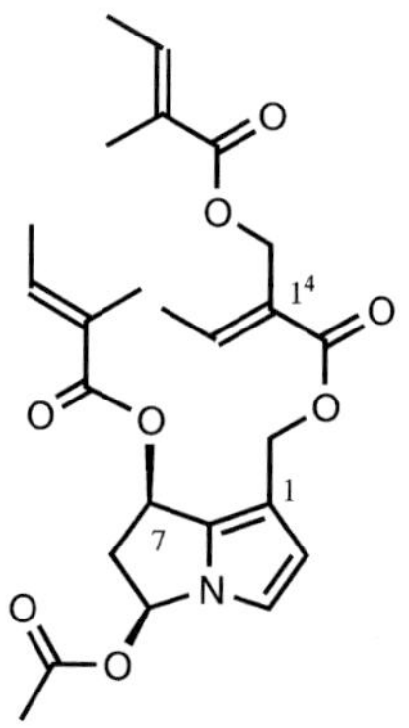   Senampeline B (**165**) 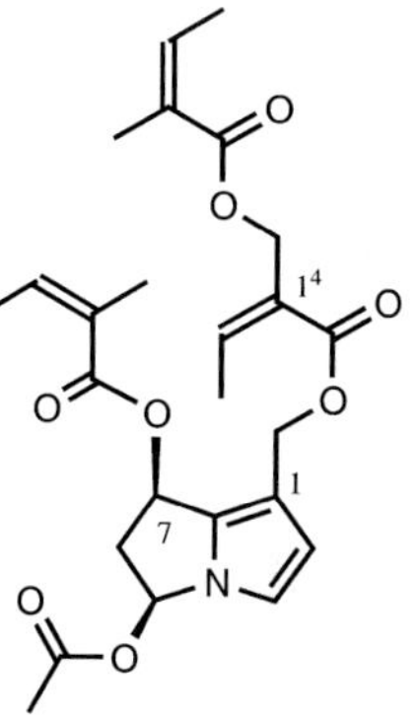   Senampeline G (**166**)

Hydrolysis of senampelines yields a neutral component (the dihydropyrrolizine moiety) which corresponds to the basic component (necine) of the related pyrrolizidine alkaloids. In all senampelines, the 5-hydroxy group of the dihydropyrrolizine skeleton is esterified with acetic acid, whereas the 7-hydroxy group is esterified either with tiglic acid (in senampelines A, B, C, and D) or with angelic acid (in senampelines E, F, and G). In addition, the primary alcohol group of senampelines A and B is esterified with (*E*)-2-(hydroxymethyl)crotonic acid, but senampeline A contains, moreover, 3-methylcrotonic acid (senecioic acid) instead of tiglic acid ((*E*)-2-methylcrotonic acid), which is present in senampelin B. Likewise, the primary alcohol group of senampelins C, D, E, F, and G is esterified with (*Z*)-2-(hydroxymethyl)crotonic acid; they differ on the third enoyl residue, which originates from senecioic acid (in senampelines C and F), tiglic acid (in senampelines D and E), or angelic acid (in senampeline G).

Another group of plant alkaloids from *Senecio* species containing a pyrrole ring comprises senaetnine (**167**) (*321*) isosenaetnine (**168**) (*322*) (15-*E*)-senaetnine (*320*) 13,19-dehydroisosenaetnine (*322, 323*) as well as (+)-pterophorine (**169**) (*319*) (−)-isopterophorine (*324*) and inaequidenine (**170**) (*321*). All these plant alkaloids contain instead of the necine base commonly present in other *Senecio* alkaloids a dihydropyrrolizinone nucleus as part of the macrocyclic diester system. Chemical syntheses of the parent 5,7a-didehydroheliotridin-3-one (= 3,8-didehydroheliotridin-5-one) have been reported (*325, 326*). With the exception of 13,19-dehydroisosenaetnine, both C7-epimers of which are known, all above alkaloids occur naturally with either 7α of 7β

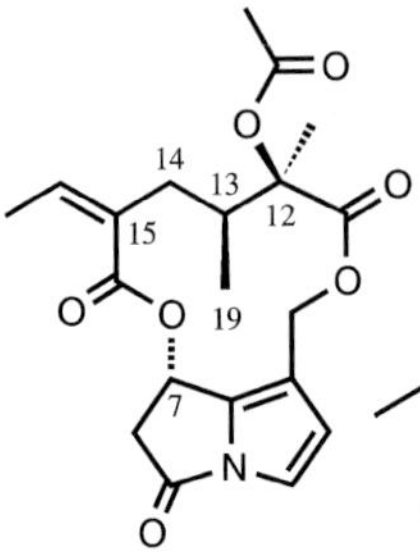

(+)-Senaetnine (**167**)

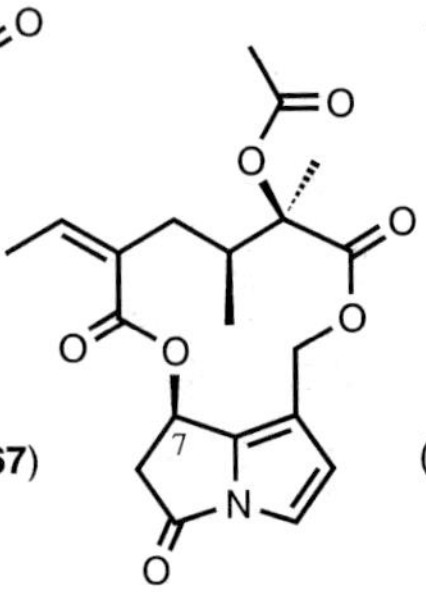

(−)-Isosenaetnine (**168**)

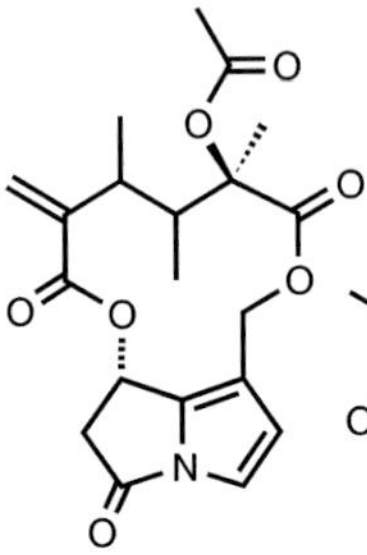

(+)-Pterophorine (**169**)

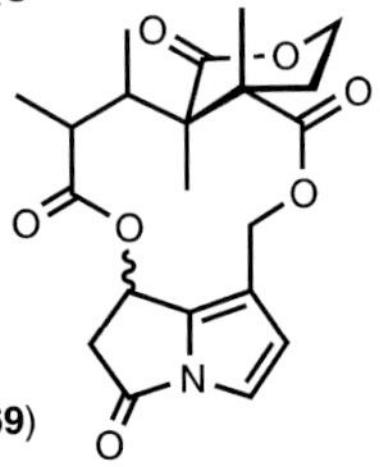

Inaequidenine (**170**)

configuration. The absolute configuration in this series, except for inaequidenine, for which arbitrary configurations at C12 and C13 are given in **170**, has been established by comparison of their CD-spectra (*323*). Biogenetically, these macrocyclic alkaloids may be derived from the senampelins by intramolecular closure of the macrolide ring through formation of the C13–C14 bond (*319*).

4.2. Pyrrole-2-Carboxylic Acid Derivatives from Plants

As in animals, naturally occurring pyrrole derivatives in plants are often esters or amides of pyrrole-2-carboxylic acid. Some pyrrole-2-carboxylic acid esters of lupinine, aphylline (10-oxosparteine) and lupanine (2-oxosparteine) occur in Leguminosae. Thus, two pyrrole-2-carboxylic acid esters of 4-(*S*)-hydroxyepilupinine (**171**, **172**) were isolated, among 43 quinolizidine alkaloids, from *Virgilia divaricata* and *V. oroboides* which occur in moist sites of the south-western and southern Cape coastal region of South Africa. Their structures were elucidated by NMR spectroscopy (*327*).

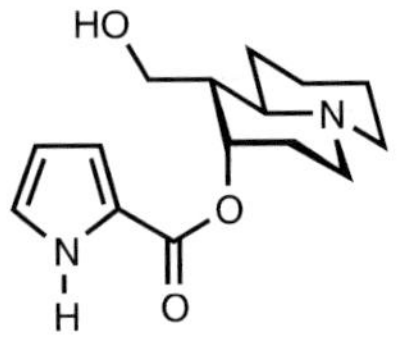

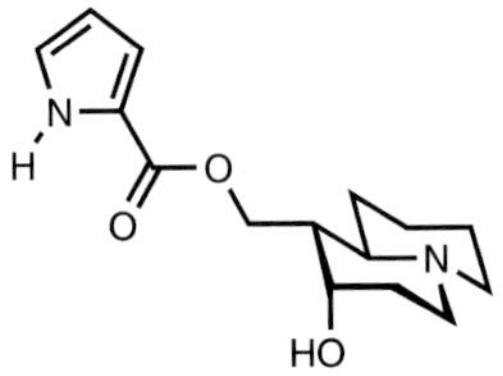

4-Hydroxyepilupinine Pyrrole-2-carboxylates

171   **172**

References, pp. 140–188

V. oroboides contains also two isomeric alkaloids *O*-(Pyrrol-2-ylcarbonyl)virgiline (**173**), the pyrrole-2-carboxylic ester of 13-hydroxyaphylline in which the OH group is axial, and calpurnine = oroboidine (**174**). Compound **173** was first reported by White to be present, along with virgiline, in the branchelets of the plant (*328*). Later, the same alkaloid and its 2,3-dehydro derivative, as a minor component, were found in the leaves of the Fijian Rubiaceae *Readea membranaceae* (*329*).

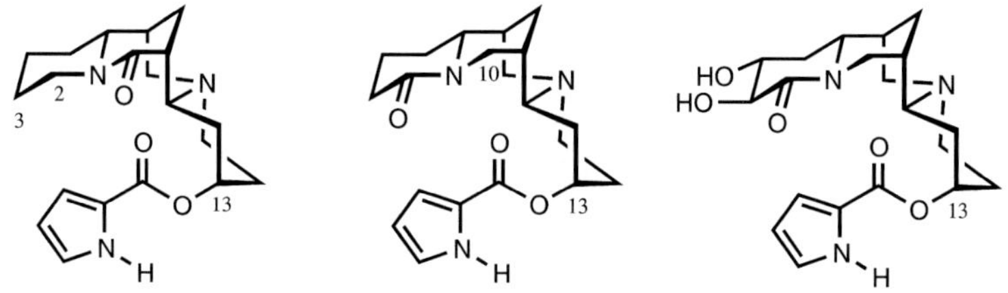

(−)-*O*-(Pyrrol-2-yl-carbonyl)virgiline (**173**)

Calpurnine (Oroboidine) (**174**)

Calpaurine (**175**)

Calpurnine (**174**), the pyrrole-2-carboxylic ester of (+)-13-hydroxylupanine, was first isolated from the aerial portion of *Calpurnia subdecandra* by Goosen (*330*). Independently, the same structure was reported as oroboidine, a name which was then abandoned in favor of calpurnine, for a base which was isolated from the seeds of the allied Leguminosae *Virgilia oroboides* (= *V. capensis*) by Gerrans and Harley-Mason (*331*) Later on, calpurnine was also obtained, together with 13-hydroxylupanine, from the roots of *Cadia ellisiana*, a toxic plant growing in Madagascar (*332, 333*) as well as, along with virgiline pyrrolecarboxylic ester, from the Ethiopian leguminous plant *Calpurnia aurea* (*334*). Whereas 13-hydroxylupanine itself has relatively weak hypotensive and antiarrhythmic effects, the corresponding 2-pyrrolylcarboxylic acid ester shows pronounced activities (*335*).

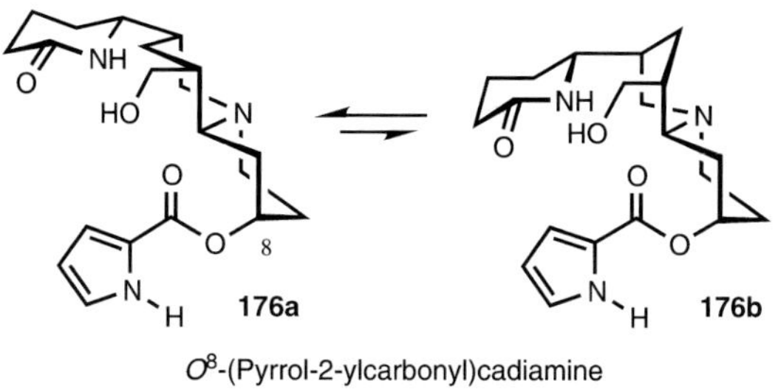

O^8-(Pyrrol-2-ylcarbonyl)cadiamine

The leaves of *C. aurea* contain also the *trans*-3,4-dihydroxy derivative of calpurnine, so-called calpaurine (**175**) (*336*) the structure of which was elucidated by spectroscopic analysis, chiefly 1D- and 2D-NMR methods (*337*) On the other hand, 10-hydroxycalpurnine and 1,10-*seco*-10-hydroxycalpurnine = O^8-pyrrol-2-ylcarbonyl)cadiamine (**176**) have been reported to occur in *Cadia purpurea*, a poisonous shub from Ethiopia, which belongs to the family of Caesalpiniaceae. Their structures were elucidated my spectroscopic methods, particularly by mass spectrometry (*338*). The less favorable conformation of the molecule (**176b**) is represented on the right in order to illustrate the structural and probably also biogenetic – relationship between O^8-(pyrrol-2-ylcarbonyl)cadiamin and calpurnine (**174**), in which the C10–N bond replaces the hydroxy group of the former.

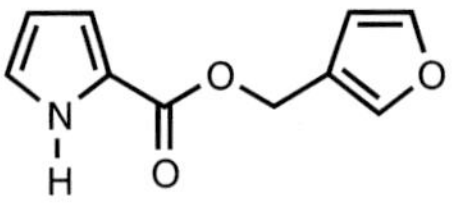

3-Furfuryl Pyrrole-2-carboxylate (**177**)

In the above pyrrolecarboxylates the alcoholic part of the ester is derived from a L-lysine metabolite; a few pyrrolecarboxylates naturally occurring in plants are known, however, in which the alcoholic part of the ester is a terpenoid. Thus, 3-furfuryl pyrrole-2-carboxylate (**177**) has been isolated from the Chinese herbal medicine tai-zi-shen (*Pseudostellaria heterophylla*), which is used as a pedriatic or geriatric tonic (*339*) Presumably, the furan ring originates from isopenthenylpyrophosphate.

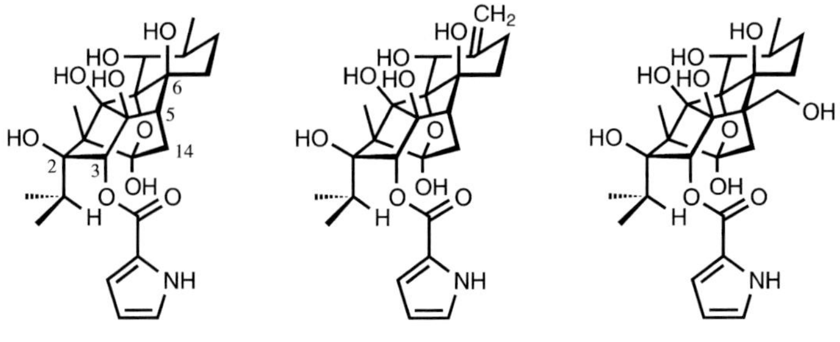

Ryanodine (**178**)   Didehydroryanodine (**179**)   Spiganthine (**180**)

The insecticidal principle ryanodine (**178**) (*340*) was isolated along with didehydroryanodine (**179**) as the principal toxic constituent, from the flacourtiaceous plant *Ryania speciosa*, which grows in Trinidad (*341, 342*). Later on, five more diterpene esters of pyrrole-α-carboxylic acid closely related to ryanodine were obtained from the same plant (*343*) On catalytical hydrogenation, **179** yields both **178** and 9-epiryanodine in a 1:9

ratio (*341, 342*). Alkaline hydrolysis of **178** gives pyrrole-α-carboxylic acid and an alcohol (ryanodol) (*344*). Ryanodine (**178**) suffers a very easy dehydratation either on mild acid treatment or on sublimation; the resulting compound anhydroryanodine (*344*), the structure of which was elucidated first (*345*), lends itself much better to degradation studies than ryanodine (*346*). The structures of anhydroryanodine (*345*) and ryanodine (*344, 347*) were defined by Wiesner et al. using classical methods of chemical degradation. The X-ray diffraction analysis of the p-bromobenzyl ether of ryanodol confirmed the structure except for a minor point: the configuration of the isopropyl and tertiary hydroxyl groups at C2 had to be reversed (*348, 349*). The structure of the hydrolysis product, (+)-ryanodol, has also been subject of a total synthesis (*350, 351*).

The powdered stems of *Ryania speciosa* are used as ryania powder to control agricultural and garden pests. Both ryanodine and didehydroryanodine cause muscle contraction at very low concentration – possibly by a direct influence on the calcium release of the junctional sarcoplasmic reticulum of the heart muscle cell (*352*) – binding at the terminal cisternae of the sarcoplasmic reticulum junctions keeping the calcium-release channels open and thereby preventing the accumulation of calcium (*353, 354*).

The 5-hydroxymethyl derivative of ryanodine has been named spiganthine (**180**). It is the cardioactive principle isolated from the ethanol extract of the aerial parts of the tropical plant *Spigelia anthelmia*, which is native to Asia and tropical America (*355*).

R = OH; R' = H; X = H : Asterinin A (**181a**)

R = OH; R' = H; X = CH$_3$: Asterinin B (**181b**)

R = H; R' = OH; X = CH$_3$: Asterinin C (**181c**)

R = R' = X = H : Astin J (**181d**)

Until now, only a few amides of pyrrole-2-carboxylic acid have been isolated from plants. Three oligopeptides, asterinins A–C (**181a–c**), which are characterized by the presence of β-phenylalanine in the peptide chain, were isolated from *Aster tataricus* roots (*356*). Later on, a forth oligopeptide, astin J (**181d**) was isolated by another group from the same plant (*357*). The corresponding amide (X=NH$_2$) can be obtained by chemical transformation of the antitumor cyclic pentapeptide astin C, which has been also isolated from *A. tataricus* (*358*). In astin C, however, the pyrrole carboxylate unit is replaced by a 2,3-dichloroproline residue, which probably yields the pyrrole ring by elimination of hydrogen chloride. Likewise, asterinins A, B, and asterinin C were considered to be artefacts derived from the corresponding cyclic pentapeptides astin B and astin A, respectively (*359*, *360*).

4.3. Pyrrolinone Derivatives in Plants

Until now, the scarce number of pyrrolinone derivatives which have been found in plants is restricted to jathropham and its derivatives, and to eremophilene lactam (**182**). The latter being a sesquiterpene alkaloid which has been isolated from the rhizomes of *Petasites hybridus* (*361*).

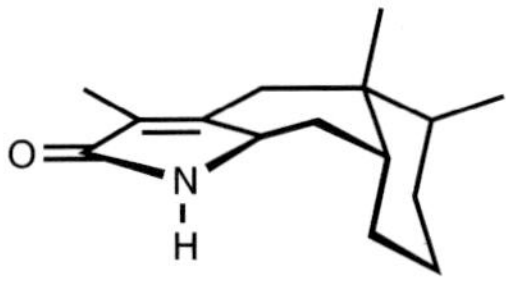

Eremophilene lactam (**182**)

Jathropham (**183**) was first isolated from *Jatropha macrorhiza* (Euphorbiaceae) as an antitumor alkaloid (*362*) and later, together with its 5-*O*-β-glucopyranoside, from *Lilium hansonii* (*363*, *364*) and other *Lilium* sp. (*365–367*) (see below). The revised structure of jathropham (*368*) – actually a tautomer of 3-methylsuccinimide – has been confirmed by the synthesis of its racemate (*369–371*)

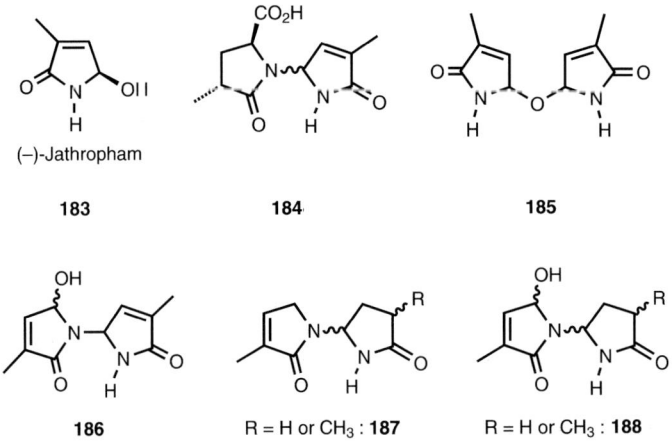

References, pp. 140–188

Along with jatropham, its racemic methyl ether and its 5-O-β-D-glucopyranoside, fresh bulbs of *L. hansonii* contain compound **184** ($[\alpha]_D^{20} = +40.9$ in methanol) as well as the jatropham dimer **185**, the achiral (*meso*) structure of which was demonstrated by single-crystal X-ray diffraction analysis (*364*). The same 5,5'-oxydi(3-methyl-3-pyrrolin-2-one), together with the 3-pyrrolin-2-one dimer **186**, had been previously isolated from bulbs of *L. candidum* by another group (*372*). In addition, *L. candidum* contains the dimeric oxo-3-pyrroline-pyrrolidin-2-one alkaloids **187** and **188** (*373, 374*). Although the optical rotations of two of the latter compounds, in methanol solution, are reported: **187** (R = CH$_3$): $[\alpha]_{546}^{20} = +248$; **188** (R = CH$_3$) : $[\alpha]_{546}^{20} = +206$, neither their relative nor their absolute configurations are known.

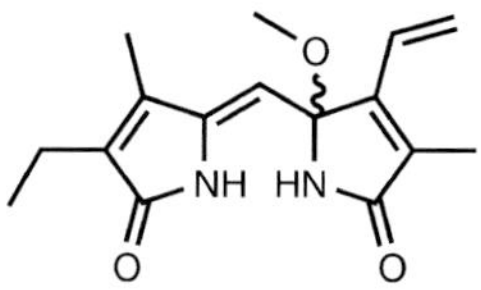

Rollipyrrole (**189**)

More recently, rollipyrrole (**189**) has been isolated from the Formosan annonaceous plant *Rollinia mucosa* (*375*). According to the analytical (mainly 1D and 2D NMR spectroscopic) data, which served to the elucidation of its structure, rollipyrrole is a propentdyopent methanol-adduct derivative. However, as propentdyopent adducts are formed *in vitro* by photo-oxidation of 1(10H)-dipyrrinones, followed by addition of a nucleophile (*e.g.* methanol) to the exocyclic C = C-bond of the primarily formed 1,10-dihydro-10H-dipyrrin-1,10-diones (so-called propentdyopents because they yield, on reduction, pigments with $\lambda_{max} \approx 525$ nm), the possibility that **189** is an isolation artefact cannot be ruled out.

5. Pyrroles from Fungi

5.1. Pyrroles from Basidiomycetes

A pyrrolylalanine derivative (**190**) has been isolated from *Clitocybe acromelalga*, a poisonous mushroom which is found in Japan only. Due to the small amount of the sample available, its structure was confirmed by chemical synthesis. A biogenetic pathway, similar to the biosynthesis of acromelic acids, with 3,4-dihydroxyphanylalanine (DOPA) as the precursor seems likely, since the 2H-pyran-2-one suggested as

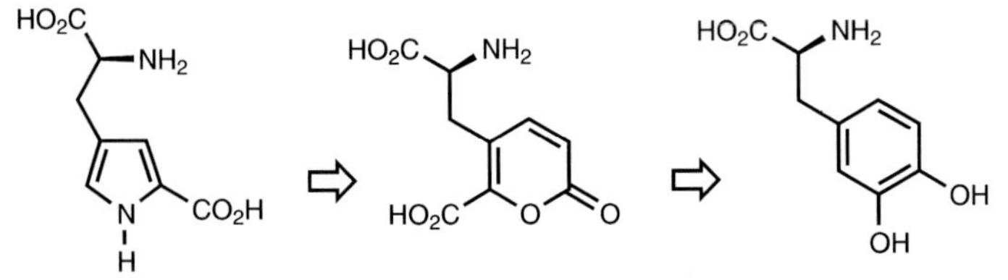

3-(2-Carboxy pyrrol-4-yl)alanine (190)

DOPA

Scheme 5.1. Pressumed biogenetic precursors of 3-(2-carboxypyrrol-4-yl)alanine (**190**)

intermediate was present among the metabolites isolated from the fungus (Scheme 5.1.) (*376*)

5.2. Pyrroles from Deuteromycetes

Verrucarin E (**191**) was isolated from *Myrothecium verrucaria*, a mold fungus which decomposes cellulose, by Tamm in 1962 (*377*). The structure of verrucarin E, which substantially differs from those of other verrucarins, which are sesquiterpenes, was demonstrated by chemical transformation into 3-acetyl-4-methylpyrrole. Synthetic verrucarin E was obtained first as a minor product (3% yield) of the reaction of 3-acetylpyrrole with formaldehyde in aqueous sodium carbonate (*378*). Later syntheses (*191, 379, 380*) as well as X-ray diffraction studies (*381*) have confirmed its structure (*cf,* Scheme 5.2.)

The biosynthetic precursor of both verrucarin E and desoxyverrucarin E (**192**), the main nitrogenous metabolite of *Eupenicillium hirayamae* (*382*) is neither proline nor δ-aminolaevulinic acid. As four molecules of ^{14}C-labeled acetate are incorporated into both pyrrole derivatives *in vivo*, their biosynthesis takes place probably by the polyketide pathway (*383*). Incorporation studies with $[1,2-^{13}C]$-acetate

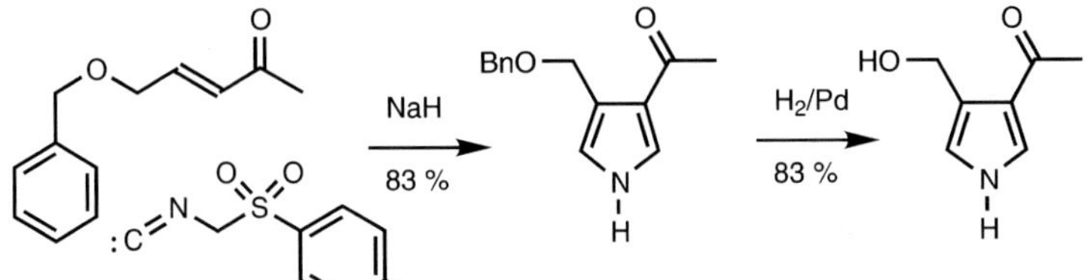

Scheme 5.2. A short chemical synthesis of verrucarin E (**191**) (*380*)

References, pp. 140–188

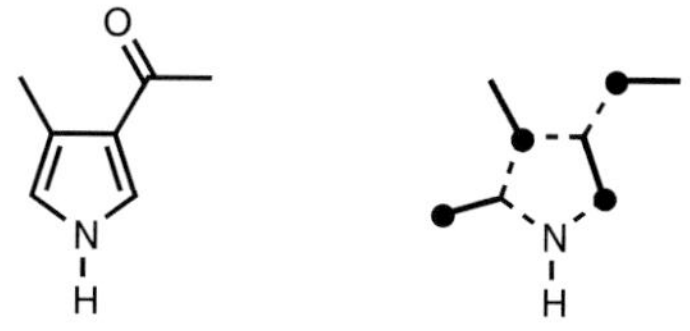

Desoxyverrucarin E (**192**)

C1 C2 ●—— = Acetate

Scheme 5.3. Biogenetic origin of the carbon atoms of verrucarin E (**191**) and deoxyverrucarin E (**192**)

have demonstrated that two acetoacetate units couple together as shown in Scheme 5.3. (*384*). One carboxylate group is lost in order to account for the molecular structure of the final product. The origin of the N-atom is unknown. Verrucarin E is a cytostatic active antibiotic.

Wallemia A (**193a**) and wallemia C (**193b**), as well as the corresponding 4-chloro derivatives, wallemia E (**193c**) and F (**193d**), are the main components of the brightly shining orange Deuteromyete *Wallemia sebi*. The previously suggested structure (*385*) and geometry of the C=C-bond adjacent to the pyrrole ring (*386*) of the wallemia pigments was later revised following to NOE studies (*387*). Wallemia A and E have been identified with synthetic samples (*388*).

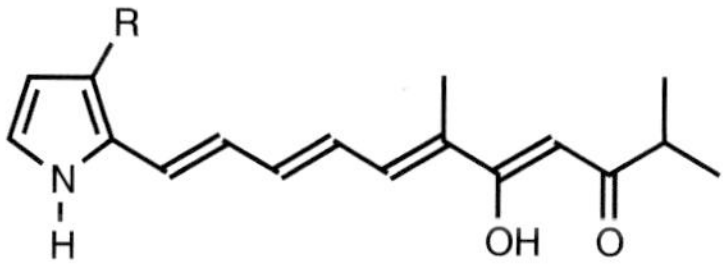

R = H : Wallemia A (**193a**)
R = Cl : Wallemia E (**193c**)

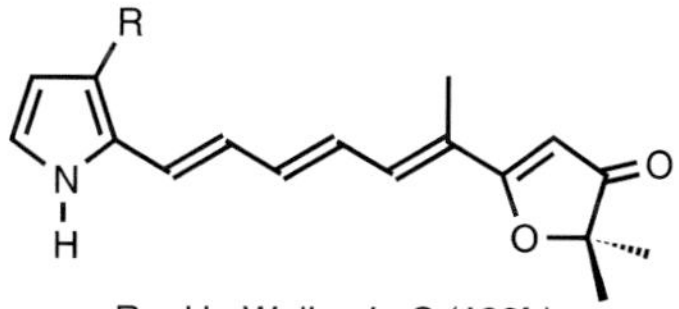

R = H : Wallemia C (**193b**)
R = Cl : Wallemia F (**193d**)

Analogous to the keronopsins (**138**) from ciliates, but with (Z)-geometry at the C7=C8 bond, is rumbrin (**194**). It was isolated from the fungus *Auxarthron umbrinum* (*389, 390*). Its structure has been elucidated by 2D NMR spectral analysis. Penochalasins A (**195a**), B (**195b**) and C (**195c**) are cytotoxic metabolites from a *Penicillium* sp. which was initially separated from the marine alga *Enteromorpha intestinalis* collected from Tanabe Bay in Japan (*391*). Penochalasins are cytochalasan derivatives which differ from each other in the structure of the isoindoline part of the molecules. Probably, penochalasins B–C derive from penochalasin A by protolytic opening of the epoxide ring present in the latter.

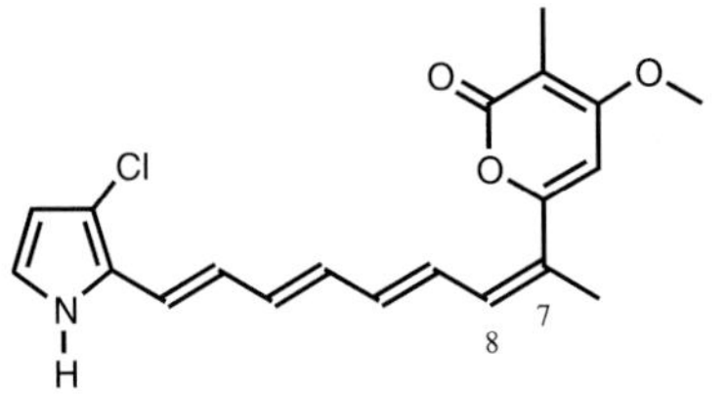

Rumbrin (**194**)

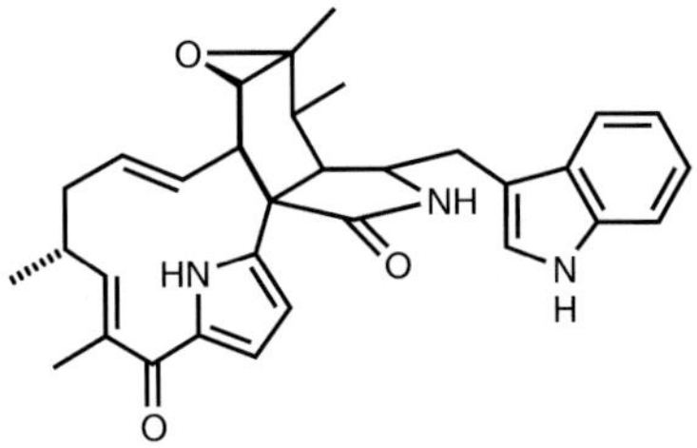

Penochalasin A (**195a**)

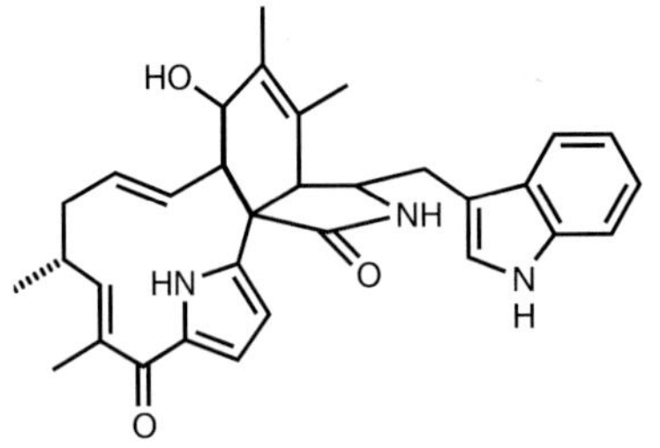

Penochalasin B (**195b**)

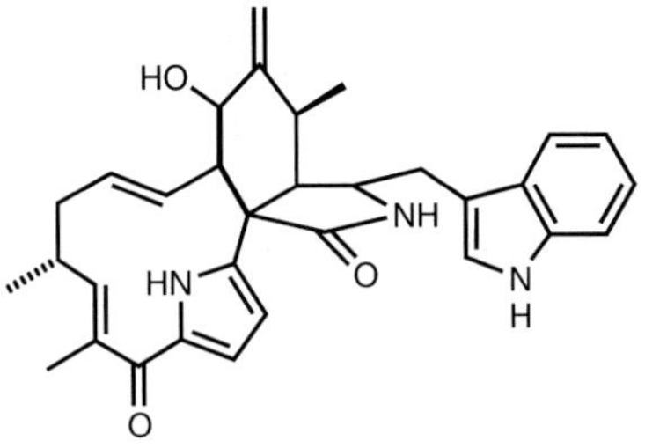

Penochalasin C (**195c**)

Macrophominol (**196**) is a pyrrolo[1,2-*a*]pyrazine derivative which has been isolated from cultures of the *Macrophomina phaseolina*, a deuteromycete which has been identified as pathogenic to several highly commercially valued crops such as black bean, water-melon, sorghum, melon, and sesame (*392*). Although the biosynthesis of macrophominol has not yet been investigated, it is most likely that the non-pyrrolic moiety of the molecule stems from L-threonine.

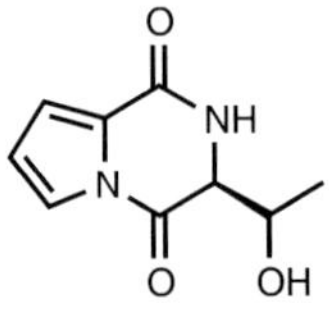

(–)-Macrophominol (**196**)

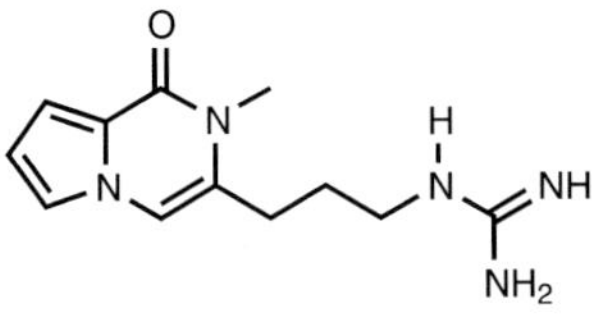

Peramine (**197**)

As macrophominol, peramine (**197**) is a pyrrolo[1,2-*a*]pyrazine derivative (actually a pyrrole-2-carboxylic acid lactam). It was isolated from perennial ryegrass (*Lolium perenne* L.), which was infected by the endoparasite *Acremomium loliae* (*393*, *394*). Endophytic fungal

References, pp. 140–188

symbiotes infect a number of grass species where their presence is associated with both insect resistance and, in some cases, with animal toxicity of the infected pasture. The identification of peramine in the mycelium of cultured *A. loliae* established the fungus as the producer of the insect feeding deterrent found in *A. loliae*-infected ryegrass. The structure of peramine has been confirmed by two independent chemical syntheses (*395, 396*). Biosynthetically, peramine appears to be derived from the cyclization of proline and arginine.

5.3. Pyrroles from Myxomycetes

Slime molds (Myxomycetes) are microorganisms on the borderline between the plant an animal kingdom. It is difficult to collect slime molds in large amounts and, therefore, only a few of the more than 500 species have been chemically investigated, so far. The dimethyl esters of lycogalic acid A, B and C (**198a–c**) (*397*) – also called lycogarubins C, B and A, respectively (*398*) – were isolated by two different groups independently from each other, from the slime mold *Lycogala epidendrum* which also contains lycogalic acid A (**198a**) and staurosporinone (**199**), a pyrrolin-2-one derivative, which had been previously isolated from a soil bacterium *Nocardiopsis* sp. K-290 (*399, 400*). The structures of these metabolites were established by 2D NMR spectroscopy and chemical degradation reactions. Moreover, the structures of lycogarubin C (dimethyl lycogalate A) and lycogarubin A (dimethyl lycogalate C) have been confirmed by chemical synthesis from methyl 3-(indol-3-yl)pyruvate (*397*) and by X-ray diffraction analysis of the corresponding *O,O,N*-trimethyl derivative (*398*), respectively. As it has been shown that lycogylic acid A (so-called chromopyrrolic acid) is derived from tryptophan in *Chromobacterium violaceum* (*401*) the reported chemical synthesis of the corresponding dimethyl ester probable imitates the biogenetic pathway of the above fungal metabolites.

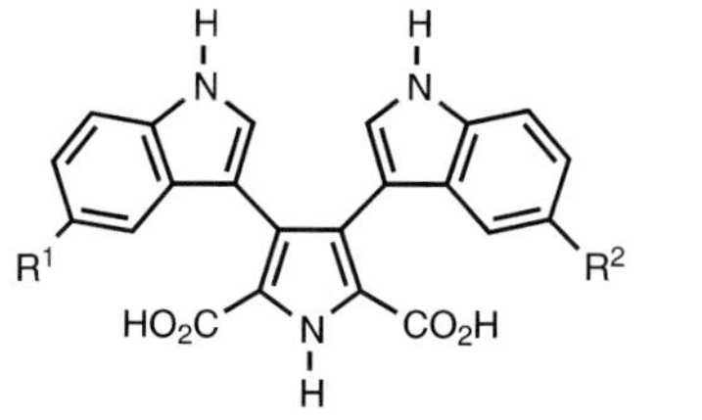

$R^1 = R^2 = H$: Lycogalic acid A (**198a**)
$R^1 = H; R^2 = OH$: Lycogalic acid B (**198b**)
$R^1 = R^2 = OH$: Lycogalic acid C (**198c**)

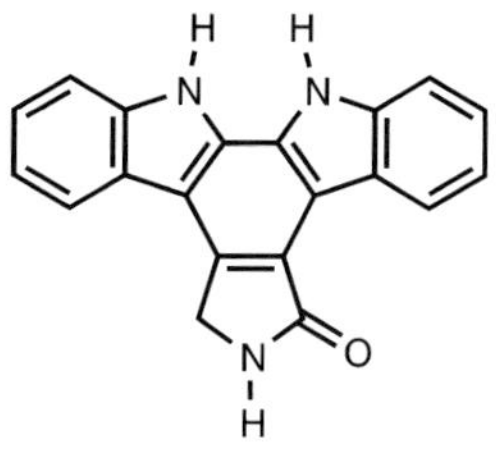

Staurosporinone (**199**)

It is interesting to note that both 3,4-*bis*(indol-3-yl)pyrrole-2,5-dicarboxylic acid derivatives and acetylenic acylglycerols have been isolated recently from *Lycogala epidendrum* (*402*), whereas in previous investigations on this species only one or the other of these types of compounds had been reported.

5.4. Pyrrolinone Derivatives in Fungi

Pyrrolizidine alkaloids are in microorganisms much less common than in plants (*cf.* Section 4.1.). Among the former phenopyrrocin (**200**), which is produced by *Penicillium* sp. (*403*), and pyrrolam (**277**) are the only pyrrolizin-3-one derivatives, which are not hydroxylated at C7a. Therefore, they may be formally considered as pyrrole derivatives.

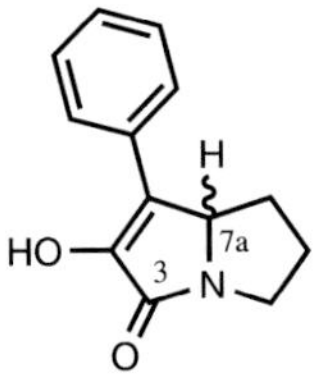

Phenopyrrozin (**200**)

5.5. Tetramic Acid Derivatives in Fungi

Derivatives of tetramic acid occur in fungi and bacteria (*cf.* Section 6.9.) much more frequently than in taxonomically higher organisms (*cf.* Section 3.2.6.). Tenuazonic acid (**201**), the simplest naturally occurring derivative of 3-acetyltetramic acid, was originally isolated from the culture filtrate of *Alternaria tenuis* by Stickings and his co-workers (*404*). It is a cytotoxic compound, which occurs in several phytopathogenic deuteromycetes (*e.g. Alternaria alternata* (*405*, *406*), *A. longipes* (*407*), and *Pyricularia oryzae* (*408–410*), among others).

The structure of tenuazonic acid was elucidated by Stickings (*411*) who obtained L-isoleucine after degradation of the mycotoxin by ozonolysis and subsequent acidic hydrolysis, thus establishing the absolute configuration as (2S,6S). On mild acidic hydrolysis, cleavage of the acetyl chain of **201** occurs.

Tenuazonic acid (**201**), which has been synthesized from L-isoleucine and diketene using Lacey's general synthesis of α-acetyltetramic acids

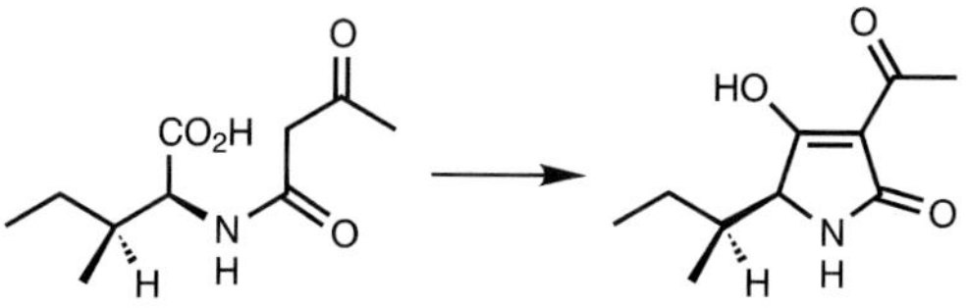

(−)-Tenuazonic acid (**201**)

Scheme 5.4. Structure and biogenetic origin of tenuazonic acid (**201**)

(*217*) (*cf.* Scheme 3.21.), was shown to be biosynthesized from two molecules of acetate and one of L-isoleucine, maintaining the chirality of the amino acid in the final product. Feeding experiments using [1-^{14}C]-labeled acetate led to the isolation of radioactive *N*-acetoacetyl-L-isoleucine and it was concluded that biosynthesis of **201** occurs *via* cyclization of the latter thus following a pathway similar to the chemical synthesis (*cf.* Scheme 5.4.) (*412–414*).

Tenuazonic acid (**201**), the growth-inhibitory action of which against human tumors growing in the embryonated egg was first reported by Gitterman and his co-workers (*415*), is a cytotoxic substance with antiviral, bactericide and insecticide activity (*416*). Tenuazonic acid inhibits the incorporation *in vivo* and *in vitro* of amino acids into proteins (*417*). Studies with cell-free systems have indicated that the inhibition by **201** of protein synthesis was due essentially to the inhibition of protein biosynthesis at the ribosome level (*418–421*) thus originating different diseases in plants (*422*) like blast disease (*409*) and brown-spot disease of tobacco and rice plants (*407*).

Laccarin (**202**) is an alkaloid which has been isolated from the Japanese mushroom karebakitsunetake (*Laccaria vinaceoavellanea*) (*423*). Laccarin was found to have a similar chromophore to that of tenuazonic acid, thus suggesting an analogous pathway in their biogenesis. Laccarin, which did not show a cytotoxicity against P388 mouse leukemia nor inhibitory activity against platelet aggregation, showed, however, a moderate inhibition on phosphodiesterase activity.

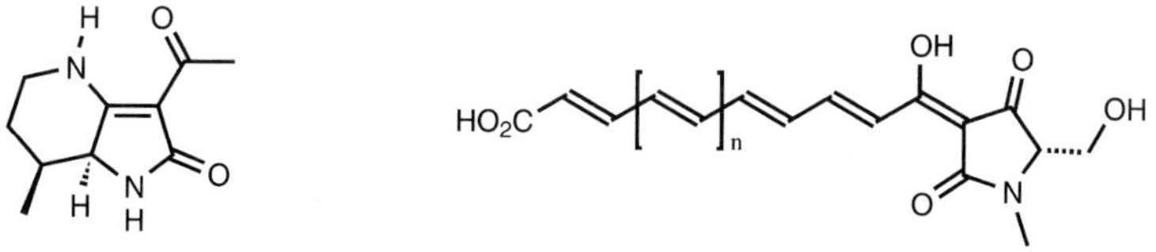

(+)-Laccarin (**202**) n = 2: Physarorubinic acid (**203a**)

n = 1: Physarorubinic acid B (**203b**)

Physarorubinic acid (**203a**) is a plasmoidal pigment, which was first isolated and characterized by Nowak and Steffan from the yellow-colored microplasmodia of the slime mold *Physarum polycephalum* (*424*). The structure was elucidated by means of spectroscopic methods. the stereochemistry was proven by synthesis of the tetramic unit and application of circular dichroism.

As well as containing an *N*-methylserine-derived acyltetramic acid terminus, **203a** contains a fully conjugated *all-trans*-pentaene chain attached to a terminal carboxylic acid function. The structure of physarorubinic acid has been confirmed by total synthesis from (*E*)-3-iodoacrylic acid (*425*). A lower homolog of physarorubinic acid, containing a polyenic chain of nine C-atoms, has been isolated from the slime mold *Physarum polycephalum*. It was named physarorubinic acid B (**203b**) (*426*)

Fuligorubin A (**204**)

Fuligorubin A (**204**) is a red pigment which has been isolated from the yellow plasmoids of the slime mold *Fuligo septica* (*427*). Fuligorubin, which occurs in the organism as a stable calcium complex, acts presumably as a photoreceptor ($\lambda_{max} = 450$ nm) playing a role on phototaxis and sporulation of the myxomycete. However, as acyltetramic acids exhibit remarkable antibiotic and cytotoxic activities and some of them act as tremorgenic mycotoxins, the possibility that fuligorubins and related metabolites may protect the vulnerable plasmodia of slime molds against attack of microorganisms has to be considered as well as their possible role as photoreceptors or metal chelating agents. As the tetramic acid derivative which was obtained by condensation of diketene and dimethyl L-glutamate and decahydrofuligorubin A (obtained on catalytical hydrogenation of the natural compound) display CD curves of opposite sign, it was concluded that the absolute configuration of the stereogenic C5-atom of fuligorubin A is (*R*). The total synthesis developed for fuligorubin (*428*) suggests a biosynthetic pathway in which D-glutamic acid condensed with a heptaketide chain.

Some from L-tyrosine biogenetically derived acyltetramic acid analogs to fugilorubin (**205–207**) are also responsible for the orange-yellow color of plasmodia from *Leocarpus fragilis*, a common slime mold which forms insect egg-like fruiting bodies which are found in autumn attached to dead conifer needles, grass, or similar substrates (*429*).

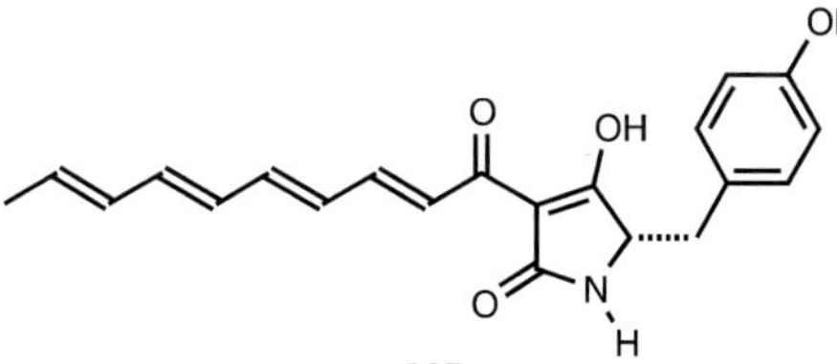

205

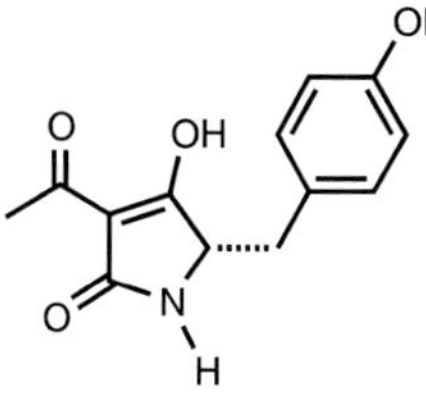

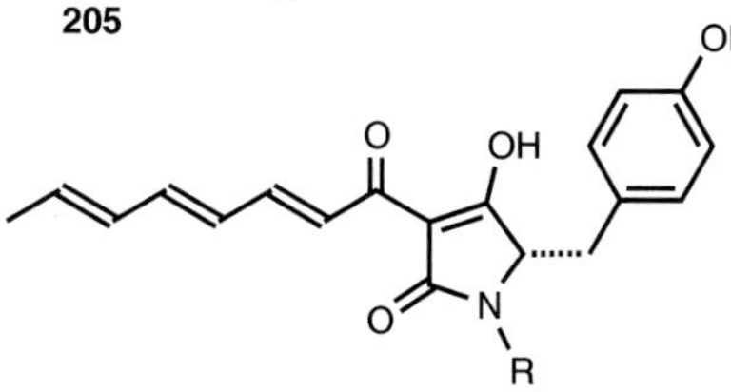

206

R = H or CH₃ : **207**

Erythroskyrine (**208**), an antibiotic active polyenoyltetramic acid derivative and the principal pigment of *Penicillium islandicum*, was first isolated by Howard and Raistrick in 1954 (*430*). Reisolation of the mycotoxin in 1964 by Shoji *et al.* (*431, 432*) and later, in 1975, by Ueno *et al.* (*433*) allowed partial structure determination, but it was not until 1988 that an unambiguous absolute and relative stereochemical assignment of erythroskyrine was obtained by Beutler and co-workers (*434*). The first total synthesis of the antibiotic was achieved in 1999 (*435*). Erythroskyrine is a mycotoxin ($LD_{50} = 60$ mg/kg) which exhibits antibiotic action against several *Staphylococcus* species. Feeding studies using sodium acetate and diethyl malonate both of which were labeled with ^{14}C at C1 and C2 showed that C9 to C24 are derived from malonate while C25 and C26 are acetate derived. The uptake of ^{14}C-carboxy-labeled L-valine indicated the origin of the tetramic acid ring. Attempts to determine the biosynthetic origin of the *N*-methyl group as well as C2 and C3 in the tetramic acid moiety were, however, inconclusive (*436*).

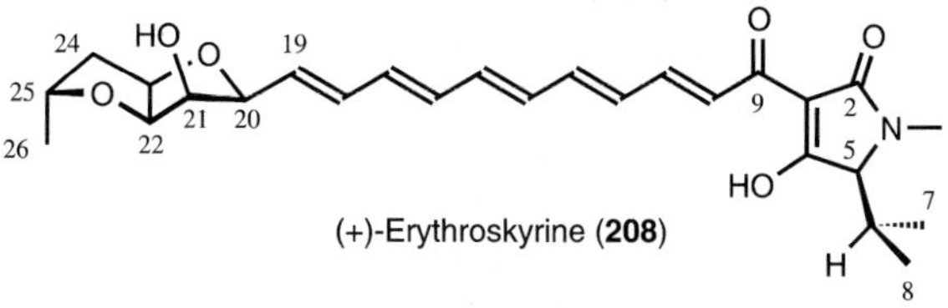

(+)-Erythroskyrine (**208**)

Aflastatin A (**209a**) (*437, 438*) and aflastatin B (**209b**) (*438*), as the minor component, which are produced by *Aspergillus parasiticus* inhibit the biosynthesis of aflatoxins without essentially affecting the growth of its producer. Their structures have been elucidated both by NMR spectroscopy and chemical degradation. Their biosynthesis have been investigated using ^{13}C-labeled substrates. Incorporation experiments using ^{13}C-labeled acetate, propionate, glucose and glycolate suggested that most of the C_2 and C_3 units involved in the alkyl chain moiety of aflastatin A were biosynthesized from acetic and propionic acids, but five C_2 units in the alkyl chain originated from glycolic acid. According to an expected polyketide pathway, C2 and C3 of the tetramic acid moiety originate from acetate, whereas the origin of the C_3 unit consisting of the C4, C5 and C6 atoms is presumably alanine biosynthesized *via* pyruvate. Since aflastatin B (**209b**), the *N*-demetyl derivative of aflastatin A (**209a**), was obtained as a minor component, *N*-methylation of **209b** might be the final step in the biosynthetic pathway of aflastatin A.

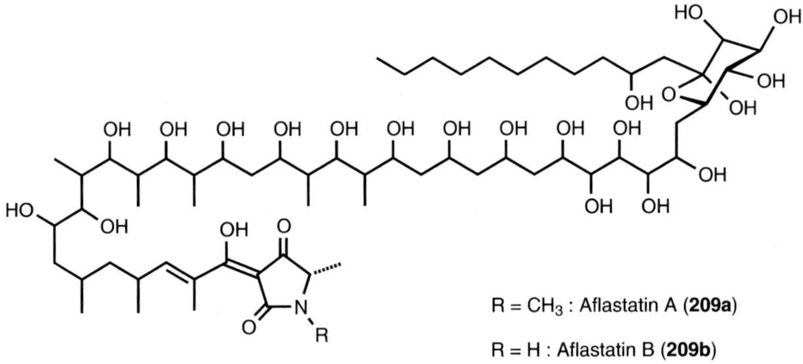

R = CH$_3$: Aflastatin A (**209a**)

R = H : Aflastatin B (**209b**)

Polycephalin B (**210a**) and polycephalin C (**210b**) have been isolated from plasmodia of the myxomycete (slime mold) *Physarum polycephalum* (*426*). Young plasmodia, which are of remarkable yellow color, live inside decaying trees and move away from the light, while older plasmodia, which have stopped growing, move towards the light and sporulate. Photoreceptors in the UV-A or blue light range are responsible for these phenomena. The absolute configuration of polycephalin C (**210b**) at C5 of the tetramic acid subunits was established, after catalytical hydrogenation of the pigment, by comparison of the CD spectrum with that of decahydrophysarorubinic acid (s. above) as well as that of synthetically prepared 3-acetyl-5-hydroxymethyl-*N*-methyltetramic acid with (5*S*) configuration. Accordingly, polycephalin C should have (*S*) configuration at both tetramic acid residues. ^{1}H-NMR spectroscopic studies established the *trans* relative configuration of the triene chains at the cyclohexene ring by comparison with appropriate *cis*- and

References, pp. 140–188

trans-substituted model compounds. Finally, the absolute configuration at both asymmetric C-atoms of the cyclohexene ring of polycephalin C was established to be (R) by chemical synthesis of the natural compound (*439*). Polycephalin B (**210a**) is a *N*-demethylated polycephalin C. However, which of the two tetramic acid moieties is *N*-methylated has not yet been determined. Polycephalin C is very sensitive to light. Possibly, the biosynthesis of the polycephalins takes place by condensation of physarorubinic acid (**203a**) with physarorubinic acid B (**203b**).

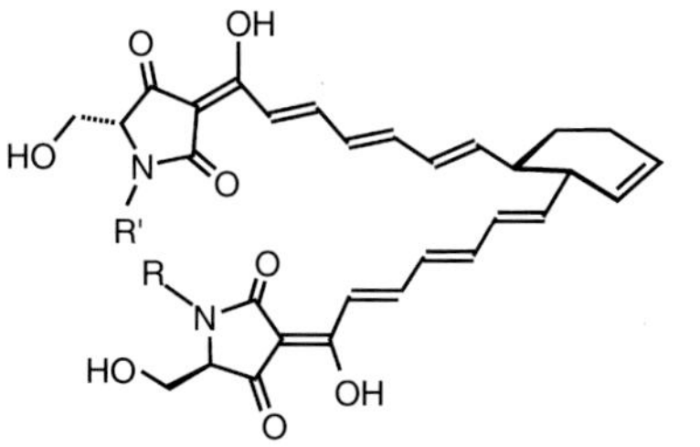

R/R' = H; R'/R = CH$_3$: Polycephalin B (**210a**)
R = R' = CH$_3$: Polycephalin C (**210b**)

The fungal metabolite equisetin (**211**), isolated from the white mold *Fusarium equiseti*, is suspected to be a promoter of chronic environmental diseases including certain leukemias. Equisetin contains an acyltetramic acid moiety, derivable from L-*N*-methylserine (*440, 441*). The total synthesis of equisetin has been accomplished in a manner that establishes its relative stereochemistry (*442*). The tetramic acid ring of the *Fusarium* metabolite was assigned as having the absolute configuration of equisetin derived from the direct biosynthetic incorporation of the amino acid L-serine. A close structural relationship exists between equisetin and the antibiotic LL-49F233α (**212a**), which is produced by a taxonomically unclassified fungus. It was found to exhibit activity against some antibiotic-resistant bacteria (*443*). A further acyltetramic acid metabolite containing an octahydronaphthalene moiety is antibiotic CJ-17,572 (**212b**); it has been recently isolated from a

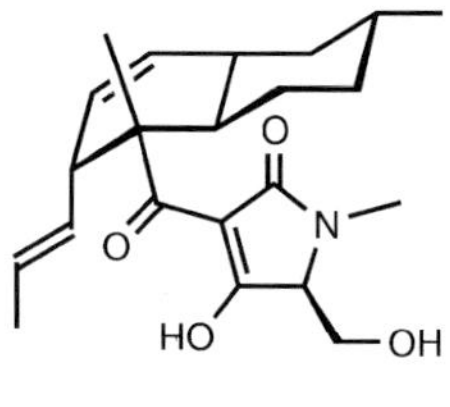

(–)-Equisetin (**211**) 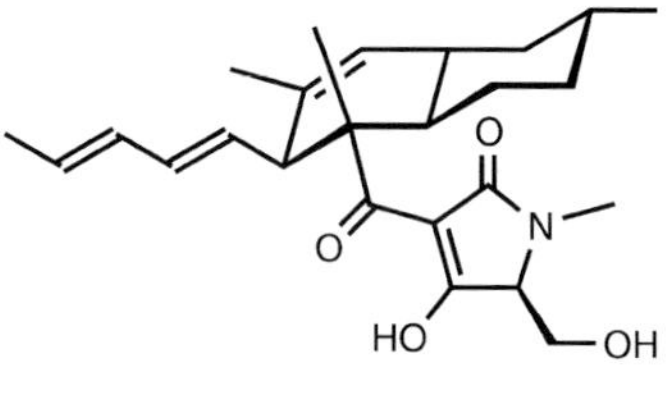   LL-49F233α (**212a**) 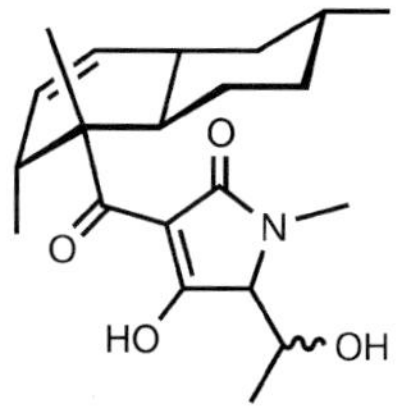   (+)-CJ-17,572 (**212b**)

fungus of the genus *Pezicula* (*444*). The spatial view of the molecule given in formula **212b** is, however, arbitrary, since until now neither the absolute nor the relative configuration of this antibiotic has been elucidated.

α-Cyclopiazonic acid (**213**) is a mycotoxin which shows neurotoxic activity ($LD_{50} = 2$–6 mg/kg). It occurs in numerous *Penicillium* sp. (*e.g. P. cyclopium, P. camemberti, P. viridicatum*) and *Aspergillus* sp. (*e.g. A. flavus, A. oryzae, A. versicolor*), which grow on stored cereal products and other foodtuffs (*e.g.* beans, maize flour, wheat and sausages) (*445*). Maize-meal was used for the large-scale cultivation of the most toxic *Penicillium cyclopium* strain and the toxic principles were extracted quantitatively with chloroform-methanol. From the fraction soluble in sodium hydrogen carbonate solution, α-cyclopiazonic acid was isolated by chromatography on formamide-impregnated cellulose and ion-exchange columns (*445*). α-Cyclopiazonic acid was found to be the main cause of the toxicity of the fungus.

α-Cyclopiazonic acid (**213**)   (−)-β-Cyclopiazonic acid (**214**)

The relative configuration of α-cyclopiazonic acid was established by ^{1}H-NMR spectroscopy and confirmed both by chemical synthesis (*446, 447*) and by X-ray diffraction analysis (*448*). The molecule crystallizes as an *exo*-enol rather than the *endo*-enol tautomer (*cf.* Section 3.2.6.) and the *exo*-cyclic enolic moieties have different orientations in the two molecules in the asymmetric unit.

Radiolabelling experiments furnished evidence that cyclopiazonic acid is derived from tryptophan, an C5-unit formed from mevalonic acid and two molecules of acetic acid (*414*). Its biosynthetic precursor was identified (*449*) as β-cyclopiazonic acid = *bis*(4,11:9,10)-secodehydrocyclopiazonic acid (**214**), which along with α-cyclopiazonic acid imine has been isolated from *P. cyclopium* (*450*) and *A. versicolor* (*451*). The structure of β-cyclopiazonic acid has been confirmed by synthesis of both racemic and optically pure material (*452*). As γ,γ-dimethylallyltryptophan is not a precursor of α-cyclopiazonic acid, prenylation of the benzene ring (to form β-cyclopiazonic acid) must occur after coupling of the tryptophan and acetyltetramic acid subunits (*453*). Moreover, incorporation of (3*R*)- and (3*S*)-[3-^{3}H,3-^{14}C]-tryptophan into α-cyclopiazonic acid in *P. cyclopium* proceeds with loss of pro-(*S*)-tritium showing that, in the final cyclization step, formation of the new C–C bond occurs from the opposite side of the molecule to proton removal (*454, 455*). This result has been confirmed using isotopic label with [1,2-^{13}C$_2$]acetate (*456*). The enzyme which transforms **214** into **213** (β-cyclopiazonate oxidocyclase) has been isolated and purified from the mycellium of *P. cyclopium* by Schabort et al. (*457, 458*).

6. Pyrroles from Bacteria

Some structurally quite simple monopyrrolic derivatives have been isolated from actynomycetes; their physiological function is, however, not yet understood. Thus, cystamidin A = pyrrole-3-propanamide (**215**) was obtained from *Streptomyces* KP-1241 culture broth (*459*), and reported to be an inhibitor of human calpain, an ubiquitous intracellular calcium-activated cysteine protease. Although the structure of cystamidin A was confirmed by synthesis (*460*), the synthetic product is inactive, thus suggesting the presence of a potent inhibitor contaminant in isolated cystamidin A.

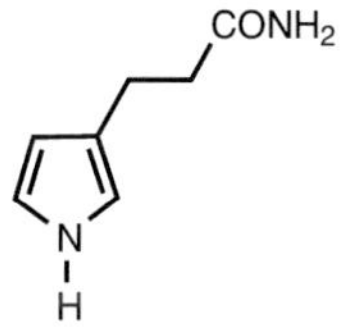

Cystamidin A (**215**)

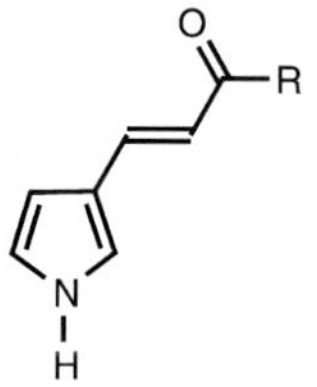

R = OH : (**216a**)
R = NH$_2$: (**216b**)

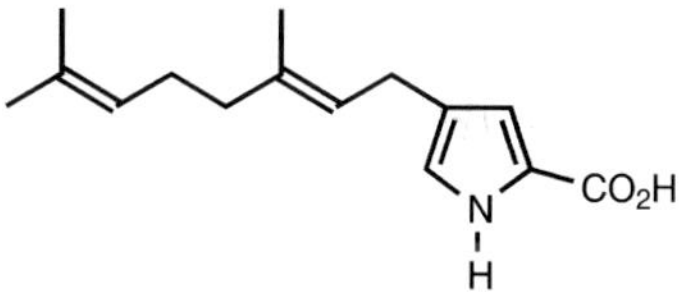

Pyrrolostatin (**217**)

Likewise, both (*E*)-pyrrole-3-acrylic acid (**216a**) and pyrrole-3-acrylamide (**216b**), which have been isolated from *Streptomyces parvulus* cultures, lack antimicrobial activity against bacteria, yeasts, or fungi. Their structures were elucidated by spectroscopic analysis and comparison with the corresponding synthetic derivatives (*461*).

Pyrrolostatin = 4-geranylpyrrole-2-carboxylic acid (**217**), isolated from the mycelium of *Streptomyces chrestomyceticus*, shows activity similar to that of vitamin E, as it is an inhibitor of lipid peroxidation in rat brain homogenate (*462*). The structure was elucidated by spectroscopic methods and later confirmed by synthesis (*463*).

6.1. Halogenated Monopyrroles from Bacteria

A number of most highly halogenated monopyrroles has been found in bacteria. Thus, tetrabromopyrrole (**218**) and hexabromo-2,2'-bipyrrole (**219**) have been isolated, along with pentabromopseudilin (**227**), from marine Chromobacteria (*464*). Tetrabromopyrrole is a very unstable compound, especially when exposed to light and oxygen, which inhibits the growth of the *Chromobacterium* itself. Most of the structures of halogenated pyrroles from bacteria are represented, however, by the pyrrolomycins a series of antibiotics, which are obtained from the fermentation broth of *Actinosporangium vitaminophilum* (*465*) a microorganism isolated from a soil sample collected at the Chikuma River of Nagano City in Japan. Pyrrolomycins can be differentiated into four different types: *i*) monocyclic pyrroles (Pyrrolomycin A), *ii*) aryl pyrroles, in which the phenyl and pyrrole rings are bound directly, (pyrrolomycin E), *iii*) benzyl derivatives, in which the phenyl and the pyrrole rings are linked through a methane bridge (pyrrolomycin B), and *iv*) benzoyl derivatives, in which the linking C-atom belongs to a carbonyl group (pyrrolomycins C, D and F). The structure of pyrrolomycin A (**220**) was established by analytical methods (*466*) and by synthesis (*467*).

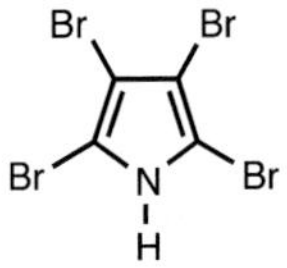

2,3,4,5-Tetrabromo-1*H*-prrole (**218**)

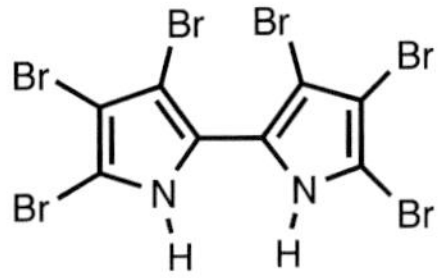

Hexabromo-2,2'-bipyrrole (**219**)

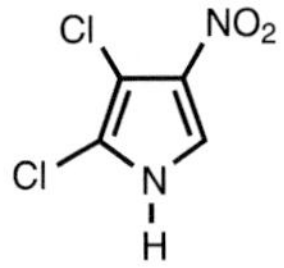

Pyrrolomycin A (**220**)

6.1.1. Halogenated Benzylpyrroles

In the series of pyrrolomycins, pyrrolomycin B (**221**) differs from the other members of antibiotics isolated from *Actinosporangium vitaminophilum* on the presence of a methane bridge between the pyrrole and

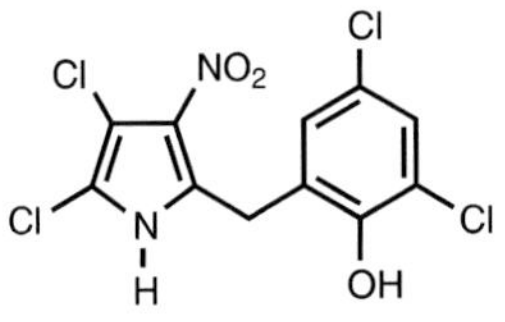

Pyrrolomycin B (**221**)

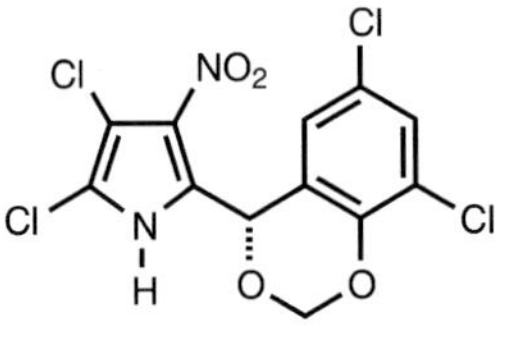

(−)-Pyrroxamycin (Dioxapyrrolomycin) (**222**)

the phenyl ring. Pyrrolomycin B has been obtained also from the culture filtrate of *Streptomyces fragilis* (*468*). The structure of **221** has been elucidated by single-crystal X-ray diffraction analysis (*469*).

Closely related to pyrrolomycin B is pyrroxamycin (*470*) which is also known in the literature as dioxapyrrolomycin (**222**) (*471*), antibiotic Al-R2081 (*472*), SS 46506A (*473*), or LL-F42248α (*474*) since its isolation from the culture broth of *Streptomyces* sp. was reported by five different laboratories independently from each other. Moreover, in 1990 Masuda *et al.* (*475*) reported the isolation of an actinomycete metabolite which they designated HS3. This compound was found to be identical with dioxapyrrolomycin. The structure of dioxapyrrolomycin, which was assigned mainly based on NMR analysis, was confirmed by X-ray diffraction analysis of the *N*-methyl derivative obtained on treatment of the antibiotic with diazomethane. Dioxapyrrolomycin ($[\alpha]_D^{25} = -110$ in ethanol) is the only chiral pyrrolomycin reported to date; its absolute configuration has been determined by the anomalous dispersion method (*471*).

In *Streptomyces fumanus* the pyrrole ring and bridging carbon atom C6 of dioxapyrrolomycin (**222**) are derived from L-proline, while the benzenoid ring is constructed from three acetate units. The process appears to be a modified polyketide pathway in which the L-proline serves as the chain starter, presumably as the coenzyme A thioester derivative. The methylenedioxy carbon atom arises from the methyl group of methionine, probably *via* S-adenosylmethionine. It is noteworthy that the nitro group of dioxapyrrolomycin is introduced into the molecule by an enzymatic process analogous to electrophlic nitration of aromatic compounds and not *via* oxidation of an amino group as in structurally related nitropyrrole antibiotics (*476*).

6.1.2. Halogenated Benzoylpyrroles

Within the group of antibiotics produced by *Actinosporangium vitaminophilum*, pyrrolomycin C (**223a**), pyrrolomycin D (**223b**), and the pyrrolomycins F (**224a–d**) are derivatives of α-benzoylpyrrole. Pyrrolomycins C and D were isolated from the toluene extracts of the culture broth, which contains also pyrrolomycins A, B, and E, by column chromatography on basic alumina (*477*). Their structures have been determined by means of spectroscopic and synthetic methods (*478*). Pyrrolomycins F, on the other hand, were produced by *Actinosporangium vitaminophilum* when bromide ion was added to the fermentation medium (*479*). Addition of other halide ions such as chloride, iodide, or fluoride showed no effect on pyrrolomycin production, affording polychlorinated pyrrolomycins A, B, C, D, and E, but no F components. The structures of all four pyrrolomycins F were determined by X-ray analysis and synthesis supported by spectroscopic analysis.

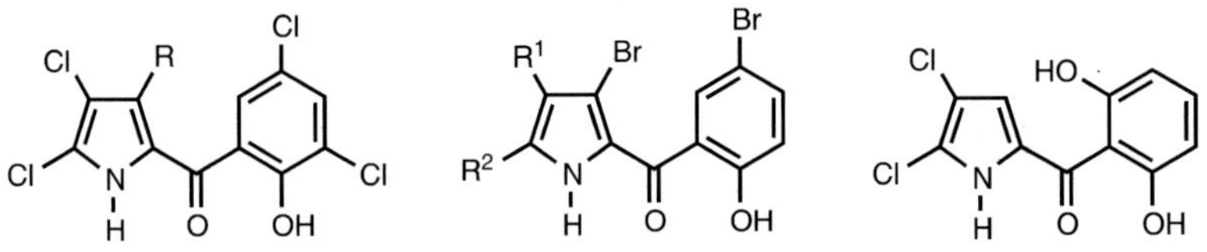

R = H : Pyrrolomycin C (**223a**) $R^1 = R^2 = Br$: Pyrrolomycin F_1 (**224a**)   Pyoluteorin (**225**)

R = Cl : Pyrrolomycin D (**223b**) $R^1 = Cl; R^2 = Br$: Pyrrolomycin F_{2a} (**224b**)

$R^1 = Br; R^2 = Cl$: Pyrrolomycin F_{2b} (**224c**)

$R^1 = R^2 = Cl$: Pyrrolomycin F_3 (**224d**)

Interestingly, the benzoyl group-containing pyrrolomycines are structurally related to pyoluteorin (**225**), the first halogenated pyrrole derivative encountered in nature. Pyoluteorin was isolated from *Pseudomonas aeruginosa* in 1958 by Takeda (*480*), who suggested also the actual structure of the molecule on the basis of analytical data (*481, 482*). The proposed structure has been confirmed repeatedly by synthesis of pyoluteorin itself (*483–490*) (*cf.* Scheme 6.1.) and analogs therefrom (*491*) as well as by X-ray diffraction analysis of its *O,O,N*-trimethyl derivative (*492*). *P. aeruginosa* is a hydrocarbon-utilizing microorganism, which produces also 3'-nitropyoluteorin and other derivatives of pyoluteorin (*493*).

The substitution pattern at the benzene ring of pyoluteorin suggests that this antibiotic is biosynthesized from a tetrakedide with proline (or

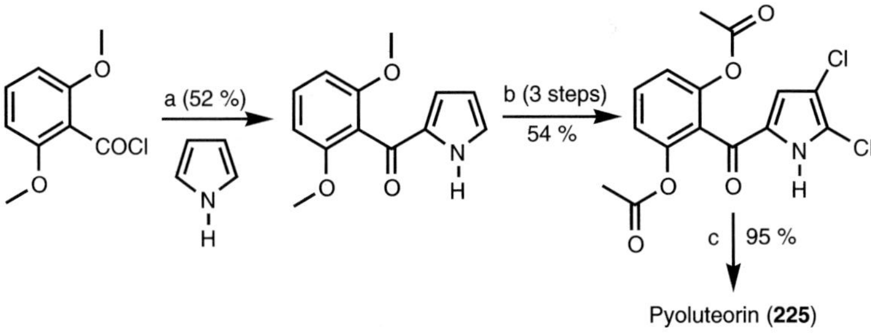

Pyoluteorin (**225**)

Scheme 6.1. A large-scale synthesis of pyoluteorin (**225**) (*490*): **a)** $SnCl_4$ in CH_2Cl_2. **b)** i: $AlBr_3$ in benzene; ii: $AcCl/(C_2H_5)_3N$ in ethyl acetate, iii: *N*-chlorosuccinimide in $CHCl_3$. **c)** HCl in methanol

References, pp. 140–188

Scheme 6.2. Biosynthetic pathway to pyoluteorin (225)

some equivalent) as the starter unit. This has been confirmed by incorporation experiments using [1,2-^{13}C$_2$]-acetate, as precursor, and analysis of the isolated labeled metabolite by ^{13}C-NMR spectroscopy (*cf.* Scheme 6.2.) (*494*).

6.1.3. Halogenated α-Arylpyrroles

Pyrrolomycin E (**226**), along with pyrrolomycins C (**223a**) and D (**223b**), has been isolated by column chromatography on basic alumina from the same culture broth of *Actinosporangium vitaminophilum*, which contains pyrrolomycins A (**220**) and B (**221**) (*477*). Structurally related to pyrrolomycin E is pentabromopseudilin (**227**), which has been isolated from *Pseudomonas bromoutilis* (*495*) and *Alteromonas luteoviolaceus* (*496*). Its structure was elucidated by X-ray diffraction analysis (*497*) and confirmed by chemical synthesis (*496, 498, 499*). Pentabromopseudilin, which is excreted into the culture medium by the same chromobacteria which produce the halogenated pyrroles **218** and **219**, was shown to be responsible for autoinhibition of the *Chromobacterium* itself, as well as for antibiotic action against other kinds of bacteria (*464*). The study of a series of pentabromopseudilin analogs suggests that the location of the phenyl group is essential for the high antibiotic and cytotoxic activity of pentabromopseudilin, since its C3 aryl analog is inactive. Contrarily, the degree of bromination of the antibiotic seems to be less important (*500–502*)

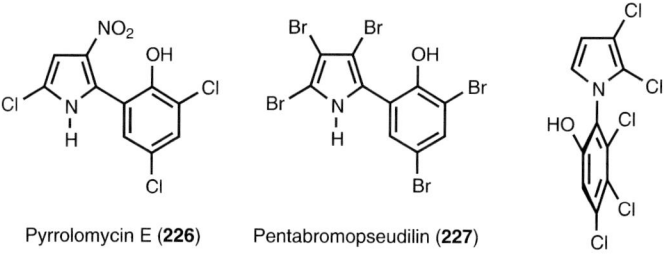

chlorine atom at C3 results also in a loss of activity. On the contrary, the nitro group on the phenyl ring is dispensable. Indeed, derivatives which lack this group are more active in a more broad spectrum of activity than pyrrolnitrin itself (*535, 536*).

Studies of the biosynthesis of pyrrolnitrin (**229**) in *Pseudomonas aureofaciens*, which were begun by Gorman and his co-workers in 1966, showed that this bacterium metabolizes exogenous tryptophan to yield pyrrolnitrin (*513, 537–539*). The C4-atom of the latter originates, as traced by ^{13}C-NMR spectroscopy (*540*), from the β-C-atom of the side-chain of trypthophan, and the indole *N*-atom of the latter becomes the amino group of aminopyrrolnitrin (**230**), which was shown to be a precursor of pyrrolnitrin (*538, 541*). As pyrrolnitrin itself is produced, along with aminopyrrolnitrin and the 3-chloro and 2,3-dichloro derivatives of the latter, by *Pseudomonas cepacia* (*516, 517*) the introduction of the halogen atoms in the pyrrole ring seems to be at the end of the biogenetic pathway of pyrrolnitrin. Thus, in the most recent work on pyrrolnitrin biosynthesis, Van Pée and co-workers propose the biosynthesis pathway depicted in Scheme 6.4. They shown that both chlorine atoms in pyrrolnitrin are introduced via NADPH-dependent enzymes, rather than conventional haloperoxidases (*542*). Most likely, arene epoxides act as intermediates, which can by opened by nucleophilic attack by Cl$^-$ before undergoing aromatization by dehydratation of the resulting chlorohydrin.

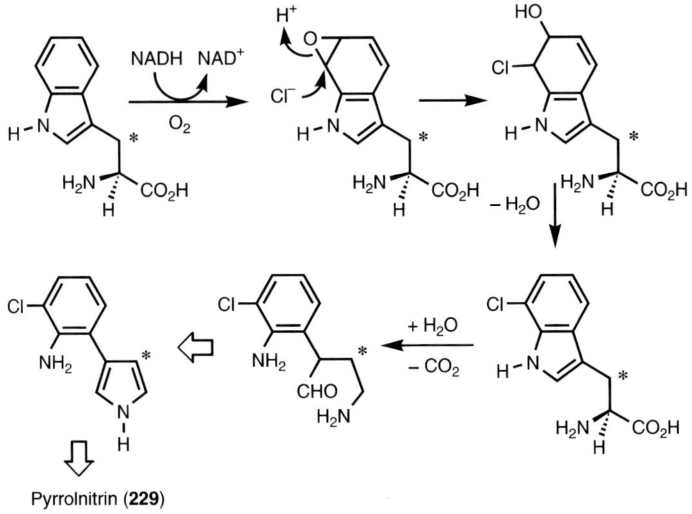

Scheme 6.4. Biosynthetic pathway to pyrrolnitrin (**229**). The labeled C-atom of L-tryptophan (marked with an asterisk in the figure) is retained at the 4-position of pyrrolnitrin

References, pp. 140–188

In a culture medium containing NH$_4$Br *Pseudomonas pyrrolnitrica* and *P. acidula* produce, among others brominated pyrrolnitrin derivatives, the dibromo analog of pyrrolnitrin (*i.e.* bromonitrin A), which is effective against *Trichophyton interdigitale* and *Penicillium chrysogenum* (*543–545*). On the other hand, some natural pyrrolnitrin analogs – *e.g.* 4'-dechloropyrrolnitrin (**231**) (*546*) isopyrrolnitrin, (**232**) (*547*) as well as oxypyrrolnitrin (**233**) (*548*) – have been isolated as minor components from *Pseudomonas* sp. cultures.

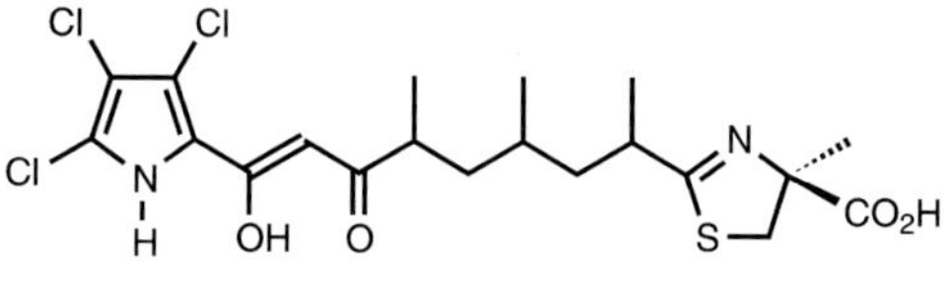

Thiazohalostatin (**234**)

A trichloropyrrole derivative, (−)-thiazohalostatin (**234**) has been isolated from *Actinomadura* sp. HQ24 by the Kirin Brewery Co. group (*549*). In order to determine the locations of the chlorine atoms, the tribromo analog was prepared by addition of KBr to the culturing medium of the actinomycete, and its ^{13}C-NMR spectrum was compared with that of **234**. Thus, from the upfield shift of the signals of the carbon atoms bound to the halogens on replacing bromine for chlorine atoms, the corresponding signals were assigned to three sp^2-hybridized C-atoms of the pyrrole ring. The stereochemistry of (−)-thiazohalostatin ($[\alpha]_D^{22} = -122$ in methanol) is not known.

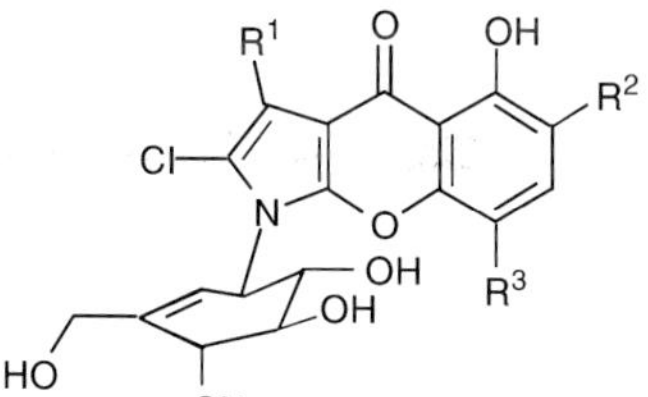

Pyralomycins 1 (**235**)

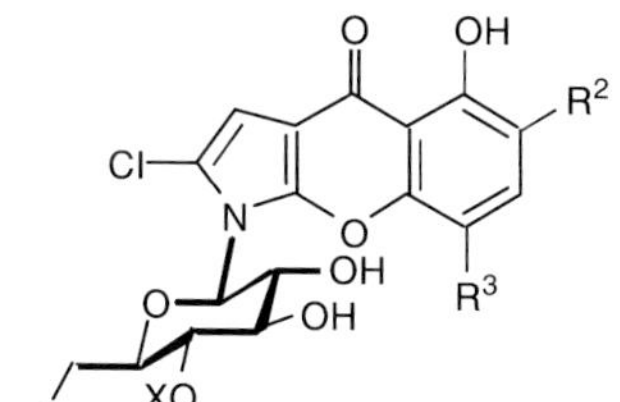

Pyralomycins 2 (**236**)

	R^1	R^2	R^3	X
a	H	Cl	CH$_3$	CH$_3$
b	H	CH$_3$	Cl	CH$_3$
c	H	Cl	CH$_3$	H
d	Cl	Cl	CH$_3$	H

	R^2	R^3	X
a	Cl	CH$_3$	CH$_3$
b	CH$_3$	Cl	CH$_3$
c	Cl	CH$_3$	H

A series of antibiotics called pyralomycins, (**235**) and (**236**), has been isolated from the culture broth of *Microtetraspora* (*Actinomadura*) *spiralis* (*550, 551*) and their structures and relative configurations were elucidated NMR spectroscopy (*552*). The absolute configurations of pyralomycin 1a and pyralomycin 2a were determined by X-ray diffraction analysis of the corresponding 7'-*O*- and 2'-*O*-*p*-bromobenzoate, respectively, using anomalous scattering of the bromine atom. The absolute configurations of pyralomycins 1b–d and 2b–c were suggested to be the same as those of pyralomycin 1a and pyralomycin 2a, respectively, by comparison of the circular dichroism spectra (*553*). Pyralomycin 2c has been obtained by chemical total synthesis (*554*).

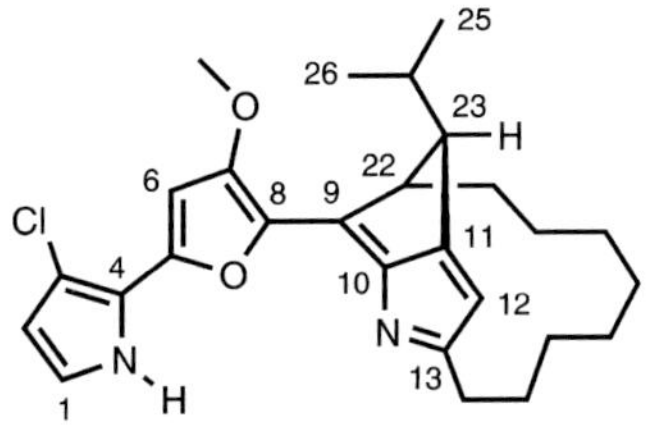

Roseophilin (**237**)

Roseophilin (**237**) is a structurally unique metabolite isolated from a culture broth of *Streptomyces griseoviridis* (*555*). The pyrrole-containing compound exhibits cytotoxicity against several human epidermoid and leukemia cell lines in the submicromolecular range. Several chemical syntheses of roseophilin have been reported, which differ on the strategy employed for the preparation of the macrotricyclic part of the molecule (atoms C9 to C26), whereas the coupling reaction of the latter with the *N*-protected furanylpyrrole moiety is common to all them (*cf.* scheme 6.5.) (*556–560*).

6.2. Pyrrole-2-carboxylates from Bacteria

As in eukaryotes, a number of important pyrrole metabolites from bacteria are esters or amides of pyrrole-2-carboxylic acid. Thus, antibiotic RP 18,631 = chlorobiocin (**238**), which is produced by several *Streptomyces* species (*S. hygroscopicus, S. roseochromogenus* var. *oscitans*, and *S. albocinerescens*), belongs together with novobiocin (**239**) and the coumermycins (see below) to a group of structurally related bacterial metabolites which contain substituted coumarins as a part of their structure. In **238** a chlorine atom and a 5-methylpyrrol-2-yl residue replace the methyl and carbamoyl group, respectively, of

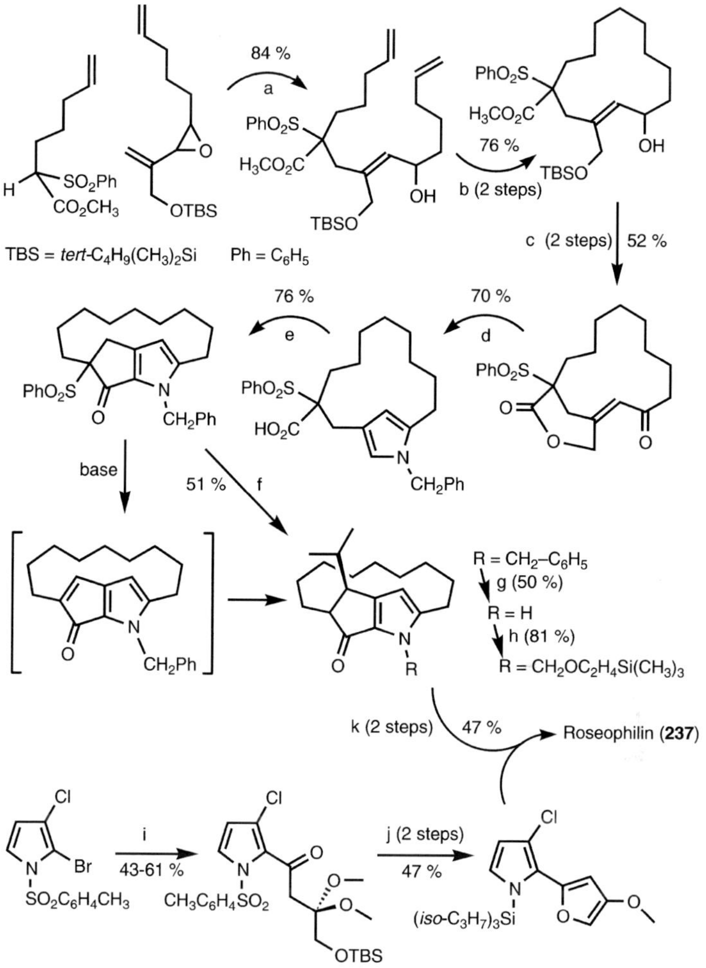

Scheme 6.5. A total synthesis of roseophilin (**237**) based on ring closing methathesis (*558*)
a) Pd[P(C₆H₅)₃]₄ in tetrahydrofuran (THF). **b)** *i*: Cl₂[P(C₆H₁₁)₃)₂Ru=CH–CH=C(C₆H₅)₂ in CH₂Cl₂; *ii*: H₂/RhClP(C₆H₅)₃ in ethanol. **c)** *i*: (C₄H₉)₄N⁺F⁻/NH₄F in THF; *ii*: 1,1,1,-Triacetoxy-1,1-dihydro-1,2.benziodoxol-3(1H)-one (Dess-Martin periodinane) in CH₂Cl₂. **d)** C₆H₅CH₂NH₂/Pd[P(C₆H₅)₃]₄ in THF. **e)** *i*: 1-(*N,N*,dimethylamino)-1-chloro-2-methyl-prop-1-ene in CH₂Cl₂; *ii*: SnCl₄ in 1,2-dichloroethane. **f)** (*iso*-C₃H₇)(CH₃)₂ZnMgCl/*tert.*-C₄H₉OK in THF. **g)** Ca in liquid NH₃, then C₅H₆N⁺ CrO₃Cl⁻ in CH₂Cl₂ **h)** KH/2-(trimethylsilyl)ethoxymethylchloride in dimethylformamide. **i)** *n*-C₄H₉Li in THF, then successively ZnCl₂, Pd[P(C₆H₅)₃]₄ and 4-[*tert*-Butyl)dimethylsilyl) oxy]-3.3-dimethoxybutanoic acid chloride. **j)** *i*: pyridinium *p*-toluenesulfonate in methanol; *ii*: K₂CO₃ in methanol, then KH and tri-*iso*-propylsilyl chloride in THF. **k)** *i*: *n*-C₄H₉Li in THF, then CeCl₃ *ii*: (C₄H₉)₄N⁺F⁻ in THF

novobiocin (**239**) (*561, 562*) the structure of which had been elucidated previously. Chlorobiocin shows greater antibacterial activity *in vitro* and *in vivo* against *Micrococcus* and *Neisseria* than the other antibiotics of this group with no increase of toxicity.

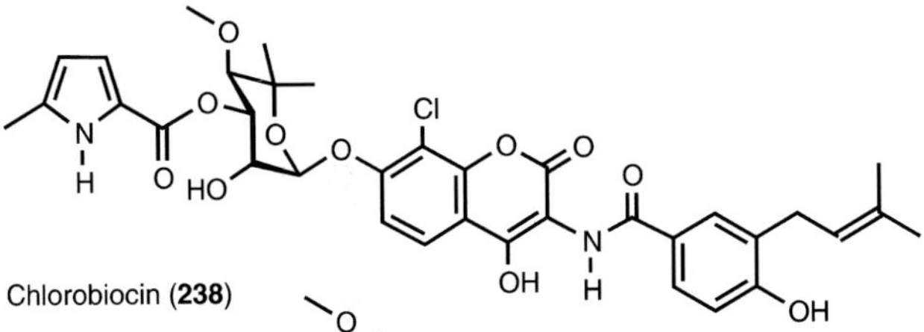

Chlorobiocin (**238**)

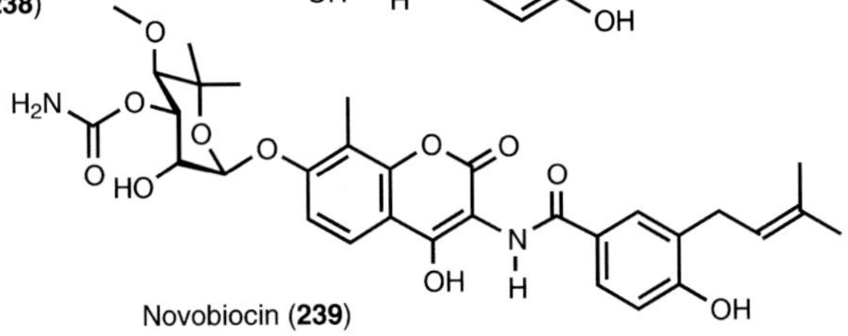

Novobiocin (**239**)

In contrast to the synthesis of coumarin in plants – where it is derived from L-phenylalanine *via* cinnamic and coumaric acids – the coumarin moiety of the above antibiotics is formed from tyrosine by oxidative cyclization. Also, L-tyrosine is utilized in the formation of the substituted 4-hydroxybenzoic acid portion of novobiocin and chlorobiocin, while the sugar moiety (called noviose) is derived from glucose by epimerization and reduction of a nucleotide diphosphate intermediate (*563*).

The complete structure of coumermycin A_1 (**240a**) (*564–566*) and coumermycin A_2 (**240b**) (*567*) both isolated from *Streptomyces rishiriensis*, were first disclosed by Kawaguchi, Okanishi, and Miyaki in 1965. In the same year, sugordomycin was isolated, as a mixture of eight antibiotics, from *S. hazeliensis* in the laboratories of the F. Hoffmann-La Roche Inc. at Nutley (New Jersey) and Basle (Switzerland) (*568–570*). Sugordomycin I and IV proved to be identical with coumermycin A_1 and A_2, respectively (*568*). The α-configuration of the glykoside linkages has been proved by X-ray diffraction analysis of a brominated degradation product of coumermycin A_1 (*571*). Partial syntheses of coumermycin A_1 and A_2 (*568, 572, 573*) as well as a total synthesis of the pyrrole-noviose-coumarin fragment bound to the central 3-methylpyrrole-2,4-dicarboxylic acid moiety of coumermycin A_1 (*574*) have been reported.

References, pp. 140–188

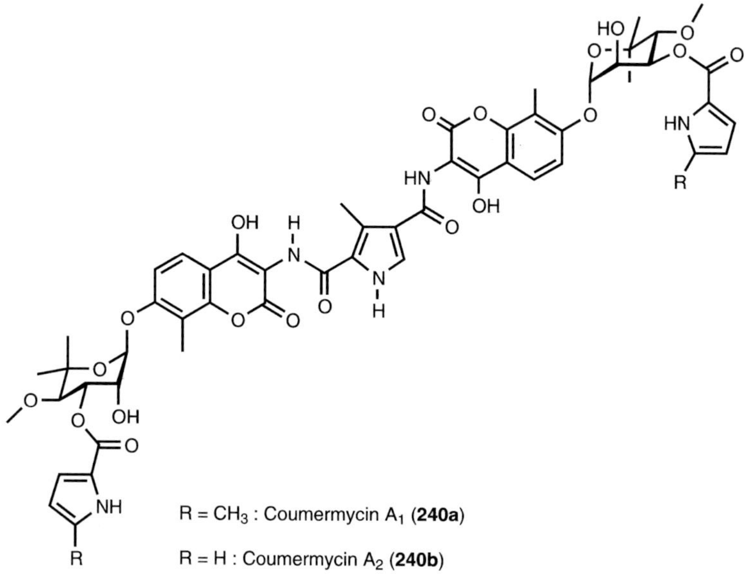

R = CH₃ : Coumermycin A₁ (**240a**)
R = H : Coumermycin A₂ (**240b**)

Coumermycin A_1 (**240a**), which was also isolated from *S. spinicoumarensis* and *S. spinichromogenes* cultures (*575*) was early recognized as a potentially useful antibiotic, which inhibits growth of *Mycobacterium tuberculosis* (*576*), *Staphylococcus aureus* (*577, 578*), *Neisseria meningitidis* (*579, 580*), *Diplococcus pneumoniae*, and *Haemophilus influenzae* (*581*). It is known that coumermicyn – as well as novobiocin (see above) – inhibits, in *Escherichia coli*, the supercoiling of DNA catalyzed by gyrase, thus preventing the replication of DNA (*582–584*).

In spite of its excellent *in vitro* activity, the insoluble character of coumermycin A_1 (**240a**) makes parental administration unsatisfactory. Poor oral absorption was also observed. Because of these shortcomings a large number of analogs have been prepared by partial syntheses, replacing the pyrrole rings at one or both ends of the molecule by monocarboxylic acids. Some of them are analogs of chlorobiocin (s. above) and they are less toxic than the natural compound (*585–587*). The structure-activity relationship of coumermycins and their analogs has been reviewed (*588, 589*).

The chemical similarity of the coumermycins and novobiocin indicates analogous biosynthetic pathways of the sugar and aminocoumarin moieties which they have in common. However, both the coumermycins and chlorobiocin are characterized by pyrrole moieties, which are not present in novobiocin. Studies by Scannell and Kong (*590*) have shown that proline serves as an efficient precursor of the pyrrole groups or the coumermycins. As expected, methionine contributed the methyl group in the methylpyrrole substituent of coumermycin A_1.

R = OH : Nargenicin A_1 (**241a**)

R = H : (+)-18-Deoxynargenicin A_1 (**241b**)

The antibioticum CP-47,444, later called nargenicin A_1 (**241a**), which is active against *Staphylococcus aureus*, was first isolated from cultures of *Nocardia argentinensis* and proved to be 9-*O*-nodusmicin pyrrole-2-carboxylate (*591, 592*).

Nodusmicin (lat. *nodus* = knot) had been isolated previously from cultures of *Saccharopolyspora hirsuta* (*593*) which produces also 18-deoxynargenicin (Antibiotic U-61,732: **241b**) (*594*). As nodusmicin, the structure of which had been elucidated by X-ray diffraction analysis (*593*), can be converted synthetically into nargenicin, the structure and relative configuration of the latter could be established by a combination of chemical and spectroscopic methods (*592*). The absolute configuration of nargenicin was subsequently determined by CD exciton chirality analysis of the derived 11-(*p*-nitrobenzoyl)-18-acetyl ester (*595*). On the other hand, the structure of (+)-18-deoxynargenicin (**241b**), which is synthetically accessible from nargenicin (*596*) has been confirmed by total synthesis (*597*).

Nodusmicin, and hence nargenicin, are members of a group of naturally occurring alicyclic polyketides containing a characteristic *cis*-fused octalin ring system, the polyketide origin of which was originally confirmed by biosynthetic studies which established that nargenicin is derived from five acetate and four propionate building blocks (*598, 599*). Thus, the formation of nargenicin can be envisaged as taking place in three distinct phases: (1) elaboration of a branched-chain, partially saturated polyketide fatty acid (2) cyclization of the latter with generation of the macrolide and octaline ring systems, and (3) late stage oxidations, methylation, and esterification with pyrrolecarboxylic acid. Complementary experiments with $^{18}O_2$ established the origins of the various oxygen atoms and ruled out plausible epoxyolefin cyclization mechanisms, thus suggesting that the characteristic octalin ring system is generated by an enzyme-catalyzed intramolecular Diels-Alder reaction (*600*).

References, pp. 140–188

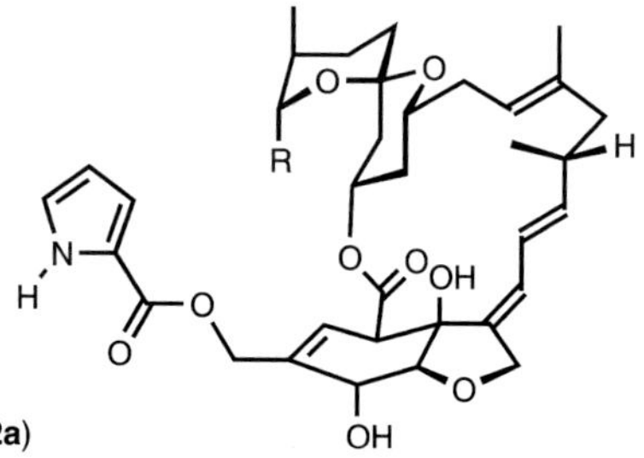

R = CH$_3$: (+)-Milbemycin α_9 (C$_1$) (**242a**)

R = C$_2$H$_5$: (+)-Milbemycin α_{10} (C$_2$) (**242b**)

Milbemycins, (**242a**) and (**242b**), are 16-membered macrolide antibiotics which are produced in submerged cultures of *Streptomyces hygroscopicus* subsp. *aureolacrimosus*. From the culture broth 13 milbemycins were purified to homogeneity by column and thin-layer chromatography in silica gel and alumina. Two of the antibiotics (milbemycin α_9 and milbemycin α_{10}) were characterized as esters of pyrrole-2-carboxylic acid. Milbemycins, which are structurally closely related to the avermectins, isolated from the culture broth of *S. avermitilis*, possess insecticidal and acaricidal activity (*601*).

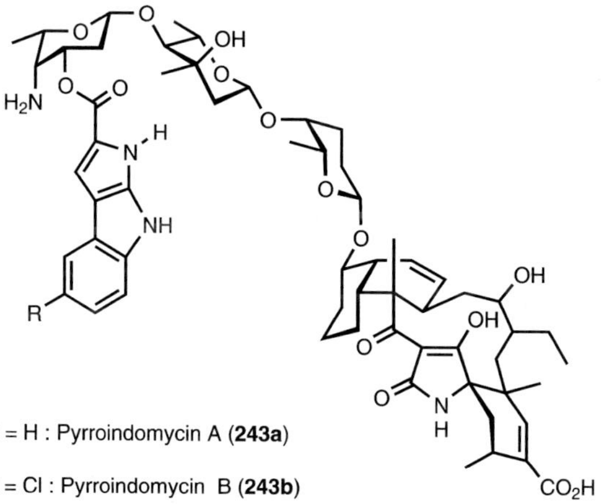

R = H : Pyrroindomycin A (**243a**)

R = Cl : Pyrroindomycin B (**243b**)

Pyrroindomycin A (**243a**) and pyrroindomycin B (**243b**) were found to be two principal components of the antibiotic complex isolated from fermentations of culture LL-42D005, which is a strain of *Streptomyces rugisporus*. Their structures, which recall those of pyrrolosporin A (**249**), were determined by using 1- and 2-D NMR, mass spectrometry,

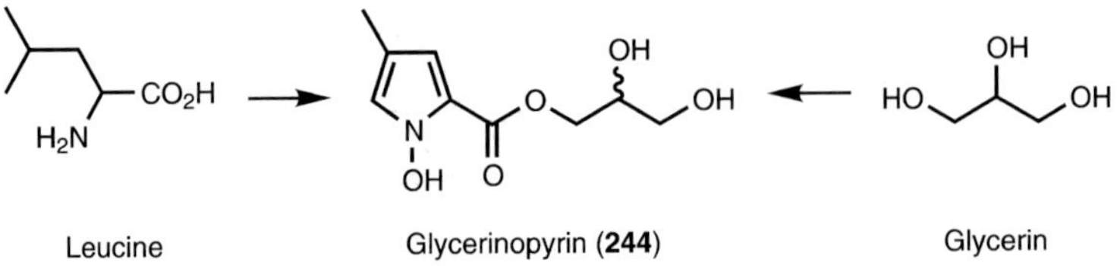

Scheme 6.6. Biogenetic precursors of glycerinopyrin (**244**)

and chemical degradations. The pyrroindomycins contain a tetramic acid nucleus and are structurally related to other antibiotics containing a tetronic acid moiety and a macrocyclic ring (*602*).

These two compounds demonstrated excellent *in vivo* activity against Gram-positive bacteria. It is believed that the mode of action is interference with the integrity of the bacterial membrane (*603*). Although the pyrroindolizins are active in vitro against clinical isolates, they demonstrate limited *in vivo* activity and a low therapeutic index.

The unusual *N*-hydroxypyrrole metabolite glycerinopyrin (**244**) has been isolated, as a mixture of 85% (*S*) and 15% (*R*) enantiomers at the side-chain stereogenic center, from a Sri Lankan strain of *Streptomyces violaceus* (*604*). Expectedly, the side-chain is biosynthesized from glycerin. The origin of the pyrrole ring, however, takes place by an unprecedented pathway in which leucine is the immediate precursor (*cf.* Scheme 6.6.) (*605*). The same pathway may lead to the formation of **9**, the pheromone of *Atta texana*, which (like **244**) is a derivative of 3-methylpyrrole-2-carboxylic acid.

6.3. Pyrrole-2-carboxamides from Bacteria

6.3.1. The Distamycin Group

(+)-Kikumycin A (**245a**) and (+)-kikumycin B (**245b**) are basic antiviral antibiotics isolated from the culture filtrate of *Streptomyces phaeochromogenes* (*606*). Their structures were established by chemical and spectroscopic studies including high resolution mass spectrometry (*607, 608*). Owing to the bathochromic effect of their C12=C13-bond kikumycins absorb at longer wavelenghts ($\lambda_{max} = 357$ nm in 0.1 N NaOH) than the structurally related antibiotics netropsin (**246**), distamycin A (**247**) and anthelvencins (**248**).

References, pp. 140–188

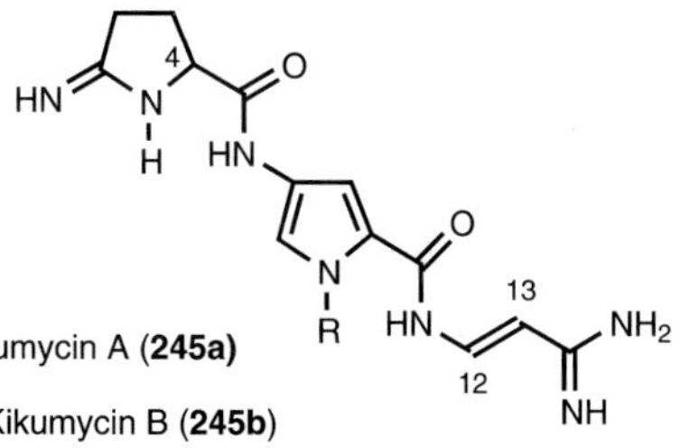

R = H : Kikumycin A (**245a**)
R = CH₃ : Kikumycin B (**245b**)

The structure of the corresponding 12,13-dihydro derivative of kikumycin B (dihydrokikumycin B) was confirmed by synthesis of both enantiomers (*609*). As the structurally related antibiotics netropsin, distamycin A, and anthelvencins (see below) kikumycins are specific inhibitors of the DNA-polymerase.

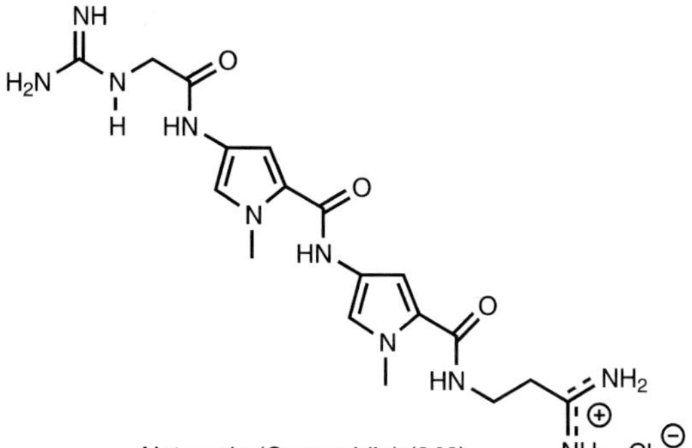

Netropsin (Congocidin) (**246**)

Netropsin (**246**), the first oligopeptide derived from *N*-methyl-4-aminopyrrole-2-carboxylic acid which was obtained from actinomycetes cultures, was isolated in 1951 by Finlay from *Streptomyces netropsis* (*610*). Later on, sinamomycin (*611*) and the so-called antibiotics T 1384 (*612*) and 2814 A (*613*) were isolated in other laboratories and identified with netropsin. Initial efforts to elucidate the structure of this antibiotic were undertaken by van Tamelen and his co-workers, who determined the correct elemental composition ($C_{18}H_{26}N_{10}O_3$) of the compound (*614*). The structure of netropsin suggested by Waller *et al.* (*615*) was revised (*616, 617*) when its identity with congocidin (*618*), the structure of which had been confirmed by synthesis (*619, 620*) was proved (*cf.* Scheme 6.7.). After the structure of netropsin was known, a series of closely related oligopeptide antibiotics were isolated from Streptomycetes, namely distamycin A, anthelvencin, and the kikumycins.

Until now, the biosynthesis of netropsin (**246**) has not been elucidated in detail. Incubation experiments with ^{14}C-labeled substrates show that the

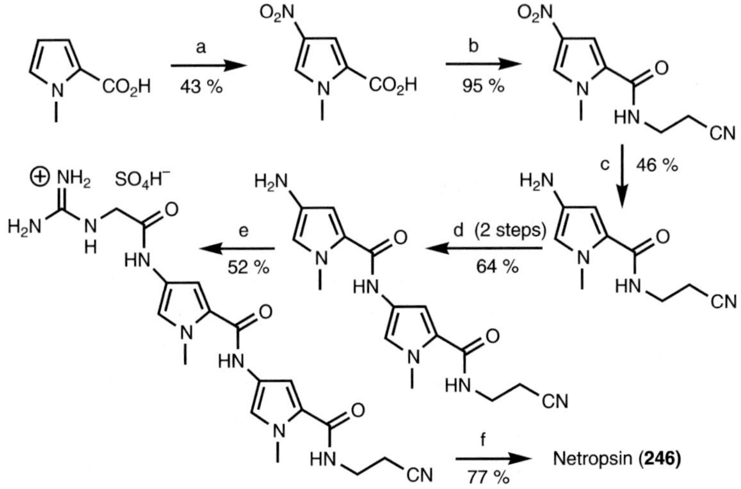

Scheme 6.7. An improved total synthesis of netropsin (**246**) and distamycin A (**247**) (*620*): **a)** HNO₃ in acetanhydride; **b)** SOCl₂, then 3-aminopropionitrile, **c)** H₂ on Pd/C in methanol; **d)** *i:* 1-methyl-4-nitropyrrole-2-acyl chloride and ethyldiisopropylamine (Hunig's base) in tetrahydrofuran; *ii:* H₂ on Pd/C in methanol; **e)** guanidine-acetic acid hydrochloride and dicyclohexylcarbodiimide in dimethylsulfoxide; **f)** HCl in ethanol.
Repetition of steps **d)** before **e)** leads to distamycin A (**247**)

α-C-atom of glycine is incorporated both in the pyrrole rings and the glyocyamyl residue. The guanidine residue, on the contrary, proceeds from arginine, and is probably incorporated by action of a transamidinase (*621*).

Netropsin (**246**) is active against Gram-positive and Gram-negative microorganisms. The same antibiotic had been named congocidin (see above) because of its activity against *Tripanosoma congolense*, a spirochaete which is particularly resistant to other antibiotics.

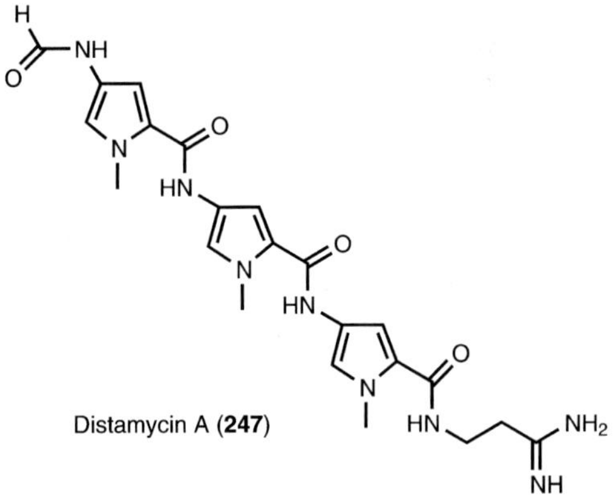

Distamycin A (**247**) is a specific inhibitor of the DNA-polymerase which was isolated from *Streptomyces distallicus* in 1967 by Arcamone *et al.* (*622*), who established its structure by chemical synthesis (*623, 624*). The interest of distamycin A as a potential antiviral drug with favorable therapeutic indices brought about the development of both improved modifications of the original procedure (*625, 626*) and of more recent synthetic approaches (*620, 627*). Moreover, a large number of analogs of both **246** and **247** have been prepared, and their antibiotic properties have been extensively investigated (*628–634*). Netropsin (**246**) and distamycin A (**247**) are antiviral antitumor antibiotics that, although too toxic for clinical use, have received extensive study as prototype lexitropsins (*i.e.* information reading oligopeptides) possessing base-specific yet non-intercalative DNA-binding properties (*635–637*). Both antibiotics exert their biological activities by blocking the template function of DNA by binding to (A-T)$_n$ sequences (*638–645*).

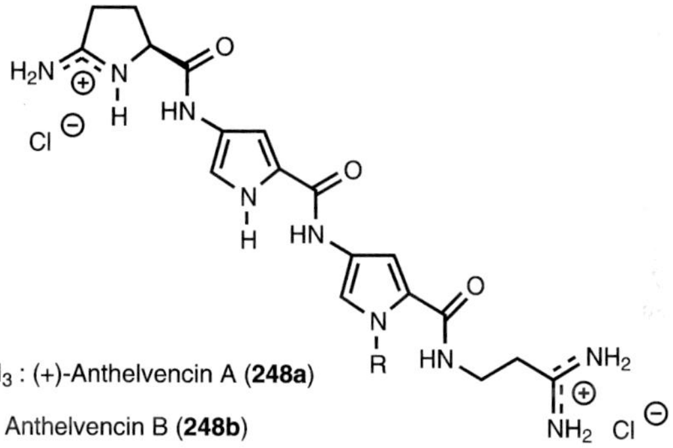

R = CH$_3$: (+)-Anthelvencin A (**248a**)

R = H : Anthelvencin B (**248b**)

Anthelvencin A (**248a**) and anthelvencin B (**248b**) have been isolated from culture filtrates of an strain of *Streptomyces venezuelae* (*646*). The relationship of the hydrolysis products of the anthelvencins with the known degradation products of netropsin revealed the similar structures of these antibiotics. Data from nuclear magnetic resonance studies indicated, however, that anthelvencins A and B differ by the number of *N*-methylpyrrole units in their molecules. The ^{1}H-NMR spectrum of **248a**, the predominant component in the culture filtrates, revealed only one *N*-methyl group in the molecule, whereas **248b** has no *N*-methyl group. The antibiotic properties of the anthelvencins resemble those reported for netropsin and distamycin A. Anthelvencin A inhibits a broad spectrum of microorganisms *in vitro* and is effective in controlling nematode infections in mice and swine. The antimicrobial activity of anthelvencin B (**248b**) is approximately half that of anthelvencin A

(**248a**). As other antibiotics of this group, anthelvencin A is a specific inhibitor of the DNA-polymerase (*647*).

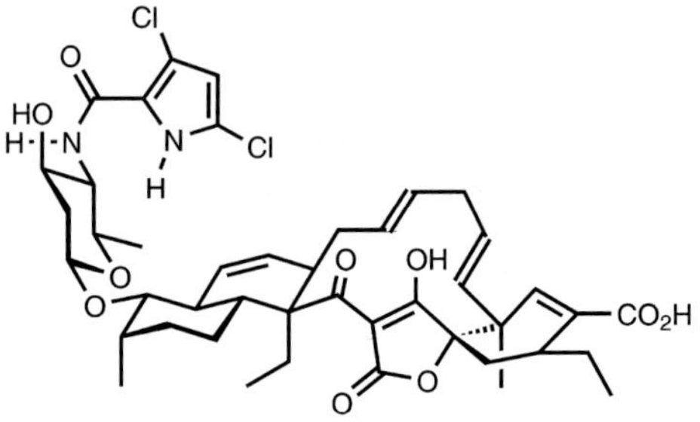

Pyrrolosporin A (**249**)

Pyrrolosporin A (**249**) is a new macrolide antitumor antibiotic possessing an unusual spiro-α-acyltetronic acid substructure and a glycosidic moiety (designated as pyrrolosamine) consisting in a 4-amino-2,4,6-trideoxyhexopyranose, which is *N*-acylated with 3,5-dichloropyrrole-2-carboxylic acid. The antibiotic was obtained as a white crystalline solid (mp. 235°C) from the fermentation broth of an streptomycete (*Micromonospora* sp.) isolated from a soil sample collected at Puerto Viejo (Peru) by vacuum liquid chromatography, crystallization and reverse phase HPLC (*648*). The structure of **249**, which recall that of the pyrroindomycins (**243**), has been determined by a combination of spectroscopic methods (NMR, MS, etc.), X-ray diffraction analysis and chemical degradation studies (*649*). However, the absolute configuration of pyrrolosporin A, which in methanol solution is weakly dextrorotatory ($[\alpha_D^{26}] = +2.8$), has not yet been established.

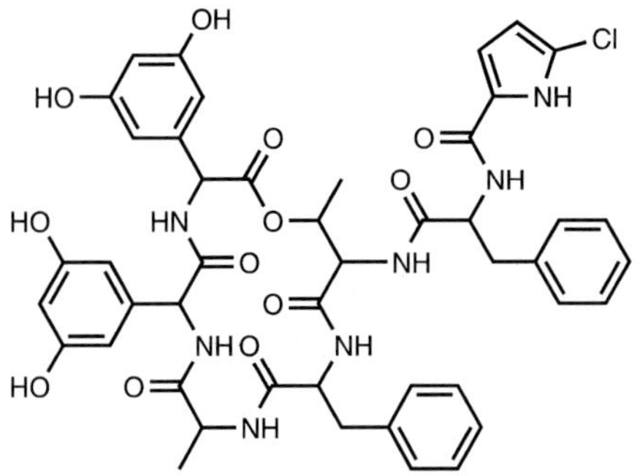

S 5185 RP (**250**)

References, pp. 140–188

The presence of a chlorinated pyrrole-2-carboxamide moiety has also been demonstrated in an oligopeptide-macrolactone described in the patent literature. The so-called antibiotic S 5185 RP (**250**) is an antibiotic, which was obtained by cultivation of a *Streptomyces* strain in the laboratories of Rhone-Poulenc Sante. For this compound immuno-suppressive properties have been reported (*650*).

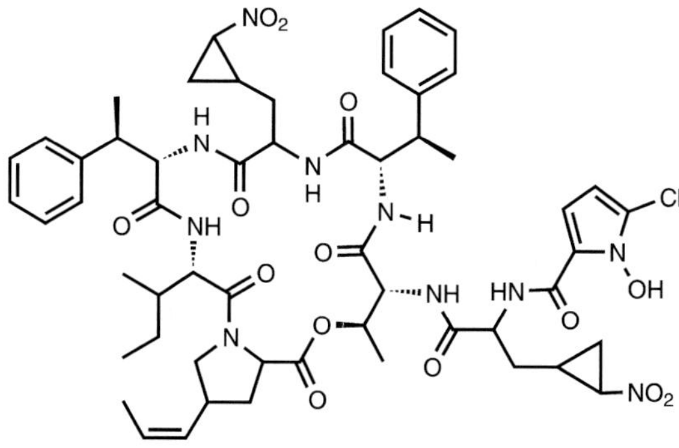

Hormaomycin (**251**)

Besides the above mentioned antibiotic S 5185 RP, hormaomycin (**251**), another peptide-macrolactone containing a 5-chloropyrrole-2-carboxamide residue, has been isolated from *Streptomyces griseoflavus*. Hormaomycin belongs to a group of intercellular hormones, which influences both the formation of aerial mycelium and the production of antibiotics in *S. griseoflavus* as well as in some other streptomycetes (*651*). Hormaomycin is also active against some Gram-positive bacteria, particularly *Arthrobacter oxydans* and *A. crystallopoietes*. The constituent amino acids of the antibiotic were delivered by acidic hydrolysis and assigned by high resolution GC/MS analysis in combination with extended 2D-NMR spectroscopy experiments (*652*). From the latter, it became plausible that the *N*-terminus of the peptide chain is acylated by a chlorine containing derivative of pyrrole-2-carboxylic acid, which, however, could not be detected in the hydrolysates of the peptide.

The unusual *N*-hydroxy-1*H*-pyrrole substructure encountered in glycerinopyrin (**244**) and hormaomycin (**251**) is present also in chromoxymycin (**252**), an antitumor antibiotic produced by *Streptomyces rubropurpureus*. Its structure was elucidated (*653*) mainly on the basis of ^{1}H- and ^{13}C-NMR spectral analysis, by comparison with the already known structure of hedamycin (**253**), which contains a similar chromophore consisting of a 4*H*-anthra[1,2-*b*]pyrone ring system (*654*) Hedamycin, however, differs from **252** not only on the configuration of all but one of the stereogenic centers of the desoxysugar moiety at C10 but also on the lack of the pyrrole-2-carboxamide residue. Most likely, therefore, hedamycin (**253**), which is also produced by *Streptomyces* species, is a biogenetic precursor of chromoxymycin (**252**), the pyrrole moiety of which is introduced in a later stage of its biosynthesis.

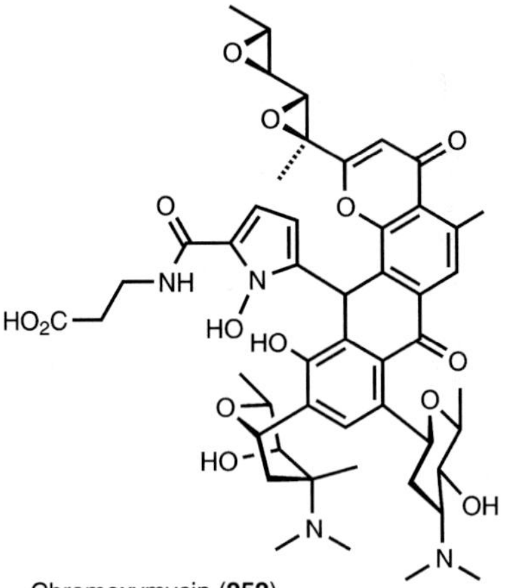

Chromoxymycin (**252**)

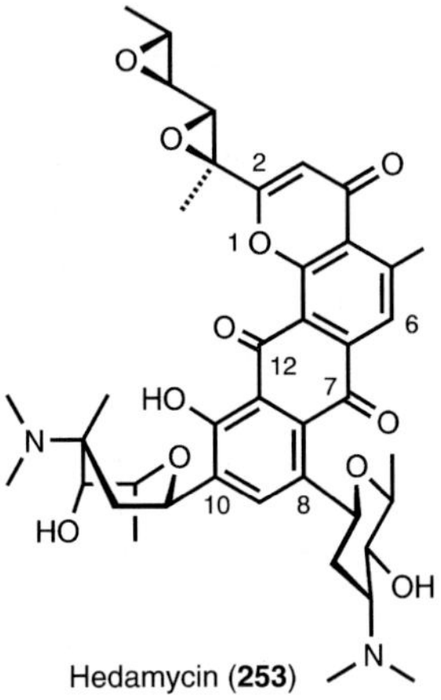

Hedamycin (**253**)

6.4. Pyrrol-2-carbacyl Derivatives from Bacteria

Antibiotic A 23 187 = calcimycin (**254a**) is a divalent calcium selective ionophore which was isolated from cultures of *Streptomyces chartreusensis* as the mixed magnesium–calcium salt (*655*). Its structure was determined by X-ray diffraction analysis on the corresponding crystalline free acid, the absolute configuration of which was first elucidated by chemical and physical methods (*656*) and subsequently confirmed by different chemical total syntheses (*657–669*).

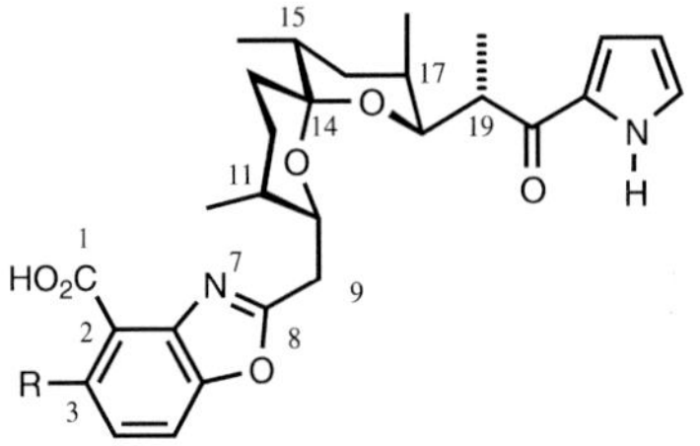

R = NH-CH$_3$: (–)-Calcimycin (**254a**)
R = H : Cezomycin (**254b**)
R = OH : Antibiotic AC 7230 (**254c**)

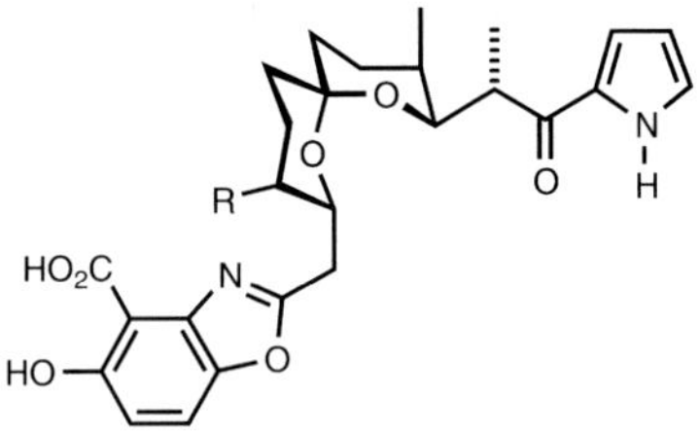

R = H: Routiennocin (Antibiotic CP-61,405) (**255a**)
R = CH$_3$: Antibiotic X-14885 A (**255b**)

References, pp. 140–188

Calcimycin (**254a**) is one of a very small group of ionophore antibiotics able to transport alkaline-earth cations across biological membranes. The structure of **254a** reveals, therefore, some novel features uncharacteristic of the known polyether ionophore antibiotics. A unique benzoxazole ring system along with an α-ketopyrrole are bridged by a spiro ketal of the oleane (1,7-dioxaspiro[5.5] undecane) type. Since its first isolation in 1972, calcimycin – one of the most frequently quoted chemicals in the biochemical literature – has been widely largely used as a tool for investigating the Ca^{++} metabolism in a variety of tissues, cells and subcellular organelles (*670*). Thermodynamic (*671*) as well as kinetic (*672*) spectrophotometric (*673–675*), potentiometric (*674*), and electrochemical (*676*) studies of its association with alkaline and/or alkaline-earth cations have been carried out. These and other studies (*677*) indicated that cation complexation by the calcimycin molecule requires the reorientation of one or both aromatic moieties with respect to the adjacent asymmetric centers. Crystallographic studies have shown that the benzoxazole ring provides two coordinating sites in the 2:1 complexes investigated with calcium (*678*) and magnesium (*679*). Rate constants for the different steps in the diffusion process have been studied in model phospholipid membranes (*680*). The Ca^{++} selectivity of calcimycin *vs.* Mg^{++} has predominantly a kinetic origin, with faster formation and dissociation rates of the calcium complexes.

Structurally related to calcimycin are cezomycin = desmethylaminocalcimycin (**254b**) and antibiotic AC 7230 (**254c**). The latter has been isolated from *Dactylosporangium* sp. (*681*). Cezomycin, the structure of which has been confirmed by chemical synthesis (*669*), has been obtained from a culture medium of *Streptomyces chartreusensis* in which addition of L-tryptophan inhibited production of calcimycin and led to the formation of **254b** (*682*). According to this observation, it is plausible that the benzoxazole moiety of calcimycin is formed from 3-hydroxyanthranilic acid, itself formed from tryptophan *via* the classic metabolic pathway involving 3-hydroxykynurenin (*682*). Thus, supplementary tryptophan in the medium is metabolized not only to give 3-hydroxyanthranilic acid but also large amounts of anthranilic acid, which may hinder further enzymatic transformations of 3-hydroxyanthranilic by competitive inhibition and act itself as substrate of the biosynthesis of the benzoxazole ring system of cezomycin.

Routiennocin = antibiotic CP-61,405 (**255a**) and Antibiotic X-14885A (**255b**) are two further ionophores closely related to calcimycin, which have been isolated from *Streptomyces* sp. Antibiotic X-14885, the structure of which was elucidated by X-ray diffraction analysis (*683*), differs from calcimycin (**254a**) by the lack of the methyl group at C15 and the replacement of a hydroxy group for the methylamino function of the benzoxazole part of the molecule. Particularly the latter replacement has a drastic effect in the acid-base properties of X-14885A and on the acid-catalyzed dissociation pathway of its complexes (*684, 685*). Routiennocin (**255a**) was isolated from a microbial fermentation of *Streptomyces routiennii* (*686*); it differs from antibiotic X-14885A (**255b**) by the absence of the methyl group on C11. Its structure has been confirmed by total synthesis (*687*).

After calcimycin (**254a**), the second carboxylic acid ionophore which became known was antibiotic X-14 547 A, which was later called indanomycin (**256**) by Ley and his associates (*688*). Indanomycin was isolated in 1978 along with pyrrole-2-carboxylic acid (*cf.* Ref. 9), as the mayor metabolites, by silica gel chromatography of the crude methanol extract from cells of *Streptomyces antibioticus*, in the laboratories of Hoffmann La-Roche in Nutley (NJ) (*689–691*). Its complete structure was elucidated by X-ray diffraction analysis on the corresponding (*R*)-(+)-1-amino-1-(4-bromophenyl)ethane salt, which was used as internal reference for the determination of the absolute configuration of the antibiotic (*689*). The structure of indanomycin has been confirmed by several chemical total syntheses (*688, 692–696*). A complete assignment of the ^{1}H- and ^{13}C-NMR signals of indanomycin has been carried out by Lallemand *et al.* (*697*). At 70°C, in acetonitrile solution containing lithium tetrafluoroborate, indanomycin rearranges stereospecifically to afford a derivative of indano[5,4-*f*]indole (*cf.* Scheme 6.8.) (*698*). Remarkably, indanomycin (**256**) is capable of complexing and tranporting divalent as well as monovalent and trivalent metal ions, even though it has only a single tetrahydropyranyl residue (*699*). It has been concluded that indanomycin acts primarily as a K$^+$ carrier in the mitochondrial membrane (*700*).

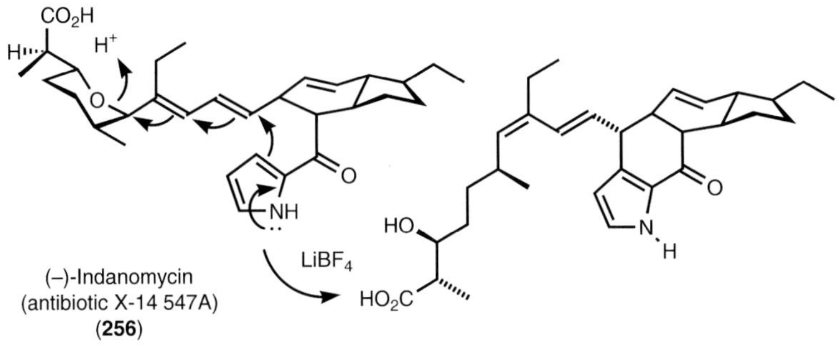

Scheme 6.8. Intramolecular cyclization of indanomycin (**256**) catalyzed by BF$_4^-$ ions

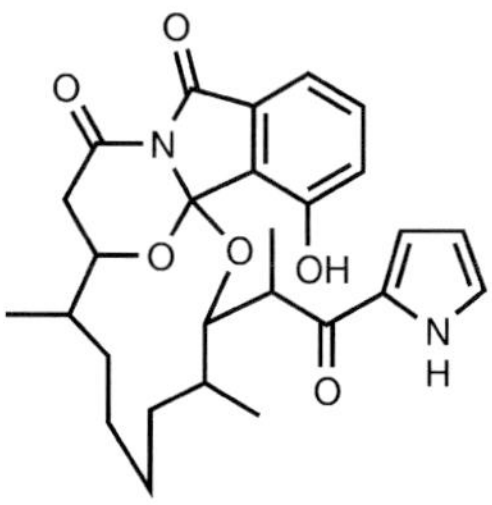

Frankiamide (**257**)

Most recently, a new pyrrole-2-carbacyl derivative, frankiamide (**257**), has been isolated from an strain of the symbiotic Actinomycete *Frankia*, that forms nitrogen-fixing root nodules in plants such as alder (*Alnus* sp.) and *Casuaria* sp. (*701*). Although structure **257** has been demonstrated by extensive ^{1}H-, ^{13}C- and ^{15}N-NMR as well as MS measurements, the configuration of the six stereogenic carbon atoms present in the molecule is not yet known. Besides antimicrobial activity, frankiamide exhibits significant inhibition of Ca^{++} fluxes in clonal rat pituitary cells.

6.5. α-Pyridylpyrroles from Bacteria

From the culture medium of *Pseudomonas roseus fluorescens*, some pyridyl-substituted amino acids have been isolated, so-called proferrosamines, which readily cyclize to the corresponding 1-pyrroline derivatives (*cf.* Scheme 6.9.). The latter form red ferrous complexes, which are named ferrosamines. In the presence of acids, elimination of a water molecule from proferrorosamine B (**258**) readily leads to the corresponding pyrrole derivative, anhydroproferrorosamine B (**259**), which is strongly fluorescent. Since this transformation occurs very easily it is possible that anhydroproferrorosamine B is not a metabolite but rather an artefact (*702*). The biosynthesis of proferrorosamines is supposedly related to the metabolism of tryptophane. Preliminary results of the localization of ^{14}C-labeling show that the pyrrolinecarboxylic acid moiety of proferrorosamin A (the analog of proferrorosamine B devoid of the glycolic acid side chain) is built with the carbon atoms of glycerol while the pyridine skeleton seems to be formed mainly from L-asparagine (*703*).

One of the most intensively investigated pyrrole metabolites from bacteria is methoxatin (**260**), which is commonly known as PQQ (pyrroloquinoline quinone). It was first isolated, purified and

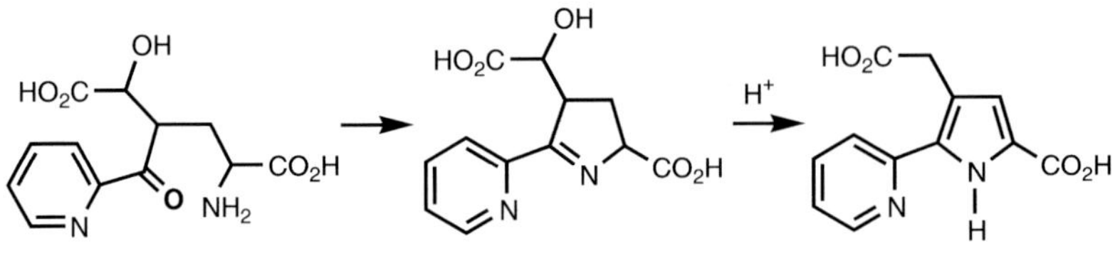

Proferrorosamine B (**258**)   Anhydroproferrorosamine B (**259**)

Scheme 6.9. Biogenetic precursors of anhydroproferrorosamine (**259**)

characterized as the prosthetic group of methanol dehydrogenase, an enzyme which replaces flavine- and nicotinamide-dependent enzymes in the soil bacterium *Pseudomonas* sp. (*704*) and other bacteria (*705, 706*). The structure of this unusual orthoquinone derivative was elucidated by X-ray diffraction analysis by Kennard and her colleagues, who named it methoxatin (*705*). From that time on, different chemical synthesis have been developed for the preparation of PQQ (*cf.* Scheme 6.10.) (*707–711*). For a complete description of methanol dehydrogenase from many bacterial sources the reader is referred to a review containing most references prior to 1985 (*712*) and to more recent reviews for full accounts of subsequent work (*713–715*).

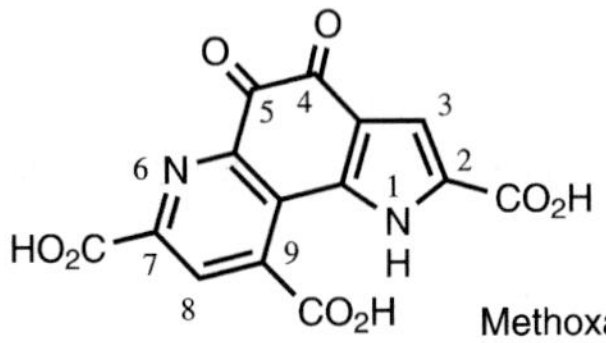

Methoxatin (Coenzyme PQQ) (**260**)

So far, the only enzymes containing **260** are bacterial dehydrogenases, which catalyze oxidation of alcohols, aldehydes, and aldose sugars (*e.g.* D-glucose) among other substrates. (for reviews see Ref. *716–719*). In all these enzymes, which are located in the periplasm of bacteria, PQQ is tightly, but not covalently bound to the apoprotein. Moreover, all these enzymes have a divalent cation at the active site, which is essential for the function of the enzyme, and the overall structures of their catalytic subunits appear to be very similar. As a coenzyme present in acetic acid bacteria (*e.g. Acetobacter aceti* and *A. pasteurianus*) (*720*) PQQ leaks into the medium and contributes to the yellow tinge of vinegar.

Extensive studies of the chemical reactivity of isolated PQQ (**260**) demonstrated the formation of a wide variety of adducts at the C5 carbonyl group of the molecule, many of which are potentially relevant

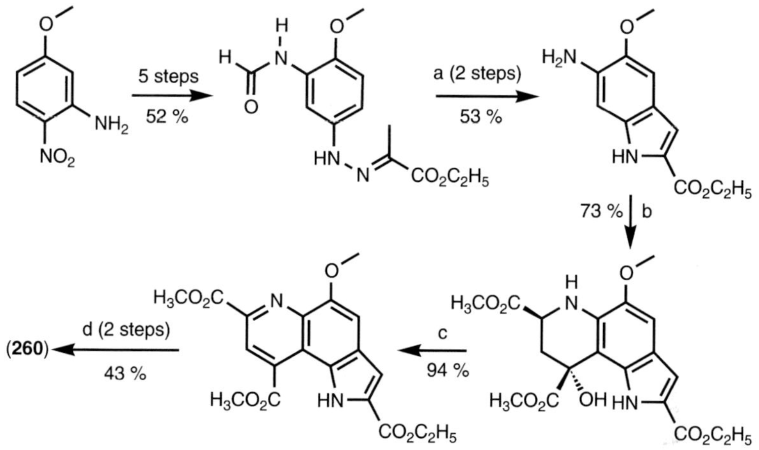

Scheme 6.10. A kg scale synthesis of methoxatin (**260**) (*711*): **a**) *i*: formic acid, 80°C, *ii*: HCl in acetone; **b**) 2-Dimethyl 2-oxoglutaconate in CH_2Cl_2; **c**) Cu(II) acetate in CH_2Cl_2/ gaseous HCl and air. **d**) *i*: Ce(IV) ammonium nitrate in acetonitrile; *ii*: LiOH in aqueous tetrahydrofuran

to the assay and mechanism of the mode of action of the enzyme (*721–723*). The reaction mechanism of methanol dehydrogenase has been reviewed extensively (*724, 725*). Steady-state kinetics is consistent with reduction of **260** by substrate and release of product, followed by two sequential single electron transfers to the cytochrome c, during which the PQQH2 is oxidized back to the quinone by way of the free-radical semiquinone (*726, 727*). The large ($k_D/k_H = 6$) deuterium isotope effect observed during the reductive phase of the reaction is consistent with cleavage of the H–C-bond as the rate-determining step of the enzymatic reaction (*728*). Scheme 6.11 illustrates a possible mechanism for the reductive half-reaction of methanol dehydrogenase, in which aspartate initiates the reaction by abstraction of a proton from the alcohol substrate (*729*). In this mechanism it is proposed that a Ca^{2+} ion acts as a Lewis acid by way of its coordination to the C5 carbonyl oxygen of **261**, thus providing the electrophilic C5 for attack by an oxyanion nucleophile. It is also possible, however, that Ca^{2+} coordinates to the substrate oxygen atom (*730, 731*).

Although little is known about the pathways for biosynthesis of **260**, an important piece of evidence has been provided by the demonstration that the nitrogen atoms and most of the carbon atoms are incorporated into the structure from tyrosine and glutamate (*732–734*). There is some evidence that the biosynthesis of PQQ occurs on a polypeptide consisting of 24–29 amino acids, which contains tyrosine and glutamate (*735*).

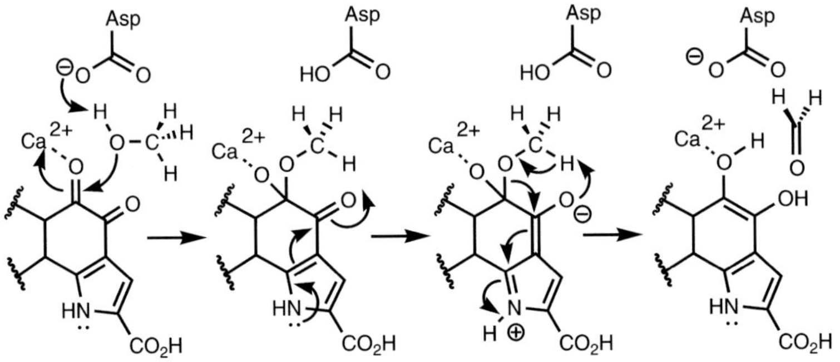

Scheme 6.11. A possible mechanism for methanol dehydrogenase with involvement of the electron pair on the nitrogen atom of the pyrrole ring of **260** (partial structure)

6.6. Other Monopyrrole Derivatives from Bacteria

Pyrrolosine (**261**), is a nucleoside analog which was isolated from the culture broth filtrate of the actinomycete *Streptomyces albus* within the scope of a systematic study on microbial products capable of arresting the development of starfish ambryos specifically at the early blastula strage. Although the analytical data of pyrrolosine suggested a structure identical with that of 9-deazainosine, a C-nucleoside analog, which had been previously obtained by chemical synthesis, both compounds show different UV spectra and retention times in HPLC analysis. Thus, as the structure of pyrrolosine – disregarding its absolute configuration, which has not been yet elucidated – was confirmed by X-ray diffraction analysis, the incorrectly called "9-deazainosine", must be a different compound (*736*).

Contrarily to synthetic "9-deazainosine", which does not interfere with development of star fish embryos, pyrrolosine inhibits RNA synthesis of starfish (*Asterina pectinifera*) embryos at blastulation and halted embryonic development just after completion of blastulation.

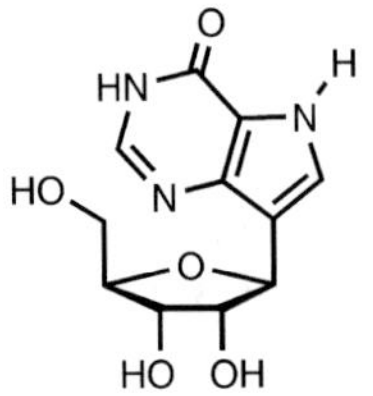

Pyrrolosine (**261**)

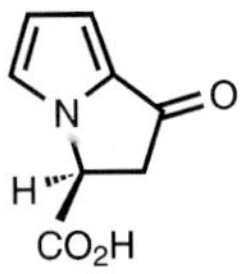

262

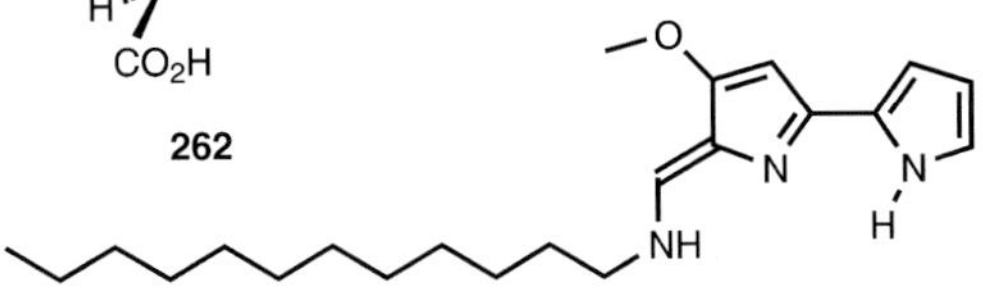

BE-18591 (**263**)

References, pp. 140–188

Besides pyrrolosine, two more pyrrole derivatives have been isolated from actinomycetes, the structures of which are analogous to those of pyrrole metabolites found in eukaryotes. Thus, an unusual pyrrole metabolite, (−)-1-oxo-2,3-dihydro-1*H*-pyrrolizin-3-carboxylic acid (**262**), an analog of the plant alkaloid loroquin (**161**) was isolated from the culture filtrate in the course of a screening of the β-lactam antibiotics generated by fermentation of the mycobacterium *Streptomyces olivaceus*. Its structure was elucidated by spectroscopic analysis and comparison with an optical active synthetic preparation, which was obtained from L-aspartic acid. Thus, the absolute configuration at C2 was established to be (*S*). In contrast to the olivanic acids, the new metabolite shoved no significant antibacterial activity or β-lactamase inhibitory properties (*737*).

On the other hand, a *N*-dodecylbipyrrole derivative (so called BE-18591: **263**) analogous to the tambjamines (**133–135**) has been isolated from a terrestrial *Streptomyces* sp. (*738*). As for the tambjamines, the biosynthesis of compound **263** may be related to that of the prodigiosins, which occur also in actinomycetes (see below).

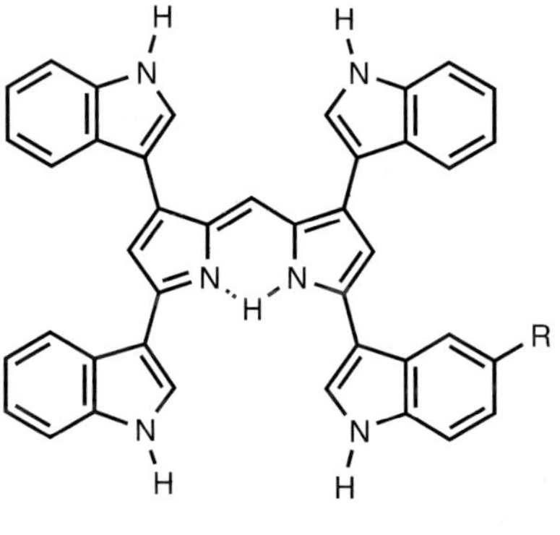

R = OH : Chromoviridan (**264a**)

R = H : Deoxychromoviridan (**264b**)

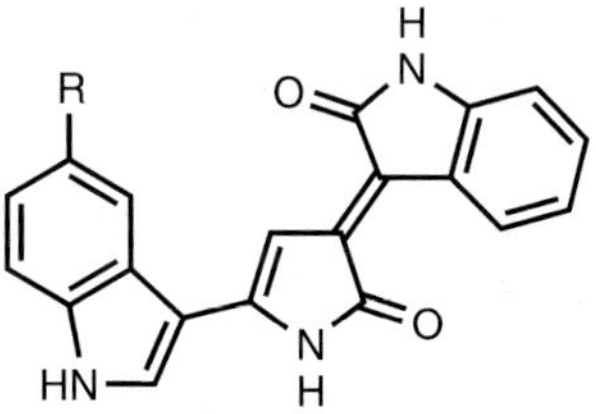

R = H : Violacein (**265**)

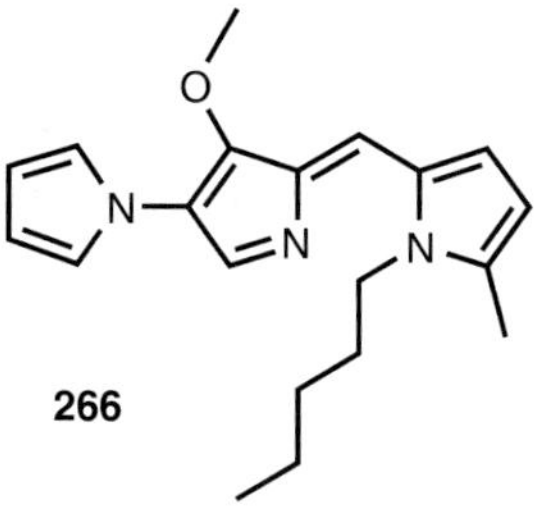

266

Chromoviridan (**264a**) and deoxychromoviridan (**264b**) are green pigments produced in cell-free cultures of *Chromobacterium violaceum*. Their structures were elucidated by FAB-mass spectrometry and different kinds of 2D-NMR spectroscopy (*739*). Together with compound **136** and the red pigment **266**, which has been isolated recently from extracts of an alkalophilic *Micrococcus* sp. (*740*), the chromoviridans are the unique so far known natural occurring dipyrrine derivatives which do not belong to the prodigiosine family (*cf.* Section 6.7.). Labeling experiments showed that **264a–b** are formed on biosynthetic pathways similar to that of violacein (**265**), which is a metabolite of tryptophan. Feeding experiments using different ^{13}C-labeled substrates demonstrated that the methene bridge of the dipyrrin moiety of the chromoviridins proceeds from L-serine, thus indicating that serine hydroxymethyl-transferase and methylene tetrahydrofolate are involved in their biosynthesis (*739*).

6.7. Prodigiosins

6.7.1. Prodigiosins from Eubacteria

The red pigment, which in the middle age was considered to be blood arising from the profanation of Hosts (*741*) – an erroneous interpretation for which many innocents paid with their live – was called *prodigiosin* by Kraft (*742*). Comprehensive reviews on the chemistry of this interesting natural compound containing most references prior to 1967 have been published by Hearn (*743*) and Feofilowa (*744*).

Prodigiosin was isolated for the first time in 1929 by Wrede (*745*) from cultures of the bacterium *Serratia marcescens* (formerly *Bacillus prodigiosus*), but it is produced also by other terrestrial and marine bacteria (*746*) such as *Pseudomonas magnesiorubra* (*747*), *Vibrio psychoerythreus* (*748*), *Beneckea gazogenes* (*749*), and *Streptomyces variegatus* (*750*), as well as *Cladophora* sp. Prodigiosin-like pigments occur both in marine bacteria and in actinomycetes (see below).

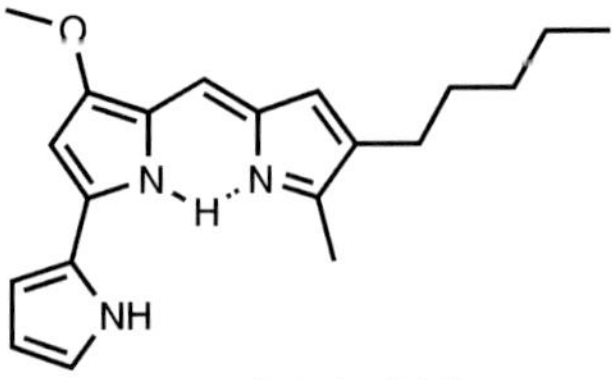

Prodigiosin (**267**)

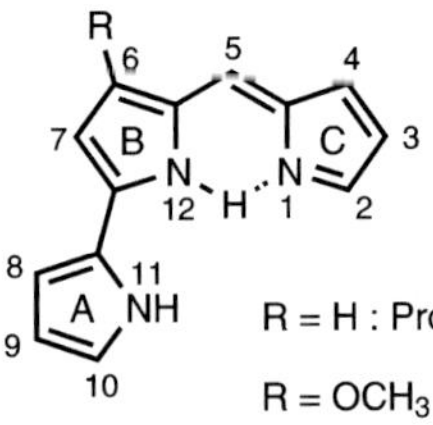

R = H : Prodigiosene (**268a**)
R = OCH$_3$: Prodiginine (**268b**)

On the basis of the structures of the products obtained by chemical degradation, Wrede and Rothhaas (*751*) suggested in 1933 the correct structure of prodigiosin (**267**). One year later, however, the same authors favored – without any experimental justification – an alternative structure with a 5-(pyrrol-2-yl)pyrromethene chromophore (*752*), which was not confirmed by later work (*753*). A decisive hint for the elucidation of the actual structure furnished the isolation of a biosynthetic precursor of prodigiosin from the culture medium of a *S. marcescens* mutant (labeled as 9-3-3) (*754, 755*). The suggested structure of this intermediate (3-methoxy-2,2′-bipyrrole-2-carbaldehyde), which is also a biogenetic precursor of the tambiamins (**133–135**), was confirmed by chemical synthesis by Rapoport and Holden (*756*), who synthesized also prodigiosin for the first time. Later syntheses have been reported by Hearn (*757*) Boger (*758*) and Wasserman (*759*) (*cf.* Scheme 6.12.).

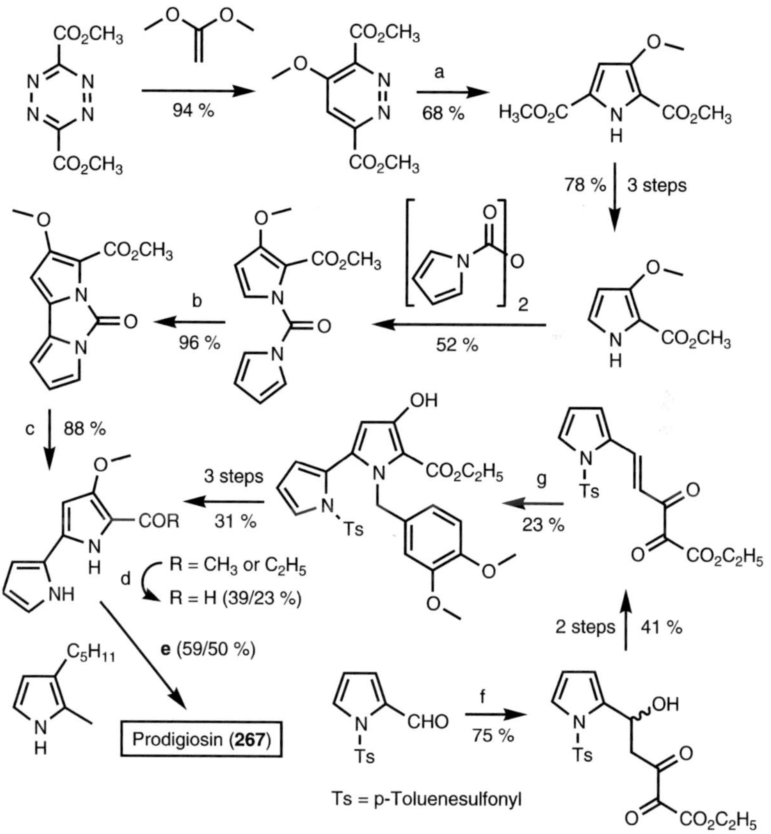

Scheme 6.12. Two recent total synthesis of prodigiosin (**267**) (*cf.* Ref. *758* and Ref. *759*). a) Zn in acetic acid. b) polymer-supported Pd(II) acetate in acetic acid. c) LiOCH$_3$ in methanol. d) i: H$_2$N–NH$_2$; ii: p-toluenesulfonyl chloride in pyridine; iii: Na$_2$CO$_3$ in ethylene glycol (*cf.* Ref. *756*). e) catalytic HCl in methanol. (*cf.* Ref. *756*). f) H$_2$C=C(OLi)CH=C(OLi)OC$_2$H$_5$ in tetrahydrofuran. g) 3,4-dimethoxybenzylamine in glacial acetic acid

Prodigiosin (**267**) has been fully characterized by IR (*753, 760*), NMR (*755*), and mass spectrometry (*761*). The chromophore of all prodigiosins is called *prodigiosene* (**268a**) (*757*), whereas 6-methoxyprodigiosene is called *prodiginine* (**268b**) (*762*). According to their dipyrrin structure, the color of prodigiosins depends on the proton concentration. Thus, in acidic medium, prodigiosins are deep red ($\lambda_{max} = 530–541$ nm) whereas in alkaline solutions their color turn to orange-yellow ($\lambda_{max} = 460–468$ nm) (*763*) and the intensity of the absorption band diminishes to the half. Actually, color change is caused not only by the protonation of the chromophore (*764*) but also by intermolecular self-association (*765*). In the living organisms, prodigiosin is located in the cell membrane, probably as integral part of a high-molecular complex with glycoproteins (*766–768*), which in contrast to the free chromophore, does not fluoresce (*769*).

Prodigiosin shows both antibiotic (*770*) and fungistatic properties (*771*). However, although several reports on the antimicrobial, cytotoxic (*772*) and antimalaria activity (*773, 774*) of this compound can be found in the literature, its toxicity for humans prevents its use as a chemotherapeutic agent. *In vitro*, prodigiosin binds calf thymus DNA by intercalation, but less specifically than tambjamine E (**134b**) (*261*).

An interesting property of the prodigiosenes is their easy disproportionation to 5.5′-di(pyrrol-2-yl) pyrromethenes (*757*). This reaction may explain the presence of blue or violet pigments – *e.g.* serratin (*753*) – which are often isolated along with prodigiosins from natural sources. Not all of these pigments are however artefacts. For instance, the brilliant blue pigment **136**, which is produced enzymatically by a mutant strain of the bacterium *Serratia marcescens* (*775*), has been isolated also from some marine invertebrates.

The biosynthesis of prodigiosin (**267**) has been mainly investigated in *Serratia marcescens* (*776–779*). Preliminary studies carried out with precursors bearing radioisotope labels shoved that glutamate, proline, and ornithine (*780, 781*), as well as aspartate and alanine (*782*), and methionine (*783, 784*) (but not 5-aminolaevulinic acid) are involved in the biosynthesis of the pyrrole rings of the molecule. However, only with the use of ^{13}C-labeled of precursors and development of ^{13}C-NMR procedures to analyze their distribution in the metabolite has the biogenesis of prodigiosin been elucidated. In a series of communications Wasserman and his co-workers have shown that proline contributes ring A and C5 of the ring B, while acetate supplies C3 and C4 of ring B, as well as C3, C4, C5, and the C3 alkyl substituent of ring C. The methene carbon atom originates with the hydroxymethyl group of serine, and the adjacent C2 of the ring B probably comes from C2 of this amino acid since C2 of glycine was a good precursor. In ring C, C2 and the adjacent methyl substituent were derived from alanine *(785–788)*.

Experiments with mutant strains of *S. marcescens* have indicated that the final stage in prodigiosin biosynthesis is the coupling of a dipyrrylaldehyde intermediate, containing rings A and B and the methene bridge atom, with a monopyrrole derivative consisting of ring C and its substituents. Each intermediate is synthesized independently, and it is likely that the pathway leading to the dipyryaldehyde in the actinomycetes (see below) is the same as than in *S. marcescens*. A plausible mechanism is suggested in Scheme 6.13.

As experiments with ^{14}C-labeled methionine suggest that one or both of the methyl groups of prodigiosin may be derived from methionine *(784)*, it has been speculate that an hydroxybipyrrole may be an intermediate in the biosynthesis of prodigiosin (*c.f.* Scheme 6.13.). In support of this speculation, a white mutant of *S. marcescens* excretes a

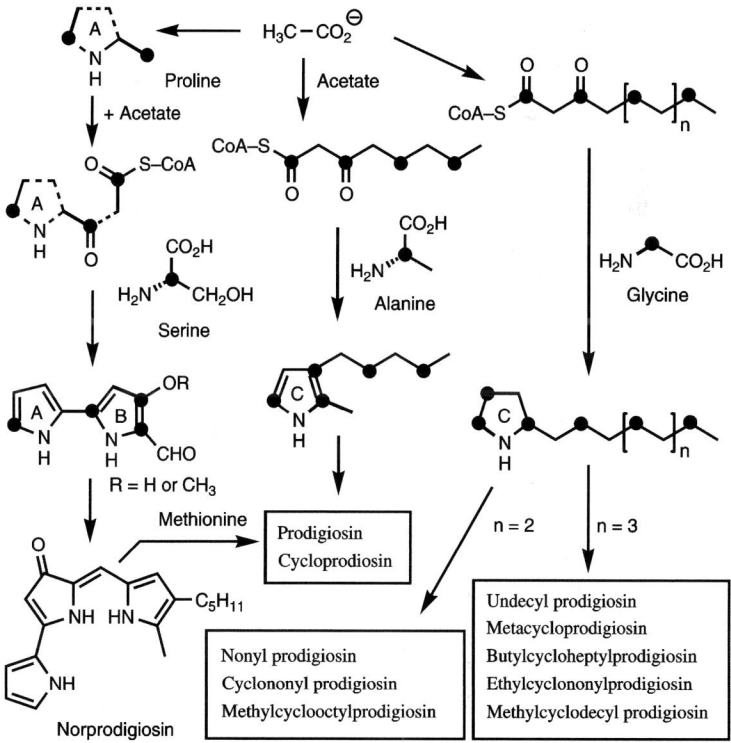

Scheme 6.13. Biosynthesis of prodigiosin (**267**) and its analogs

methoxybipyrrole (*777*) that will combine with 2-methyl-3-amylpyrrole to form prodigiosin (*778*). On the other hand, norprodigiosin (2-methyl-3-pentyl-6-hydroxyprodigiosen) has been isolated from a mutant of *S. marcescens*, after γ-irradiation (*789*) so that an alternative pathway in the biosynthesis of prodigiosin and its analogs may proceed through methylatation of norprodigiosin, as the final step. In solution, the preferred tautomer of norprodigiosin is the keto form instead of the corresponding hydroxypyrrole (*790*).

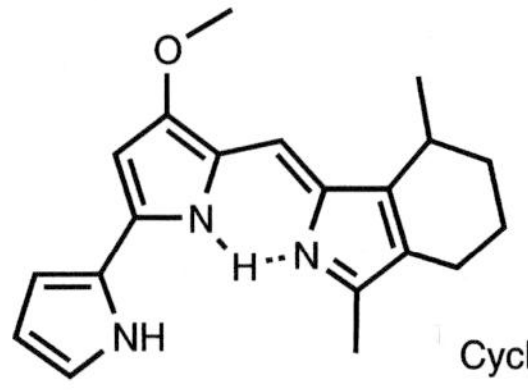

Cycloprodigiosin (**269**)

As mentioned before, prodigiosin occurs in both terrestrial and marine bacteria. In most cases, however, the pigments are hardly separable mixtures of prodigiosin with other analogs, which have not been recognized as such in earlier work on this field. Thus, cycloprodigiosin (**269**) has been isolated from the facultatively anaerobic red marine bacterium *Beneckea gazogenes*, in which it occurs in a hardly separable mixture with prodigiosin (*749*). The same pigment had been previously obtained from another marine bacterium, *Alteromonas rubra*, which is commonly found in Mediterranean coastal waters during the autumn months, but its structure was erroneously characterized as the corresponding ethylcyclopentano isomer (*791*). The structure of cycloprodigiosin (*792*) has been confirmed by synthesis of the racemic compound (*793*).

6.7.2. *Prodiginines from Actinomycetes*

Actinomycetes are filamentous bacteria which occur both as soil inhabiting saprophytes and as disease-producing parasites. The most common actinomycetes belong to the genus *Streptomyces*. Although bright-red pigments related to undecylprodiginine were presumably isolated from actinomycetes as early as 1947 (*794*), their relationship to prodigiosin was not clarified until later (*795*). At present, more than 95 microorganisms, mainly Actinomycetes, are known, which produce prodigiosin-like pigments (*796, 797*).

A noteworthy feature of prodigiosin analogs from the actinomycetes is the pattern of alkyl substitution in ring C. Whereas prodigiosin and its congeners from eubacteria possess a methyl substituent at C2 and

Undecylprodiginine (= Prodigiosin-25 C) (**270**)   Nonylprodiginine (**271**)

differ from one another in the length of the alkyl chain attached to C3, the actinomycete prodigiosin analogs are unsubstituted at C3 but have an extended alkyl chain 9 or 11 carbon atoms long at C2. Thus, prodigiosin 25 C (**270**) (*798, 799*) – also called undecylprodigiosin (*800*) or undecylprodiginine – was first characterized from two species of *Streptomyces*, and more recently from *Streptomyces coelicolor* (*801*). Undecylprodigiosin occurs also in *Actinomadura* (*802*) and *Streptoverticillium* species, e.g. *Streptoverticillium rubrireticuli*, an organism frequently incriminated in pink staining of polyvinyl chloride (*803*). The lower homolog, nonylprodiginine (**271**), on the contrary, has been found only in *Actinomadura* species (*762*). The structure of undecylprodigiosin has been confirmed by total synthesis (*799, 804, 805*).

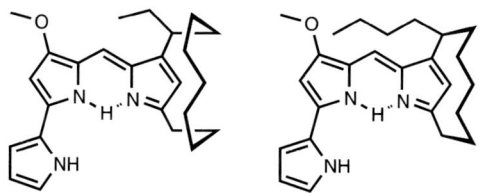

Metacycloprodigiosin (**272**)   Butylcycloheptylprodigiosin (**273**)

Variations are introduced into the structure of actinomycete prodigiosin analogs by attachment of the 2-alkyl chain of undecyl- and nonylprodigine at a second point (C4 or C10) of the prodiginine skeleton. Thus, (−)-metacycloprodigiosin = ethyl-2,4-cyclononylprodiginine (**272**), which is probably identical with streptorubrin A (*797, 806*), was isolated first, together with undecylprodigiosin, from *Streptomyces longisporus ruber* (*807, 808*). Metacycloprodigiosin, the structure of which has been confirmed by chemical synthesis (*809, 810*) (*cf*. Scheme 6.14.), is an optical active compound with a remarkable high optical activity ($[\alpha]_D^{20} = -2370$ in methanol containing 0.5% KOH). Another prodigiosin-like pigment, called vitamycin A, has been isolated from *Actinomyces aureoverticillatus* (*811*). However, as the spectroscopic

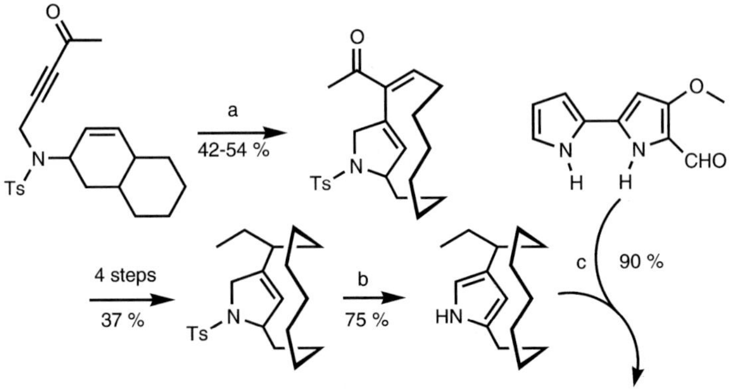

Scheme 6.14. A total synthesis of metacycloprodigiosin (**272**) using a Pt(II) or acid catalyzed enyne metathesis reaction (*810*): **a)** PtCl$_2$ (5%) or BF$_3 \cdot$Et$_2$O in toluene; **b)** excess KAPA (potassium 3-aminopropylamide) in 1,3-diaminopropane. **c)** conc. HCl in ethanol

data reported for this compound (*812*) closely match those of metacycloprodigiosin (*796*), both metabolites are probably identical and, therefore, the structure proposed for vitamycin A should be revised.

An analog of metacycloprodigiosin is butylcycloheptylprodigiosin (**273**), a pink actinomycete metabolite, which accompanies undecylprodigiosin, in *Streptomyces* sp.Y-42 (*774*), *Streptoverticillium rubrireticuli* (*803*), and *Streptomyces coelicolor* (*801*). Butylcycloheptylprodigiosin, which had been first incorrectly characterized as a 2,3-cycloalkyl derivative (*774, 796, 797, 801, 803*) is probably identical with streptorubrin B (*797, 806*). Its structure (*813*) has been confirmed by chemical synthesis (*810*). Both metacycloprodigiosin and butylcycloheptylprodigiosin, as well as ethylcyclononylprodiginine and methylcyclodecylprodiginine (see below) are probably derived from undecylprodigiosin by enzymatic oxidative cyclization of the C2-side-chain.

Actinomadura species produce a series of cyclic prodigine derivatives with the alkyl chain attached at C5 of ring A. These include cyclononylprodiginine (**274a**) (*814, 815*), and methylcyclooctylprodiginine (**275**) (*815*), which were isolated from *A. madurae*, as well as ethylcyclononylprodiginine (**274b**) (*815*) and methylcyclodecylprodiginine (**276**) (*802, 815*), which were obtained from *A. pelletieri*. The latter two have also been found in a *Streptoverticillium* species (*797*). Probably, both **274a** and **275** are biogenetically derived from nonylprodiginine (**271**) by cyclization of the C2-side-chain, whereas **274b** and **276** correspondingly originate from undecylprodigiosin (**270**).

References, pp. 140–188

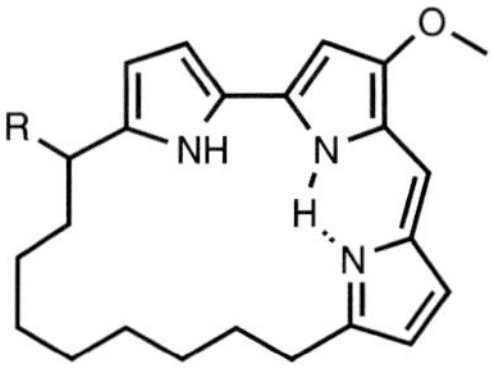

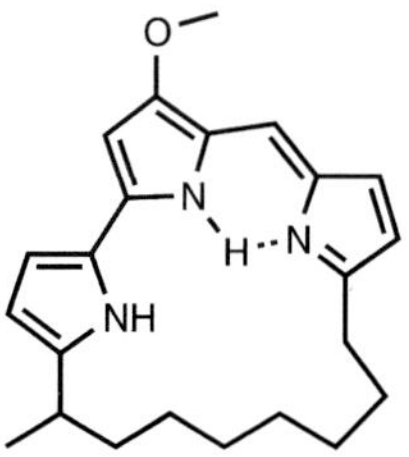

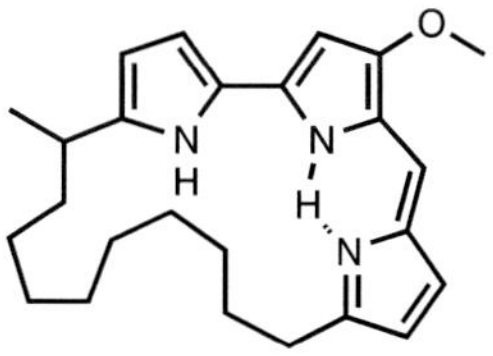

Methylcyclooctylprodiginine (**275**)

R = H : Cyclononylprodiginine (**274a**)
R = C₂H₅ : Ethylcyclononylprodiginine (**274b**)

Methylcyclodecylprodiginine (**276**)

It is likely that the biosynthetic pathway leading to the dipyrrylaldehyde precursor of the rings A and B in the actinomycetes is the same as than in *S. marcescens*. Actinomycetes and *S. marcescens* clearly differ, however, in the route by which the monopyrrole intermediate of ring C is formed, although here, too, some common elements can be seen (*cf.* Scheme 6.13.) (*816*). Thus, the biosynthesis of metacycloprodigiosin by *Streptomyces longisporus ruber* is similar to that of prodigiosin by *Serratia marcescens*, except for the incorporation of glycine in ring C and the involvement of a polyacetate chain which forms the alkyl chain of undecylprodigiosin (*vide supra*) as the precursor of metacycloprodigiosin (*817*). In the actinomycete pigment, carbons C2, C3, C4 and the alkyl substituent of undecylprodigiosin (**270**) and metacycloprodigiosin (**272**) came from acetate whereas C5 was derived from the methylene group of glycine. In each of the metabolites the incorporation of label from 1-[^{13}C]- and 2-[^{13}C]-acetate suggested that a polyketide chain of 14 carbon atoms contributed to the formation of ring C and its attached alkyl substituent (*817*).

As prodiginine derivatives from actinomycetes, particularly undecylprodigiosin (**270**) and metacycloprodigiosin (**272**), are not cytotoxic at doses which inhibit T-cell proliferation without affecting the *in vivo* functions of helper T and B cells (*818*), they themselves or their analogs (*cf.* Ref. *810*) may deserve interest – in contrast to prodigiosin (see above) – as chemotherapeutics.

6.8. Hydroxy Pyrroles from Bacteria

Pyrrolam A (**277**) is an alkaloid from *Streptomyces olivaceus* (*819*). It belongs together with phenopyrrocin (**200**) to a family of pyrrolizidones, the most of which bear a hydroxy group at C7a. Pyrrolam has been obtained by total synthesis from (*R*)-proline as the starting material (*820*).

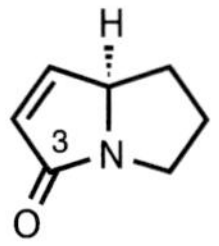

(−)-Pyrrolam A (**277**)

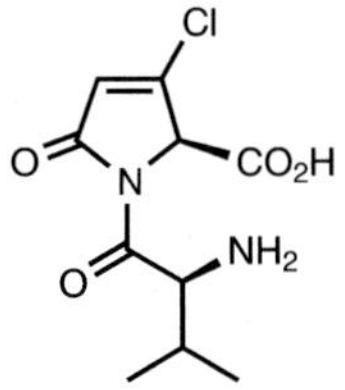

Antibiotic FR-900148 (**278**)

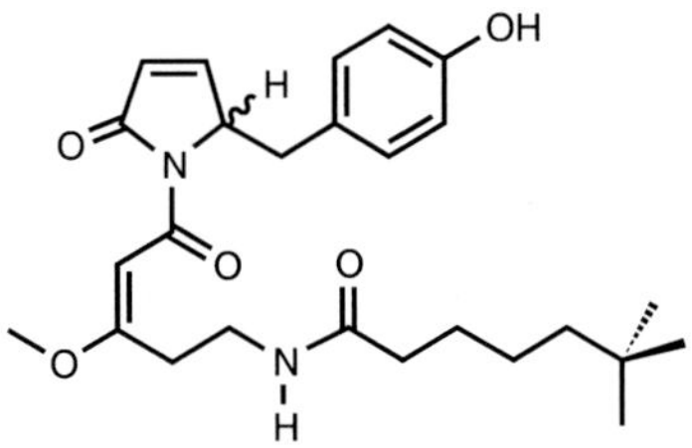

Ypaoamide (**279**)

A chlorine-containing 3-pyrroline-2-one derivative of L-valine was isolated from *Streptomyces xanthocidicus* and named Antibioticum FR-900148 (**278**) (*821*). Interestingly, on treatment with dilute acid (pH = 2.0) the antibiotic is transformed into the corresponding 4-pyrroline-2-one tautomer, which is biologically inactive (*822*). Probably, antibioticum FR-900148, which has been also isolated along with (*R*)-(*Z*)-4-amino-3-chloro-2-pentenedioic acid from *Streptomyces viridogenes* cultures (*823*), is biosynthesized from L-glutamic acid and L-valine.

Ypaoamide (**279**) is a broadly acting feeding deterrent by both the yellow-banded parrotfish (*Scarus schlegeli*) and the urchin (*Echinometra mathaei*) from the marine cyanobacterium *Lyngbya majuscula*, which became notorious following its massive blossom at Ypao beach, a popular tourist site in Guam (Mariana islands), in May 1994. Its structure was determined spectroscopically by interpretation of 2D-NMR experiments, and by comparison with model compounds (*824*).

The same marine cyanophyte (*Lyngbya majuscula*), which produces ypaoamide (see above) is the source of a series of biologically active lipopentapeptides containing a 5-methyl-3-pyrrolin-2-one residue, and of some tetramic acid derivatives as well (see below). Thus, majusculamide D (**280a**) and deoxymajusculamide D (**280b**) are two cytotoxins from *L. majuscula* collected in the lagoon of Enewetak Atoll in the Marshall Islands (*825*). Both **280a** and **280b** are structurally related to microcolin A (**281a**) and microcolin B (**281b**), respectively, which are two very potent immunosuppressive agents from a Venezuelan specimen of the same cyanobacterium (*826*). The structure and absolute configuration of microcolin A at all chiral C-atoms has been secured ba chemical synthesis (*827*).

References, pp. 140–188

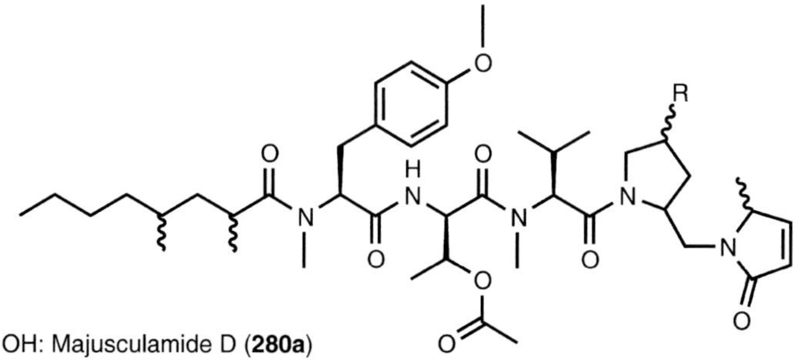

R = OH: Majusculamide D (**280a**)
R = H: Deoxymajusculamide D (**280b**)

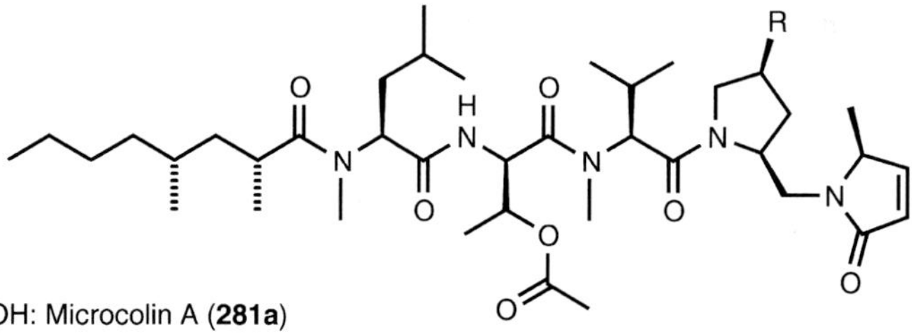

R = OH: Microcolin A (**281a**)
R = H: Microcolin B (**281b**)

6.9. Tetramic Acid Derivatives from Prokaryotes

Lyngbya majuscula, collected in 1979 at Kahala Beach, Oahu (Hawaiian Islands) was also the source of the tetramic acid derivative malyngamide A (**282**) (*828*), and of the pukeleimides (see below). Twenty years later, isomalyngamide A, the Z-isomer of malyngamide A about the chloromethylene group, has been isolated from the same Hawaiian cyanobacterium instead of **282** (*829*). On the other hand, malyngamide Q (**283a**) and malyngamide R (**283b**), in which the chloromethylene group has also Z-geometry, have been isolated from a shallow-water Madagascan

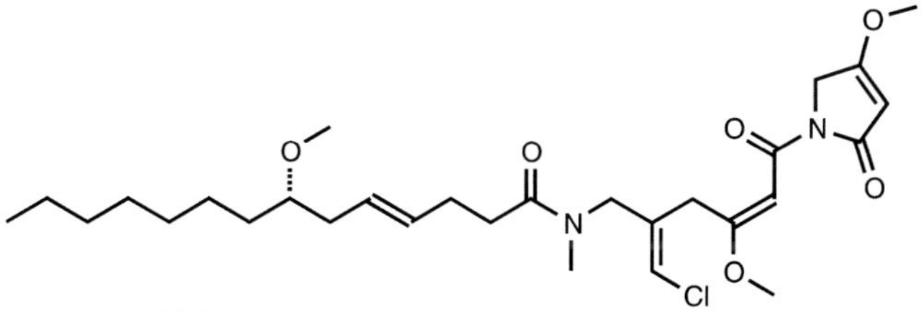

(−)-Malyngamide A (**282**)

R = H: Malyngamide Q (**283a**)

R = CH$_3$: Malyngamide R (**283b**)

Lyngbya majuscula (*830*). The structure and absolute configuration of malyngamide was elucidated by a combination of spectroscopic analysis and chemical degradation procedures (*828*). Thus, alkaline hydrolysis of **282** produced tetramic acid methyl ether and (*E*,7*S*)-7-hydroxy-4-tetradecenoic acid, the absolute configuration of which was already known. The structure of malyngamide A recalls that of the spongal metabolite dysidin (**113**). As for the latter, the biological activity of **282** is not known.

In the pukelimides (**284–286**), the side chain of the fragment of malyngamide (**282**) which results after cleavage of the 7-hydroxy-4-tetradecenoic acid residue is cyclized to a five-membered lactame ring. Thus, pukeleimides (named after the Hawaiian word *pukele* which means to gather thickly in the water) belong to a rare family of natural occurring 5-ylidenepyrrol-2(5*H*)ones produced besides malingamide A by the toxic, shallow-water variety of the marine cyanobacterium *Lyngbya majuscula*, collected in Kahala Beach, Oahu (*831, 832*). However, none of the pukeleimides are toxic to mice. Pukeleimide C (**285b**), which was isolated as a racemate, was crystallized, and a X-ray diffraction structure was obtained (*831*). The other pukeleimides were separated chomatographycally and then identified using NMR spectroscopy. Pukeleimides B and F are the (*Z*)-stereoisomers of pukeleimide A (**284a**) and G (**284b**), respectively. Pukeleimide A (**284a**) has been obtained by chemical synthesis (*833*).

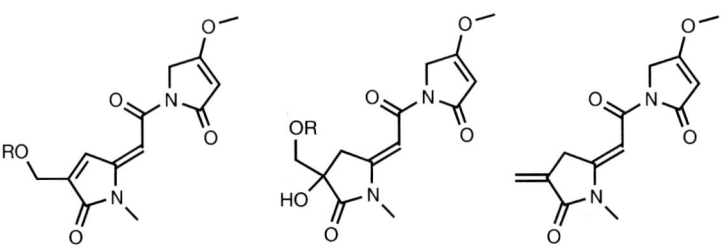

R = H : Pukeleimide A (**284a**)   R = H : Pukeleimide D (**285a**)   Pukeleimide E (**286**)
R = CH$_3$: Pukeleimide G (**284b**)   R = CH$_3$: Pukeleimide C (**285b**)

References, pp. 140–188

The tetramic acid derivative (+)-althiomycin (**287**) was first isolated from *Streptomyces althioticus* by Umezawa and his collaborators in 1957 (*834*), and later identified with matamycin, which was isolated in 1959 from *Streptomyces matensis* by Sensi *et al.* (*835*) and more recently from the myxobacterium *Cystobacter fuscus* among other *Myxococcus* sp. (*836*).

Althiomycin (**287**)   Malonomycin (**288**)

An initial investigation determined the partial peptide nature of the molecule and identified cysteine, thiazole-4-carboxylic acid and 4-methoxy-3-pyrrolin-2-one among the hydrolysis products. However, the structure suggested originally for acetylalthiomycin (*837*) was not consistent with chemical and spectroscopic data. Later work, which were carried out through spectroscopic analysis and degradation studies by Bycroft and Pinchin (*838*) as well as by Umezawa *et al.* (*839*), lead to the structure elucidation of althiomycin, which was confirmed in the last group by X-ray diffraction studies on single crystals of a dehydration product of althiomycin. A later X-ray diffraction study carried out with the methyl ester obtained on cleavage of the 4-methoxy-3-pyrrolin-2-one moiety in refluxing methanol settled the (*E*)-geometry of the aldoxime group as represented in formula **287** (*840*). However, the configuration of the C-atom bearing the hydroxymethyl group is still unknown because althiomycin is always isolated as a mixture of epimers at this stereogenic center. It is still unclear whether the natural compound is a mixture or if epimerization occurs during the isolation procedure. Althiomycin has been obtained by chemical total synthesis starting from D-cysteine (*841*, *842*).

Biosynthetically, althiomycin (**287**) appears to be derived from the pentapeptide H_2N-gly-cys-ser-cys-gly-OH by post-transcriptional modification, although this pathway has not been proven (*838*). The organization of these residues into the tricyclic structure of althiomycin gives the molecule an arc-like structure in three dimensions that suggests it could bind to the major grove of double helical *r*RNA. In fact, **287** is a naturally occurring antibiotic active against Gram-positive and

Gram-negative bacteria. Its mechanism of action involves inhibition of protein synthesis at the peptidyltransferase stage, thus supporting the hypothesis that it might bind specifically to a region of helical rRNA located close to the peptidyltransferase active site in rRNA (*843*).

6.10. Derivatives of 3-Acetyltetramic Acid from Actinomycetes

The antiprotozoal compound malonomycin (**288**) – formerly known as antibiotic K16 – was isolated from *Streptomyces rimosus*. The determination of its structure, which was carried out by van der Baan and co-workers (*844*), was complicated by the easy cleavage of the side chain on mild acidic hydrolysis, a reaction which is characteristic for acyl tetramic acid derivatives The structure of the chromophore was confirmed by ozonolysis of the corresponding *bis*-2,4-dinitrophenylhydrazone. However, the configuration at the asymmetric C5-atom of the chromophre – presumably (*S*) – remained unsolved. As the cyclization of the appropriate methyl *N*-acetoacetyl-2,3-diaminopropanoate derivative under the conditions of Lacey's synthesis of β-acyltetramic acids (*cf.* Scheme 3.21.) failed, the total synthesis of **288** was accomplished by a convergent strategy in which the crucial step is a very mild coupling of the dipeptide side chain to the 3-position of the pyrrolidin-2,4-dione ring (Scheme 6.15.) (*845*). Biosynthetic studies indicate that the aminomethyl substituent at C5 originates from L-2,3-diaminopropionic acid (*845*). A more conspicuous feature of the structure of malonomycin is the presence of the unique aminomalonic acid moiety, incorporated into

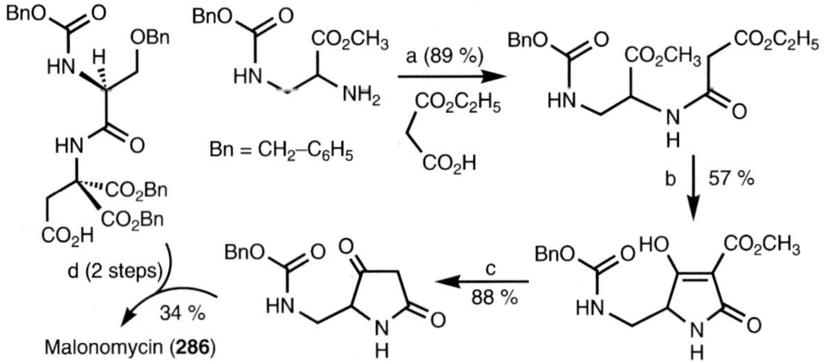

Scheme 6.15. A convergent synthesis of malonomycin (**286**) (*845*): **a)** Triethylamine/ *N*-ethoxycarbonyl-2-ethoxy-1,2-dihydroquinoline (EEDQ) in CH_2Cl_2. **b)** $NaOCH_3$ in methanol. **c)** H_2O in nitromethane. **d)** *i*: 1-hydroxybenzotriazole (HOBT)/dicyclohexylcarbodiimide (DCC)/triethylamine in acetonitrile; *ii*: H_2/Pd-C in methanol

the acyl substituent at C3; this part of the molecule is of paramount importance for the biological activity of malonomycin as all activity is lost on decarboxylation. The biosynthetic origin of this moiety is still unknown.

Alteramide A (**289**) is a macrocyclic lactam with a dienone and a dienoyltetramic acid substructure which, in solution under daylight, are susceptible to an unique intramolecular photochemical [4 + 4] cycloaddition to generate a hexacyclo derivative containing a cyclooctadiene ring. Alteramide A is produced by a bacterium *Alteromonas* sp. simbiotically associated with the marine sponge *Halichondria okadai* collected in Nagai, Kanagawa (*846*). The isolated sample contained ca. 20% of a minor component, alteramide B, which is difficult to separate by HPLC. Ateramide B has the same ring system as the alteramide A without the hydroxyl group at C25.

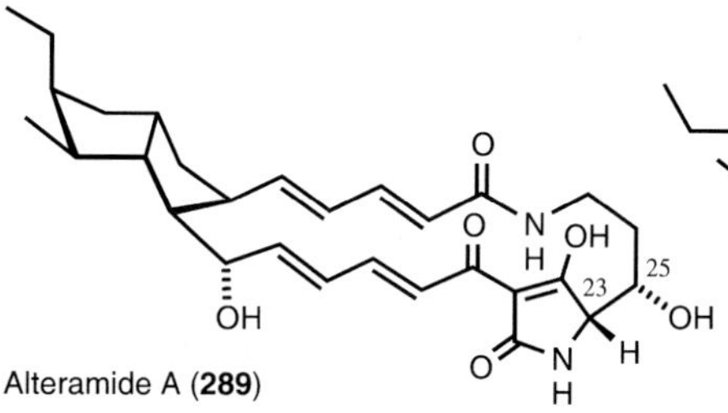

Alteramide A (**289**)

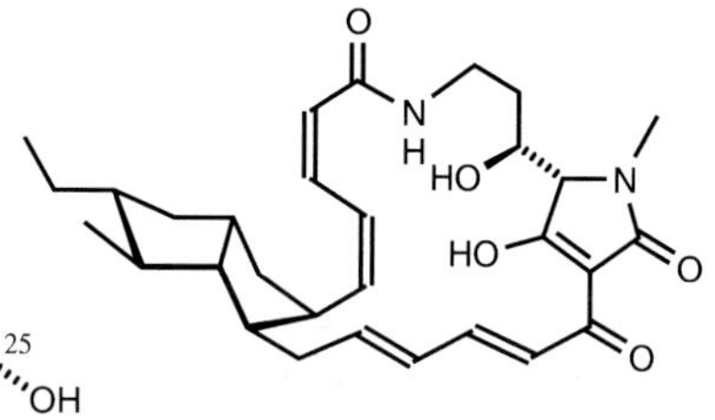

Aburatubolactam A (**290**)

The structure of alteramide A was determined by spectroscopic methods (HRFABMS, ^{1}H- and ^{13}C-NMR spectroscopy including HMQC, COSY, HMBC, NOESY techniques). On oxidative work up after ozonolysis of alteramide A, the breakdown product is L-β-hydroxyornithine hence establishing the configuration of both C23 and C25 as (*S*).

Closely related to alteramide A is the structure of aburatubolactam A (**290**), which has been isolated from a Strepomyces sp. (SCRC-A20) found on an unidenfied marine mollusk (*847*). Its structure was determined by X-ray diffraction analysis.

Both the antifungal metabolite capsimycin (**291**) (*848*) – also known as antibiotic N-461 – and the structurally related antibiotic ikarugamycin (**292**) (*849*), which possesses specific antiprotozoal activity, were isolated from *Streptomyces* strains. Ikarugamycin has been isolated from the culture broth of a variety of *Streptomyces phaeochromogenes* (*849*). The structure and configuration of the metabolite was established on the basis of chemical reactions, specifically of ozonolysis and oxidative degradation (*850*). Several chemical syntheses of the *as*-indacene unit of **292**

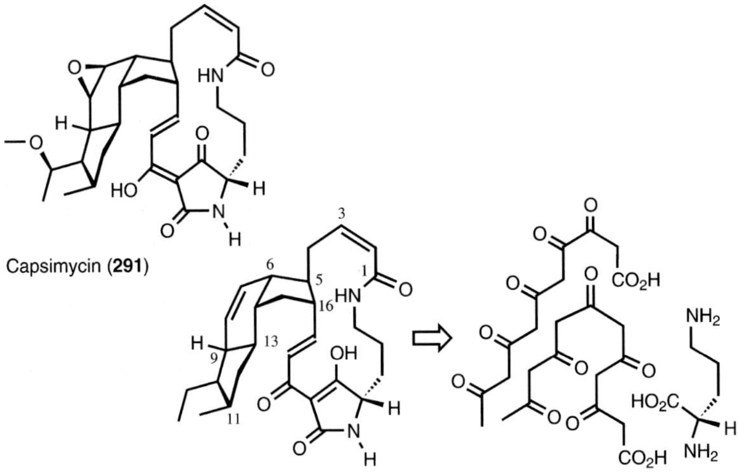

Scheme 6.16. Possible biogenetic origin of ikarugamycin (**292**) from L-ornithine and two hexaacetate chains

(*851–853*) and of the antibiotic itself (*854–858*) have been achieved. The absolute configuration of ikarugamycin (**292**) and capsimycin (**291**) has been assigned through X-ray diffraction analysis of the p-nitrobenzoate of a synthetic derivative used as intermediate of the decahydro-*as*-indacene moiety of the antibiotics (*853*).

As it has been proved that other natural acyltetramic acids, such as tenuazonic acid (**201**), can be biosynthesized from an amino acid and a polyacetate, it appears likely that both **291** and **292** can also be derived from L-ornithine and two hexaacetate chains as outlined in Scheme 6.16. Although the biosynthetic pathway of the *as*-hydrindacene skeleton remains a matter of conjecture, it was suggested that it arises biogenetically via an intramolecular Diels-Alder reaction between the butadiene part (consisting of atoms C6 to C9) and a double bond (C13 = C14) in a hypothetical intermediate formed from the two hexaacetate chains.

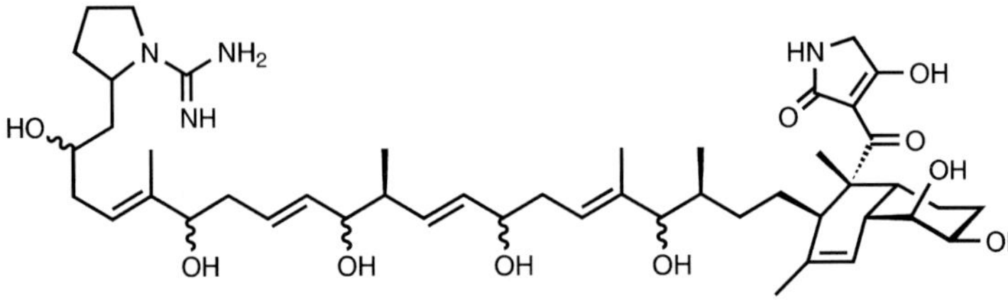

Lydicamycin (**293**)

References, pp. 140–188

Lydicamycin (**293**), which has been isolated from *Streptomyces lydicus* (*859*), possesses a skeleton containing tetramic acid and amidinopyrrolidine moieties. The structure and relative stereochemistry of the bicyclic portion of the molecule were elucidated by NMR spectral analysis including a variety of 2D techniques (*860*). The absolute configuration and the stereochemistry of the polyol side chain remained however unknown. Lydicamycin represents a new class of antibiotics which appears to be predominantly polyketide-derived with the amidinopyrrolidine moiety presumably originating from an amino acid.

α-Lipomycin (**294**), isolated from *Streptomyces aureofaciens* (*861*), and oleficin (**295**), obtained from a strain related to *S. parvulus* (*862*), are structurally related derivatives of acetyltetramic acid. β-Lipomycin, the aglycone of **294**, occurs also in nature. The structure of α-lipomycin was elucidated first through spectroscopic analysis of the products of chemical degradation of the natural compound. This indicated the presence of acyltetramic acid, *all-trans*-tetradecapentaene units and a β-anomeric linkage to a carbohydrate moiety, which was identified as the known deoxy sugar D-digitoxose through comparison with authentic material (*863*, *864*).

As L-*N*-methylglutamic acid was obtained *via* ozonolysis and subsequent acidic hydrolysis of the aglycon, β-lipomycin, the (*S*) configuration was assigned to C5. The configuration of the remaining two asymmetric C-atoms remains unknown as any asymmetry is lost upon oxidative degradation due to loss of carbon dioxide from the ensuing β-hydroxy acid. The structure of oleficin (**295**) was deduced in a similar manner (*865*) – the initial report (*866*) being erroneous.

A similar procedure as above was employed to ascertain the structure of altamycin A, an antibiotic obtained from *Actinomyces pneumonicus* var. *altamicus* by Shenin and his co-workers (*867*), which proved to be identical with α-lipomycin (*868*). As in the case of oleficin and the

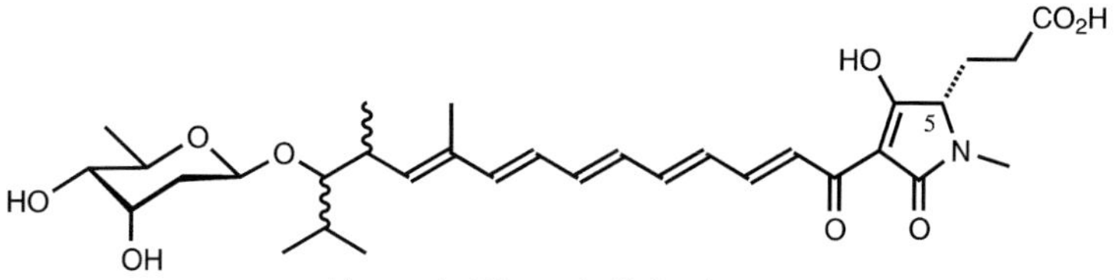

α-Lipomycin (Altamycin A) (**294**)

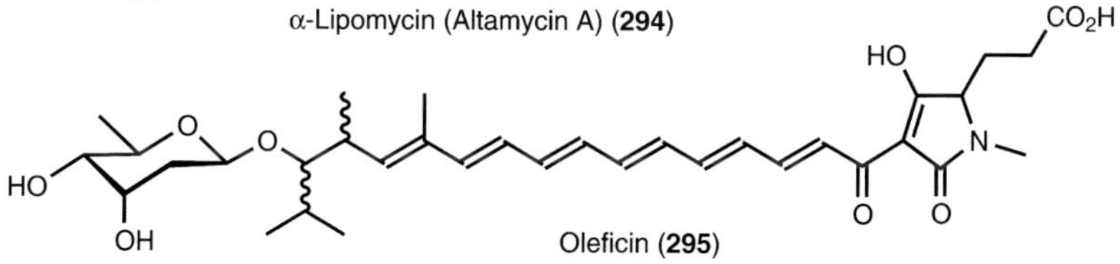

Oleficin (**295**)

lipomycins, the configuration of the asymmetric centers at the terminus of the olefinic chain could not be assigned.

A group of dienoyltetramic acid antibiotics has stimulated considerable interest on account of their biological activity as potent inhibitors of terminal DNA transferase and bacterial RNA polymerase enzymes in cell-free systems (*869*). Two of them, tirandalydigin (**297**) and streptolydigin (**298**) are derivatives of streptolic acid (**296**), which is obtained on periodate degradation of sodium streptolydigin (*870*). Streptolydigin (**298**) was the first of these tetramic acid derivatives to be isolated. It was obtained from culture filtrates of the actinomycete *Streptomyces lydicus* (*871*), and its structure was elucidated by Rinehart's group who used a combination of chemical methods and spectroscopic analysis (*872–875*). Tirandalydigin (**297**), on the other hand, was isolated from the fermentation broth of *Streptomyces* sp. AB-1006A-9 (*876*). Structural assignment was carried out using two-dimensional NMR spectroscopic studies and comparison with related known systems (*877*).

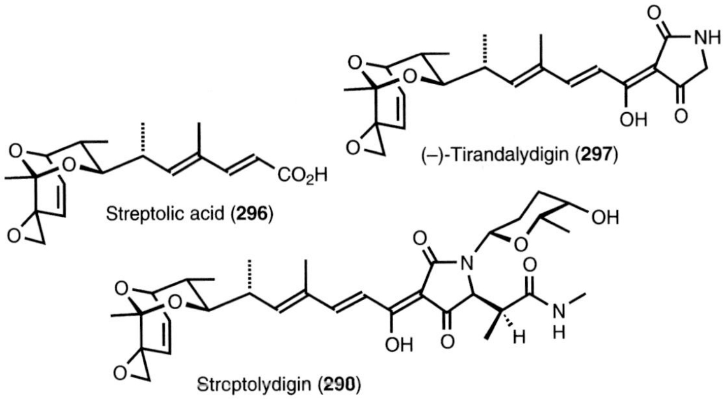

Streptolic acid (**296**)

(−)-Tirandalydigin (**297**)

Streptolydigin (**298**)

A related antibiotic, tirandamycin A (**299a**) (*878*) was isolated from the culture broth of *Streptomyces tirandis* (*879*). The biological activity was comparable to that of **298**. Later on, tirandamycin B (**299b**) was isolated, together with tyrandimycin A, from *Streptomyces flaveolus* (*880*). Structurally it only differs from tirandamycin A by the presence of an additional hydroxy group on the methyl function adjacent to the oxirane moiety. Comparison of the ^{1}H- and ^{13}C-NMR spectra of both compounds proved their absolute configurations to be identical.

The structure and absolute configuration of the tyrandamycins has been confirmed by X-ray diffraction analysis of the p-bromophenacyl

References, pp. 140–188

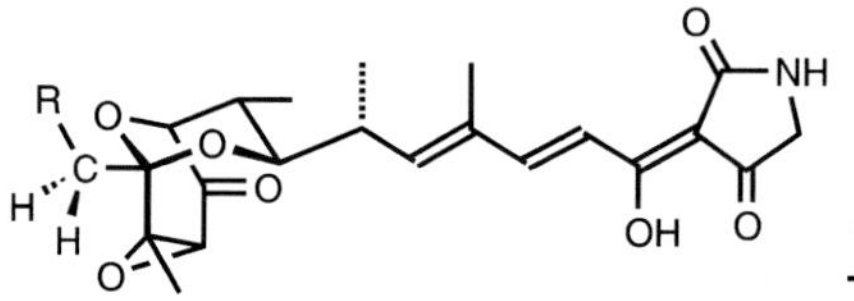

R = H : Tirandamycin A (**299a**)
R = OH : Tirandamycin B (**299b**)

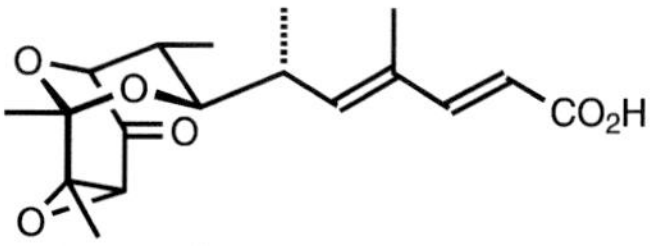

Tirandamycic acid (**300**)

ester of tirandamycic acid (**300**) (*881*), which is obtained on periodate degradation of sodium tirandamycin. As **300** and streptolic acid (**296**) – the corresponding degradation product of sodium streptolydigin (*vide supra*) – can be converted to a common derivative retaining the stereochemistry of both acids, the absolute configuration of streptolydigin was elucidated at the same time. As feeding experiments using ^{14}C-labeled substrates indicate that acetate, propionate, D-glucose and glutamic acid are all incorporated into streptolydigin (**298**) (*882*) the structures of tirandamycic acid and streptolic acid are well accounted for biogenetically by a combination of the propionate and acetate pathways.

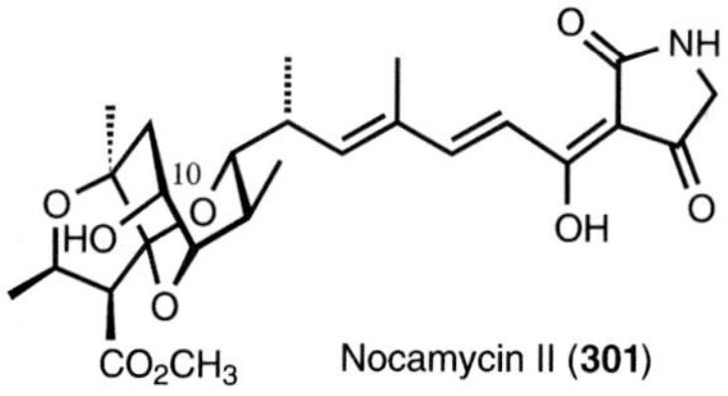

Nocamycin II (**301**)

The isolation of a further member in this series, nocamycin, from the fermentation broth of *Nocardiopsis syringae* was reported by Brazhnikova *et al.* (*883*). The original structure assignment for nocamycin I (*884*) was incorrect and it was revised later to the actual structures of nocamycin II (**301**) and the corresponding C10-ketone (nocamycin I) (*885*). Thus, nocamycin I proved to be identical with Bu-2313 B, an antibiotic previously isolated from an unidentified oligosporic Actinomycete strain by a Japanese group (*886, 887*), who reported also a partial synthesis (*886, 888*). The absolute configurations of Bu-2313 B and its *N*-methyl derivative (known as Bu-2313 A) was elucidated by X-ray diffraction analysis of the p-bromophenacyl ester of the acid which was obtained upon periodate oxidation of both antibiotics (*889*). Presumably, a common biosynthetic pathway exists to all dienoyltetramic acids (*882*).

Two more α-acyltetramic acid derivatives, catacandin A and B, have been isolated from the bacterium *Lysobacter gummosus* (*890*). The whole structure of these isomeric antifungal antibiotics is so far unknown. A similar partial structure seems to be common to fuligorubin A (**204**), a metabolite of the slime mold *Fuligo septica* and to the catacandins. However, the UV absorption ($\lambda_{max} = 258$ nm) of catacandin A ($C_{29}H_{38}N_2O_6$) points out to a shorter system of conjugated C=C-bonds than that of fuligorubin A.

References

1. Baeyer A, Emmerling A (1870) Reduction des Isatins zu Indigoblau. Chem Ber **3**: 514
2. (a) Bell CA, Lapper E (1877) Über die trockene Destillation der Ammoniumsalze der Zuckersäure. Chem Ber **10**: 1961; (b) Bell CA (1880) Einwirkung von Zinkstaub auf Succinimid. Chem Ber **13**: 877 (*cf.* Bernthsen A (1880) Über das Verhalten des Succinimids gegen Phosphorpentachlorid und gegen Zinkstaub. Chem Ber **13**: 1047
3. Battersby AR (1987) Nature's pathways to the pigments of life. Nat. Prod Rep **4**: 77
4. Sachs P (1931) Ein Fall von akuter Porphyrie mit hochgradiger Muskelatrophie. Klin Wschr **10**: 1123
5. Westall RG (1952) Isolation of phorphobilinogen from the urine of a patient with acute porphyria. Nature **170**: 614
6. Cookson GH, Rimington C (1954) Porphobilinogen. Biochem J **57**: 476
7. A comprehensive review on pyrrole alkaloids has appeared recently: Le Quesne PW, Dong Y, Blythe TA (1999) Recent Research on Pyrrole Alkaloids. In: Pelletier SW (ed) Alkaloids: Chemical & Biological Perspectives, vol 13, ch 3. Pergamon, Amsterdam, pp. 237–287
8. Jarrah MY, Thaller V (1983) Isolation and partial synthesis of 3-methoxycarbonyl-7-formyl-1-benzoxepin-5(2H)-one, the ester of a metabolite from shake cultures of the fungus *Marasmiellus ramealis* (Bull. ex Fr.) Singer. J Chem Soc Perkin Trans I, 1719
9. Corpe WA (1967) Extracellular accumulation of pyrroles in bacterial cultures. App Microbiol **11**: 145
10. Granick S, Bogorad L (1953) Porphobilinogen a monopyrrole. J Am Chem Soc **75**: 3610
11. (a) Waldenström J, Vahlquist B (1939) Studien über die Entstehung der roten Harnpigmente (Uroporphyrin und Porphobilin) bei der akuten Porphyrie aus ihrer farblosen Vorstufe (Porphobilinogen). Hoppe-Seyler's Z Physiol Chem **260**: 189; (b) Brockman PE, Gray CH (1953) Studies on porphobilinogen. Biochem J **54**: 22; (c) Mauzerall D (1960) The thermodynamic stability of porphobilinogen. J Am Chem Soc **82**: 2601
12. (a) Mauzerall D (1960) The condensation of porphobilinogen to uroporphyrinogen. J Am Chem Soc **82**: 2605; (b) Frydman RB, Reil S, Frydman B (1971) Relation between structure and reactivity of porphobilinogen and related pyrroles. Biochemistry **10**: 1154
13. (a) Gossauer A (1974) Die Chemie der Pyrrole. Springer, Berlin, p 202 ff; (b) Frydmann RB, Frydman B, Valasinas A (1979) Protoporphyrin: Synthesis and Biosynthesis of its Metabolic Intermediates. In: Dolphin D (ed) The Porphyrins. vol IV, ch 1. Academic Press, New York, p 24 ff
14. (a) Bobal P, Neier R (1997) The chemical synthesis of porphobilinogen an important intermediate of the biosynthesis of the "pigments of live". Trends Org Chem **6**: 125;

(b) Neier R (2000) A novel synthesis of porphobilinogen: Synthetic and biosynthetic studies. J Heterocycl Chem **37**: 487
15. Jackson AH, MacDonald SF (1957) A synthesis of porphobilinogen. Canad J Chem **35**: 715
16. Battersby AR, McDonald E, Wurziger HKW, James KJ (in part) (1975) Stereochemistry of biosynthesis of the vinyl groups of protoporphyrin-IX: A short synthesis of porphobilinogen. J Chem Soc Chem Commun 493
17. Battersby AR, Hunt E, McDonald E, Moron J (1973) Biosynthesis of porphyrins and related macrocycles. Part II. Synthesis of δ-amino[5-^{13}C]laevulinic acid and [11-^{13}C]porphobilinogen: Incorporation of the latter into protoporphyrin-IX. J Chem Soc Perkin Trans 1, 2917
18. Sancovich HA, Ferramola AM, Batlle AM del C, Grinstein M (1970) Preparation of porphobilinogen. Methods Enzymol **A17**: 220
19. (a) Müller G, Bezold G (1969) Gewinnung von Porphobilinogen aus δ-Aminolävulinsäure mit Zellsuspensionen von *Propiobacterium shermanii*. Z Naturforsch **24b**: 47; (b) Müller G (1972) Zur Gewinnung von Porphobilinogen aus δ-Aminolävulinsäure mittels *P. shermanii*: Bemerkenswerter Effekt einer Hitzebehandlung der Zellen. Z Naturforsch **27b**: 473
20. Gurne D, Shemin D (1973) Synthesis of the pyrrole porphobilinogen by sepharose-linked δ-aminolevulinic acid dehydratase. Science **180**: 1188
21. Scott JJ (1956) Synthesis of crystallizable porphobilinogen. Biochem J **62**: 6 P
22. Jordan PM (1991) Biosynthesis of Tetrapyrroles, ch 1 and 5. Elsevier, Amsterdam
23. (a) Irvine DG, Bayne W, Miyashita H, Majer JR (1969) Identification of kryptopyrrole in human urine and its relation to psychosis. Nature (London) **224**: 811; (b) Irvine DG, Bayne W, Majer JR (1970) Autotransfer chromatography combined with mass spectroscopy, for the characterization of pyrroles and indoles. J Chromatogr **48**: 334; (c) Sohler A, Beck R, Noval JJ (1970) Mauve factor reidentified as 2,4-dimethyl-3-ethylpyrrole and its sedative effect on the CNS. Nature (London) **228**: 1318
24. (a) Yamaguchi M, Mori Y, Nishimura N (1966) The specific fluorescent compound, f'_2, for collagen disease. Wakayama Med Rep **11**: 119 [Chem Abstr (1968) **68**: 11169]; (b) Yamaguchi M, Matsukawa S, Ura A, Koyama M, Imura T, Nishimura N (1969) Specificity of pyrrole-1,2-dicarboimide for collagen diseases. Wakayama Med Rep **13**: 169 [Chem Abstr (1970) **73**: 64176]
25. Tittlemier SA, Simon M, Jarman WM, Elliot JE, Norstrom RJ (1999) Identification of a novel $C_{10}H_6N_2Br_4Cl_2$ heterocyclic compound in seabird eggs. A bioaccumulating marine natural product? Environ Sci Technol **33**: 26
26. Gribble GW (1996) Naturally occurring organohalogen compounds – A comprehensive survey. In: Herz W, Kirby GW, Moore RE, Steglich W, Tamm C (eds) Progress in the chemistry of organic natural products, vol **68**. Springer, Wien, pp 133–141
27. Gribble GW, Blank DH, Jasinski JP (1999) Synthesis and identification of two halogenated bipyrroles present in seabird eggs. J Chem Soc Chem Commun 2195
28. Letellier G, Bouthillier LP (1956) The formation of 2-pyrrolecarboxylic acid from hydroxy-D- and allohydroxy-D-proline. Canad J Biochem and Physiol **34**: 1123
29. (a) Wolf G, Berger CRA (1958) The metabolism of hydroxyproline in the intact rat. Incorporation of hydroxyproline into protein and urinary metabolites. J Biol Chem **230**: 231; (b) Radhakrishnan AN, Meister A (1957) Conversion of hydroxyproline to pyrrole-2-carboxylic acid. J Biol Chem **226**: 559

30. Tokuyama T, Daly J, Witkop B, Karle IL, Karle J (1968) The structure of batrachotoxinin A, a novel steroidal alkaloid from Colombian arrow poison frog *Phyllobates aureotaenia*. J Am Chem Soc **90**: 1917
31. Tokuyama T, Daly J, Witkop B (1969) The structure of batrachotoxin, a steroidal alkaloid from the Colombian arrow poison frog *Phyllobates aureotaenia*, and partial synthesis of batrachotoxin and its analogs and homologs. J Am Chem Soc **91**: 3931 and references given therein
32. Witkop B (1971) New directions in the chemistry of natural products: The organic chemist as a pathfinder for biochemistry and medicine. Experientia **27**: 1121
33. Tokuyama T, Daly JW (1983) Steroidal alkaloids (batrachotoxins and 4-β-hydroxybatrachotoxins), "indole alkaloids" (calycanthine and chimonanthine) and a piperidinyldipyridine alkaloid (noranabasamine) in skin extracts from the Colombian poison-dart frog *Phylobates terribilis* (Dendrobatidae). Tetrahedron **39**: 41
34. Dumbacher JP, Beehler BM, Spande TF, Garraffo HM, Daly JW (1992) Homobatrachotoxin in the genus Pitohui: chemical defense in birds? Science (Washington) **258**: 799
35. Karle IL, Karle J (1969) The structural formula and crystal structure of the O-p-bromobenzoate derivative of batrachotoxinin A, $C_{31}H_{38}NO_6Br$, a frog venom and steroidal alkaloid. Acta Crystallogr **B25**: 428
36. Gilardi RD (1970) The absolute configuration of a steroidal substance, the O-p-bromobenzoate derivative of batrachotoxinin A. Acta Crystallogr **B26**: 440
37. Imhof R, Gössinger E, Graf W, Berner-Fenz L, Berner H, Schaufelberger R, Wehrli H (1973) Die Partialsynthese von Batrachotoxinin A. Helv Chim Acta **56**: 139
38. Daly JW (1982) Alkaloids of neotropical poison frogs (Dendrobatidae). Progr Chem Org Nat Prod **41**, Springer Wien, pp 211–234
39. Albuquerque EX, Daly JW, Witkop B (1971) Batrachotoxin: Chemistry and pharmacology. Science **172**: 995
40. (a) Tumlinson JH, Silverstein RM, Moser JC, Brownlee RG, Ruth JM (1971) Identification of the trail pheromone of a leaf-cutting ant, *Atta texana*. Nature (London) **234**: 348 (b) Tumlinson JH, Silverstein RM, Moser JC, Brownlee RG, Ruth JM (1972) A volatile trail pheromone of the leaf-cutting ant, *Atta texana*. J Insect Physiol **18**: 809
41. Riley RG, Silverstein RM, Carroll B, Carroll R (1974) Methyl 4-methylpyrrole-2-carboxylate: A volatile trail pheromone from the leaf-cutting ant, *Atta cephalotes*. J Insect Physiol **20**: 651
42. Do Nascimento RR, Morgan ED, Moreira DDO, Della Lucia TMC (1971) Trail pheromone of leaf-cutting ant *Acromyrmex subterraneus* (Forel). J Chem Ecol **20**: 1719
43. Rapoport H, Bordner J (1964) Synthesis of substituted 2,2'-bipyrroles. J Org Chem **29**: 2727
44. (a) Sonnet PE (1972) Synthesis of the trail marker of the Texas leaf-cutting ant, *Atta texana* (Buckley). J Med Chem **15**: 97; (b) Sonnet PE, Moser JC (1972) Synthetic analogs of the trail pheromone of the leaf-cutting ant, *Atta texana* (Buckley). J Agr Food Chem **6**: 1191; (c) Walizei GH, Breitmaier E (1989) Pyrrole aus 3-Alkoxyacroleinen und CH-aciden α-Aminoessigsäure-Derivaten. Synthesis 337; (d) Barton DHR, Kervagoret J, Zard SZ (1990) A useful synthesis of pyrroles from nitroolefins. Tetrahedron **46**: 7587; (e) Cornforth J, Ming-hui D (1990) Synthesis of 3-methylpyrrole via 4-methylpyrrole-2-carboxylate. A thermal oxazolone-pyrone rearrangement. J Chem Soc Perkin Trans 1, 1459; (f) Zimmer R, Collas M, Roth MM, Reißig H-U (1992) 6-Siloxy-substituted 5,6-dihydro-4H-1,2-oxazines as key building blocks

for natural products. Liebigs Ann Chem 709; (g) Xiao D, Schreier JA, Cook JH, Seybold PG, Ketcha DM (1996) Reversible Friedel-Crafts acylations of 3-alkyl-1-(phenylsulfon)pyrroles: Application to the synthesis of an ant trail pheromone. Tetrahedron Lett **37**: 1523; (h) Abbaspour Tehrani K, Borremans D, De Kimpe N (1999) Synthesis of 2-Acyl-3-chloropyroles: Application to the synthesis of the trail pheromone of the ant *Atta texana*. Tetrahedron **55**: 4133

45. Francke W, Schröder F, Walter F, Sinnwell V, Baumann H, Kaib M (1995) New alkaloids from ants: Identification and synthesis of (3R,5S,9R)-3-butyl-5-(1-oxopropyl)indolizidine and (3R,5R,9R)-3-butyl-5-(1-oxopropyl)indolizidine constituents of the poison gland secretion in *Myrmicaria eumenoides*. Liebigs Ann 965

46. Schröder F, Francke S, Francke W, Baumann H, Kaib M, Pasteels JM, Daloze D (1996) A new family of tricyclic alkaloids from *Myrmicaria* ants. Tetrahedron **52**: 13539

47. Schröder F, Francke W (1998) Synthesis of myrmicarin 217, a pyrrolo[2.1.5-cd]indolizine from ants. Tetrahedron **54**: 5259

48. Sayah B, Pelloux-Léon N, Vallée Y (2000) First synthesis of nonracemic (R)-(+)-myrmicarin 217. J Org Chem **65**: 2824

49. Schröder F, Sinnwell V, Baumann H, Kaib M (1996) Myrmicarin 430A: A new heptacyclic alkaloid from *Myrmicaria* ants. J Chem Soc Chem Commun 2139

50. Schröder F, Sinnwell V, Baumann H, Kaib M, Francke W (1997) Myrmicarin 663: A new decacyclic alkaloid from ants. Angew Chem **109**: 161; Angew Chem Int Ed Engl **36**: 77

51. Timmermans M, Braeckman J-C, Daloze D, Pasteels JM, Merlin J, Declercq J-P (1992) Exochomine, a dimeric ladybird alkaloid, isolated from *Exochromus quadripustulatus* (Coleoptera: Coccinellide). Tetrahedron Lett **33**: 1281

52. McCormick KD, Attygalle AB, Xu S-C, Svatos A, Meinwald J, Houck MA, Blankespoor CL, Eisner T (1994) Chilocorine: Heptacyclic alkaloid from a coccinellid beetle. Tetrahedron **50**: 2365

53. Shi X, Attygalle AB, Meinwald J, Houck MA, Eisner T (1995) Spirocyclic defensive alkaloid from a coccinellid beetle. Tetrahedron **51**: 8711

54. Meinwald J, Meinwald YC, Wheeler JW, Eisner T, Brower LP (1966) Major components in the exocrine secretion of a male butterfly (Lycorea). Science **151**: 583

55. Meinwald J, Meinwald YC (1966) Structure and synthesis of the major components in the hair-pencil secretion of a male butterfly *Lycorea ceres ceres* (Cramer). J Am Chem Soc **88**: 1305

56. Pliske TE, Eisner T (1969) Sex pheromone of the queen butterfly: Biology. Science **164**: 1170

57. Meinwald J, Meinwald YC, Mazzocchi PH (1969) Sex pheromone of the queen butterfly: Chemistry. Science **164**: 1174

58. Meinwald J, Thompson WR, Eisner T (1971) Pheromones. VII. African monarch. Major components of the hair-pencil secretion. Tetrahedron Lett 3485

59. Edgar JA, Culvenor CCJ, Smith LW (1971) Dihydropyrrolizine derivatives in the "hair-pencil" secretions of Danaid butterflies. Experientia **27**: 761

60. Meinwald J, Boriack CJ, Schneider D, Boppré M, Wood WF, Eisner T (1974) Volatile ketones in the hairpencil secretion of danaid butterflies (*Amauris* and *Danaus*). Experientia **30**: 721

61. Röder E, Wiedenfeld H, Bourauel T (1985) Synthese von 2,3-Dihydro-7-methyl-1H-pyrrolizin-1-on, dem Sexualpheromon Danaidon. Liebigs Ann Chem 1708

62. (a) Schneider D, Boppré M, Schneider H, Thompson WR, Boriak CJ, Petty RL, Meinwald J (1975) A pheromone precursor and its uptake in male Danais butterflies. J Comp Physiol **97**: 245; (b) Boppré M, Petty RL, Schneider D, Meinwald J (1978)

Behaviorally mediated contacts between scent organs: Another prerequisite for pheromone production in *Danaus chysippus* males (Lepidoptera). J Comp Physiol **A126**: 97
63. Conner WE, Eisner T, Vander Meer RK, Guerrero A, Meinwald J (1981) Precopulatory sexual interaction in an arctiid moth (*Utetheisa ornatrix*): Role of pheromone derived from dietary alkaloids. Behav Ecol Sociobiol **9**: 227
64. (a) Schneider D, Boppre M, Zweig J, Horsley SB, Bell TW, Meinwald J, Hansen K, Diehl EW (1982) Scent organ development in *Creatonotos* moths: Regulation by pyrrolizidine alkaloids. Science **215**: 1264; (b) Bell TW, Meinwald J (1986) Pheromones of two arctiid moths (*Creatonotos transiens* and *C. gangis*): chiral components from both sexes and achiral female components. J Chem Ecol **12**: 385
65. Tada H, Tozyo T (1988) Two bromopyrroles from a marine sponge *Agelas* sp. Chem Lett 803
66. Emrich R, Weyland H, Weber K (1990) 2,3,4-Tribromopyrrole from the marine polychaete *Polyphysia crassa*. J Nat Prod **53**: 703
67. Barrow RA, Capon RJ (1993) Brominated pyrrole carboxylic acids from an Australian marine sponge, *Axinella* sp. Nat Prod Lett **1**: 243
68. Schmitz FJ, Gunasekera SP, Lakshmi V, Tillekeratne LMV (1985) Marine natural products: Pyrrololactams from several sponges. J Nat Prod **48**: 47
69. Reddy NS, Ramesh P, Rao TP, Rao JV, Venkateswarlu Y (1999) Chemical investigation of the marine sponge *Axinalla tenuidigitata*. Indian J. Chem., Sect. B: Org Chem Incl Med Chem **38B**: 1145
70. (a) Anderson HJ, Lee S-F (1965) Pyrrole chemistry. IV. The preparation and some reactions of brominated pyrrole derivatives. Can J Chem **43**: 409; (b) Hodge P, Rickards RW (1965) The halogenation of methyl pyrrole-2-carboxylate and some related pyrroles. J Chem Soc 459
71. Forenza S, Minale L, Riccio R, Fattorusso E (1971) New bromopyrrole derivatives from the sponge *Agelas oroides*. J Chem Soc D 1129
72. König GM, Wright AD, Linden A (1998) Antiplasmodial and cytotoxic metabolites from the Maltese sponge *Agelas oroides*. Planta Med **64**: 443
73. Gunasekera SP, Cranick S, Longley RE (1989) Immunosuppressive compounds from a deep water marine sponge, *Agelas flabelliformis*. J Nat Prod **52**: 757
74. Utkina NK, Fedoreev SA, Maksimov OB (1985) Pyrrole derivatives from the marine sponge *Axinellidae*. Khim Prir Soedin 578 [Chem Abstr (1986) **104**: 145784]
75. Rinkes IJ (1941) Untersuchungen über Pyrolderivate. 5. Mitteilung. Recl Trav Chim Pays-Bas **60**: 303
76. Chevolot L, Padua S, Ravi BN, Blyth PC, Sheuer PJ (1977) Isolation of 1-methyl-4,5-dibromopyrrole-2-carboxylic acid and its 3'-(hydantoyl)propylamide (midpacamide) from a marine sponge. Heterocycles **7**: 891
77. Fathi-Afshar R, Allen TM (1988) Biologically active metabolites from *Agelas mauritania*. Canad J Chem **66**: 45
78. Fu X, Ng P-L, Schmitz FJ, Hossain MB, van der Helm D, Kelly-Borges M (1996) Makaluvic acids A and B: Novel alkaloids from the marine sponge *Zyzzya fuliginosus*. J Nat Prod **59**: 1104
79. Cafieri F, Fattorusso E, Mangoni A, Taglialatela-Scafati O, Carnuccio R (1995) A novel bromopyrrole alkaloid from the sponge *Agelas longissima* with antiserotonergic activity. Bioorg Med Chem Lett **5**: 799
80. Kobayashi J, Kanda F, Ishibashi M, Shigemori H (1991) Manzacidins A-C, novel tetrahydropyrimidine alkaloids from the Okinawan marine sponge *Hymeniacidon* sp. J Org Chem **56**: 4574

81. Namba K, Shinada T, Teramoto T, Ohfune Y (2000) Total synthesis and absolute structure of manzacidin A and C. J Am Chem Soc **122**: 10708
82. Jahn T, König GM, Wright AD, Wörheide G, Reitner J (1997) Manzacidin D: An unprecedented secondary metabolite from "living fossil" sponge *Astrosclera willeyana*. Tetrahedron Lett **38**: 3883
83. Ishida K, Ishibashi M, Shigemori H, Sasaki T, Kobayashi J (1992) Agelasine G, a new antileukemic alkaloid from the Okinawan marine sponge *Agelas* sp. Chem Pharm Bull **40**: 766
84. Capon RJ, Faulkner DJ (1984) Antimicrobial metabolites from a pacific sponge, *Agelas* sp. J Am Chem Soc **106**: 1819
85. Shoji N, Umeyama A, Teranaka M, Arihara S (1996) Four novel diterpenoids, including nakamurol A with unique thelepogane skeleton, from the marine sponge *Agelas nakamurai*. J Nat Prod **59**: 448
86. König GM, Wright AD (1994) Two new naturally occurring pyrrole derivatives from the tropical marine sponge *Agelas oroides*. Nat Prod Lett **5**: 141
87. Iwagawa T, Kaneko M, Okamura H, Nakatani M, Van Soest RWM (1998) New alkaloids from the Papua New Guninean Sponge *Agelas nakamurai*. J Nat Prod **61**: 1310
88. Umeyama A, Ito S, Yuasa E, Arihara S, Yamada T (1998) A new bromopyrrole alkaloid and the optical resolution of the racemate from the marine sponge *Homaxinella* sp. J Nat Prod **61**: 1433
89. Mancini I, Guella G, Amade P, Roussakis C, Pietra F (1997) Hanishin, a semiracemic, bioactive C_9 alkaloid of the axinellid sponge *Acanthella carteri* from the Hanish Islands. A shunt metabolite? Tetrahedron Lett **38**: 6271
90. Tsukamoto S, Kato H, Hirota H, Fusetani N (1996) Ceratinamides A and B: New antifouling dibromotyrosine derivatives from the marine sponge *Pseudoceratina purpurea*. Tetrahedron **52**: 8181
91. Tsukamoto S, Kato H, Hirota H, Fusetani N (1996) Mauritiamine, a new antifoulding oroidin dimer from the marine sponge *Agelas mauritiana*. J Nat Prod **59**: 501
92. Stempien Jr. MF, Nigrelli RF, Chib JS (1972) Isolation and synthesis of physiologically active substances from sponges of the genus *Agelas*. 164th. ACS National Meeting, New York, 21 MEDI Abstract
93. Clark WD, Corbett T, Valeriote F, Crews P (1997) Cyclocinamide A. An unusual cytotoxic halogenated hexapeptide from the marine sponge *Psammocinia*. J Am Chem Soc **119**: 9285
94. Grieco PA, Reilly M (1998) Studies related to the absolute configuration of cyclocinamide A: Total synthesis of 4(*R*),11(*R*)-cyclocinamide A. Tetrahedron Lett **39**: 8925
95. Rudi A, Stein Z, Green S, Goldberg I, Kashman Y, Benayahu Y, Schleyer M (1994) Phorbazoles A–D, novel chlorinated phenylpyrrolyloxazoles from the marine sponge *Phorbas* aff. *clathrata*. Tetrahedron Lett **35**: 2589
96. Tsukamoto S, Kato H, Hirota H, Fusetani N (1996) Pseudoceratidine: A new antifouling spermidine derivative from the marine sponge *Pseudoceratina purpurea*. Tetrahedron Lett **37**: 1439
97. (a) Ponasik JA, Kassab DJ, Ganem B (1996) Synthesis of the antifouling polyamine pseudoceratidine and its analogs: Factors influencing biocidal activity. Tetrahedron Lett **37**: 6041; (b) Ponasik JA, Conova S, Kinghorn D, Kinney WA, Rittschof D, Ganem B (1998) Pseudoceratidine, a marine natural product with antifouling activity: Synthetic and biological studies. Tetrahedron **54**: 6977

98. Behrens C, Chistoffersen MW, Gram L, Nielsen PH (1997) A convenient synthesis of pseudoceratidine and three analogs for biological evaluation. Bioorg Med Chem Lett 7: 321
99. Cafieri F, Fattorusso E, Mangoni A, Taglialatela-Scafati O (1996) Clathramides, unique bromopyrrole alkaloids from the caribbean sponge *Agelas clathrodes*. Tetrahedron 52: 13713
100. Cafieri F, Fattorusso E, Taglialatela-Scafati O (1998) Novel bromopyrrole alkaloids from the sponge *Agelas dispar*. J Nat Prod 61: 122
101. Cafieri F, Carnuccio R, Fattorusso E, Taglialatela-Scafati O, Vallefucco T (1997) Antihistaminic activity of bromopyrrole alkaloids isolated from Caribbean *Agelas* Sponges. Bioorg Med Chem Lett 7: 2283
102. Cimino G, De Stefano S, Minale L, Sodano G (1975) Metabolism in Porifera – III. Chemical patterns and the classification of the desmospongiae. Comp Biochem and Physiol B50: 279
103. Supriyono A, Schwarz B, Wray V, Witte L, Müller WEG, van Soest R, Sumaryono W, Proksch P (1995) Bioactive alkaloids from the tropical marine sponge *Axinella carteri*. Z Naturforsch 50c: 669
104. Cimino G, De Rosa S, De Stefano S, Mazzarella L, Puliti R, Sodano G (1982) Isolation and X-ray crystal structure of a novel bromocompound from two marine sponges. Tetrahedron Lett 23: 767
105. De Nanteuil G, Ahond A, Guilhem J, Poupat C, Tran Huu Dau E, Potier P, Pusset M, Pusset J, Laboute P (1985) Invertébrés marins du lagon néocalédonien. Isolement et identification des métabolites d'une nouvelle espèce de spongiaire, *Pseudaxinyssa cantharella*. Tetrahedron 41: 6019
106. Rinehart KL (1989) Biologically active marine natural products. Pure Appl Chem 61: 525
107. (a) Assmann M, Lichte E, Pawlik JR, Köck M (2000) Chemical defenses of the Caribbean sponges *Agelas wiedenmayeri* and *Agelas conifera*. Mar Ecol Progr Ser 207: 255; (b) Lindel T, Hoffmann H, Hochgürtel M, Pawlik JR (2000) Structure-activity relationship of inhibition of fish feeding by sponge-derived and synthetic pyrrole-imidazole alkaloids. J Chem Ecol 26: 1477, and references given therein
108. Garcia EE, Benjamin LE, Fryer RI (1973) Reinvestigation into the structure of oroidin, a bromopyrrole derivative from marine sponge. J Chem Soc Chem Commun 78
109. Walker RP, Faulkner DJ, Van Engen D, Clardy J (1981) Sceptrin, an antimicrobial agent from the sponge *Agelas sceptrum*. J Am Chem Soc 103: 6772
110. (a) de Nanteuil G, Ahond A, Poupat C, Thoison O, Potier P (1986) Synthèse de l'oroïdine. Bull Soc Chim Fr 813; (b) Little TL, Webber SE (1994) A simple and practical synthesis of 2-aminoimidazoles. J Org Chem 59: 7299; (c) Berrée F, Girard-Le Bleis P, Carboni B (2002) Synthesis of the marine sponge alkaloid oroidin and its analogues via Suzuki cross-coupling reactions. Tetrahedron Lett 43: 4935
111. Daninos-Zeghal S, Al Mourabit A, Ahond A, Poupat C, Potier P (1997) Synthèse de métabolites marins 2-aminoimidazoliques: Hyménidine, oroidine et kéramadine. Tetrahedron 53: 7605
112. Olofson A, Yakushijin K, Horne DA (1998) Synthesis of marine sponge alkaloids oroidin, clathrodin, and dispacamides. Preparation and transformation of 2-amino-4,5-dialkoxy-4,5-dihydroimidazolines from 2-aminoimidazoles. J Org Chem 63: 1248
113. Lindel T, Hochgürtel M (2000) Synthesis of the marine natural product oroidin and its Z-isomer. J Org Chem 65: 2806

114. Kobayashi J, Ohizumi Y, Nakamura H, Hirata Y (1986) A novel antagonist of serotonergic receptors, hymenidin, isolated from the Okinawan marine sponge *Hymeniacidon* sp. Experientia **42**: 1176
115. Morales JJ, Rodriguez AD (1991) The structure of clathrodin, a novel alkaloid isolated from the Caribbean sea sponge *Agelas clathrodes*. J Nat Prod **34**: 629
116. Nakamura H, Ohizumi Y, Kobayashi J, Hirata Y (1984) Keramadine, a novel antagonist of serotonergic receptors isolated from the Okinawan sea sponge *Agelas* sp. Tetrahedron Lett **25**: 2475
117. Lindel T, Hochgürtel M (1998) The alkyne pathway to keramadine from the marine sponge *Agelas* sp. Tetrahedron Lett **39**: 2541
118. Daninos S, Al Mourabit A, Ahond A, Zurita MB, Poupat C, Potier P (1994) Synthèse de métabolites marins 2-aminoimidazoliques: clathrodine et 3'-amino-1'-[2-aminoimidazol-4(5)-yl]-prop-2'-ène. Bull Soc Chim Fr **131**: 590
119. (a) Lindel T, Hoffmann H, Hochgürtel M (1999) Chemistry of marine pyrroleimidazole alkaloids. Bioorg Chem 8; (b) Mourabit AA, Potier P (2001) Sponge's molecular diversity through the ambivalent reactivity of 2-aminoimidazole: A universal chemical pathway to the oroidin-based pyrrole-imidazole alkaloids and their Palau'amine congeners. Eur J Org Chem 237
120. Braekman J-C, Daloze D, Stoller C, van Soest RWM (1992) Chemotaxonomy of *Agelas* (Porifera: Demospongiae). Biochem Syst Ecol **20**: 417 and references given therein
121. Wright AE, Chiles SA, Cross SS (1991) 3-Amino-1-(2-aminoimidazolyl)prop-1-ene from the marine sponges *Teichaxinella morchella* and *Ptilocaulis walpersi*. J Nat Prod **54**: 1684
122. Assmann M, Lichte E, van Soest RWM, Köck M (1999) New bromopyrrole alkaloid from the marine sponge *Agelas wiedenmayeri*. Org Lett **1**: 455
123. Cafieri F, Fattorusso E, Mangoni A, Taglialatela-Scafati O (1996) Dispacamides, antihistamine alkaloids from Caribbean *Agelas* sponges. Tetrahedron Lett **37**: 3587
124. Lindel T, Hoffmann H (1997) Synthesis of dispacamide from the marine sponge *Agelas dispar*. Tetrahedron Lett **38**: 8935
125. Fresneda PM, Molina P, Sanz MA (2001) A convergent approach to midpacamide and dispacamide pyrrole-imidazole marine alkaloids. Tetrahedron Lett **42**: 851
126. Uemoto H, Tsuda M, Kobayashi J (1999) Mukanadins A-C, new bromopyrrole alkaloids from marine sponges *Agelas nakamurai*. J Nat Prod **62**: 1581. Erratum (2000) *ibid* **63**: 1045
127. Lindel T, Hoffmann H (1997) Synthesis of *rac*-midcapamide and *spiro*-cyclization of its precursor. Liebigs Ann/Recueil **7**: 1525
128. Jiménez C, Crews P (1994) Mauritamide A and accompanying oroidin alakaloids from the sponge *Agelas mauritiana*. Tetrahedron Lett **35**: 1375
129. Fattorusso E, Taglialatela-Scafati O (2000) Two novel pyrrole-imidazole alkaloids from the Mediterranean sponge *Agelas oroides*. Tetrahedron Lett **41**: 9917
130. Kobayashi J, Inaba K, Tsuda M (1997) Tauroacidins A and B, new bromopyrrole alkaloids possessing a taurine residue from *Hymeniacidon* sponge. Tetrahedron **53**: 16679
131. Tsuda M, Uemoto H, Kobayashi J (1999) Slangenins A-C. novel bromopyrrole alkaloids from marine sponge *Agelas nakamurai*. Tetrahedron Lett **40**: 5709
132. Barrios Sosa AC, Yakushijin K, Horne DA (2000) Synthesis of slagenins A, B, and C. Org Lett **2**: 3443

133. Jiang B, Liu J-F, Zhao S-Y (2001) Enantioselective synthesis for the antipodes of slagenins B and C: Establishment of absolute stereochemistry. Org Lett **3**: 4011
134. Reddy NS, Venkateswarlu Y (2000) S-(+)-Methyl ester of hanishin from the marine sponge *Agelas ceylonica*. Biochem Syst Ecol **28**: 1035
135. Reddy NS, Yenkateswarlu Y (2000) A new bromopyrrole alkaloid from the sponge *Axinella tenuidigitata*. Indian J Chem Sect. B: Org Chem Incl Med Chem **39**: 971
136. Koul SK, Taneja SC, Agarwal VK, Dhar KL (1988) Minor amides of *Piper* species. Phytochemistry **27**: 3523
137. Cafieri F, Fattorusso LE, Mangoni A, Tagliatela-Scafati O (1995) Longamide and 3,7-dimethylisoguanidine, two novel alkaloids from the marine sponge *Agelas longissima*. Tetrahedron Lett **36**: 7893
138. Li C-J, Schmitz FJ, Kelly-Borges M (1998) A new lysine derivative and new 3-bromopyrrole carboxylic acid derivative from two marine sponges. J Nat Prod **61**: 387
139. (a) Marchais S, Al Mourabit A, Ahond A, Poupat C, Potier P (1999) Synthesis of the marine carbinolamine (±) longamide. Control of *N*-1 and *C*-3 bromopyrrole nucleophilicity. Tetrahedron Lett **40**: 5519; (b) Barrios Sosa AC, Yakushijin K, Horne DA (2000) Controlling cyclizations of 2-pyrrolecarboxamidoacetals. Facile solvation of β-amido aldehydes and revised structure of synthetic homolongamide. Tetrahedron Lett **41**: 4295
140. Banwell MG, Bray AM, Willis AC, Wong D-J (1999) First syntheses of the pyrroloketopiperazine marine natural products (±)-longamide, (±)-longamide B methyl ester and (±)-hanishin. New J Chem **23**: 687
141. Kobayashi J, Ohizumi Y, Nakamura H, Hirata Y, Wakamatsu K, Miyazawa T (1986) Hymenin, a novel α-adrenoceptor blocking agent from the Okinawan marine sponge *Hymeniacidon* sp. Experientia **42**: 1064
142. Eder C, Proksch P, Wray V, Steube K, Bringmann G, van Soest RWM, Sudarsono, Ferdinandus E, Pattisina LA, Wiryowidagdo S. Moka W (1999) New alkaloids from the indopacific sponge *Stylissa carteri*. J Nat Prod **62**: 184
143. Xu Y, Phan G, Yakushijin K, Horne DA (1994) A synthesis of (±)-hymenin. Tetrahedron Lett **35**: 351
144. (a) Xu Y, Yakushijin K, Horne DA (1997) Synthesis of $C_{11}N_5$ marine sponge alkaloids: (±)-Hymenin, stevensine, hymenialdisine and debromohymenialdisine. J Org Chem **62**: 456; (b) Barrios Sosa AC, Yakushijin K, Horne DA (2000) A practical synthesis of (Z)-debromohymenialdisine. J Org Chem **65**: 610
145. Albizati KF, Faulkner DJ (1985) Stevensine, a novel alkaloid of an unidenfied marine sponge. J Org Chem **50**: 4163
146. Andrade P, Willoughby R, Pomponi SA, Kerr RG (1999) Biosynthetic studies of the alkaloid, stevensine, in a cell culture of the marine sponge *Teichaxinella morchella*. Tetrahedron Lett **40**: 4775
147. Xu Y, Yakushijin K, Horne DA (1996) Transbromination of brominated pyrrole and imidazole derivatives: Synthesis of the $C_{11}N_5$ marine alkaloid stevensine. Tetrahedron Lett **37**: 8121
148. Pettit GR, Herald CL, Leet JE, Gupta R, Schaufelberger DE, Bates RB, Clewlow PJ, Doubek DL, Manfredi KP, Rützler K, Schmidt JM, Tackett LP, Ward FB, Bruck M, Camou F (1990) Antineoplastic agents. 168. Isolation and structure of axinohydantoin. Can J Chem **68**: 1621
149. (a) Zeng L, Fu X, Su J, De Guzman F, Schmitz FJ, Hossain MB, Van der Helm D (1991) Studies on the chemical constituents of South China sea sponge *Phacellia fusca*. Chin J Chem **9**: 136 [Chem Abstr (1991) **115**: 203706]; (b) Fu X, Zeng L, Su J,

De Guzman F, Schmitz FJ, Hossain MB, Van der Helm D (1991) A novel pyrrolactam alkaloid from South China sea sponge *Phacellia fusca* Schmidt. Chem Res Chin Univ **7**: 78 [Chem Abstr (1993) **118**: 36148]
150. Sharma GM, Buyer JS, Pomerantz MW (1980) Characterization of a yellow compound isolated from the marine sponge *Phakellia flabellata*. J Chem Soc Chem Commun 435
151. Utkina NK, Fedoreev SA, Maksimov OB (1984) Nitrogen-containing metabolites of the marine sponge *Acanthella carteri*. Khim Prir Soedin 535 [Chem Abstr (1985) **102**: 146 334]
152. Kitagawa I, Kobayashi M, Kitanaka K, Kido M, Kyogoku Y (1983) Marine natural products. XII. On the chemical constituents of the Okinawan marine sponge *Hymeniacidon aldis*. Chem Pharm Bull **31**: 2321
153. Groszek G, Kantoci D, Pettit GR (1995) The isolation and structure elucidation of debromoaxinohydantoin. Liebigs Ann 715
154. Annoura H, Tatsuoka T (1995) Total syntheses of hymenialdisine and debromohymenialdisine: Stereospecific construction of the 2-amino-4-oxo-2imidazolin-5(Z)-disubstituted ylidene ring system. Tetrahedron Lett **36**: 413
155. Mattia CA, Mazzarella L, Puliti R (1982) 4-(2-Amino-4-oxo-2-imidazolin-5-ylidene)-2-bromo-4,5,6,7-tetrahydropyrrolo[2,3-c]azepin-8-one methanol solvate: a new bromo compound from the sponge *Acanthella aurantiaca*. Acta Cryst **B38**: 2513
156. Williams DH, Faulkner DJ (1996) Isomers and tautomers of hymenialdisine and debromohymenialdisine. Nat Prod Lett **9**: 57
157. Inaba K, Sato H, Tsuda M, Kobayashi J (1998) Spongiacidins A–D, new bromopyrrole alkaloids from *Hymeniacidon* sponge. J Nat Prod **61**: 693
158. Kobayashi J, Nakamura H, Ohizumi Y (1988) α-Adrenoceptor blocking action of hymenin, a novel marine alkaloid. Experientia **44**: 86
159. (a) Kobayashi J, Ohizumi Y, Nakamura H, Hirata Y (1986) A novel antagonist of serotonergic receptors, hymenidin, isolated from the Okinawan marine sponge *Hymeniacidon* species. Experientia **42**: 1176; (b) Rosa R, Silva W, Escalona de Motta G, Rodriguez AD, Morales JJ, Ortiz M (1992) Antimuscarinic activity of a family of $C_{11}N_5$ compounds isolated from *Agelas* sponges. Experientia **42**: 885
160. Kasum B, Prager RH, Tsopelas C (1990) Dihydroindol-7(6H)-ones and 6,7-dihydropyrrolo[3,3-c]azepine-4,8-(1H,5H)dione. Aust J Chem **43**: 355
161. Prager RH, Tsopelas C (1990) Approaches to the synthesis of 5-benzylidene-2-imidazolin-4-ones. Aust J Chem **43**: 367
162. Burkholder PR, Sharma GM (1969) Antimicrobial agents from the sea. Lloydia **32**: 466
163. (a) Sharma GM, Burkholder PR (1971) Structure of dibromophakelin, a new bromine-containing alkaloid from the marine sponge *Phakellia flabellata*. J Chem Soc Chem Commun 151; (b) Sharma GM, Magdoff-Fairchild B (1977) Natural products of marine sponges. 7. The constitution of weakly basic guanidine compounds, dibromophykellin and monobromophykellin. J Org Chem **42**: 4118
164. Foley LH, Büchi G (1982) Biomimetic synthesis of dibromophakelin. J Am Chem Soc **104**: 1776
165. Fedoreev SA, Utkina NK, Il'in SG, Reshetnyak MV, Maksimov OB (1986) The structure of dibromoisophakellin from the marine sponge *Acanthella carteri*. Tetrahedron Lett **27**: 3177
166. Wiese KJ, Yakushijin K, Horne DA (2002) Synthesis of dibromophakellstatin and dibromoisophakellin Tetrahedron Lett **43**: 5135

167. Pettit GR, McNulty J, Herald DL, Doubek DL, Chapuis JC, Schmidt JM, Tackett LP, Boyd MR (1997) Antineoplastic agents. 362. Isolation and X-ray crystal structure of dibromophakellstatin from the Indian ocean sponge *Phakellia mauritiana*. J Nat Prod **60**: 180
168. Fedoreyev SA, Ilyin SG, Utkina NK, Maximov OB, Reshetnyak MV, Antipin MYu, Struchkov YuT (1989) The structure of dibromoagelaspongin – A novel bromine-containing guanidine derivative from the marine sponge *Agelas* sp. Tetrahedron **45**: 3487
169. D'Ambrosio M, Guerriero A, Debitus C, Ribes O, Pusset J, Leroy S, Pietra F (1993) Agelastatin A, a new skeleton cytotoxic alkaloid of the oroidin family. Isolation from the axinellid sponge *Agelas dendromorpha* of the Coral Sea. J Chem Soc Chem Commun 1305
170. D'Ambrosio M, Guerriero A, Chiasera G, Pietra F (1994) Conformational preferences and absolute configuration of agelastatin A, a cytotoxic alakaloid of the axinellid sponge *Agelas dendromorpha* of the Coral Sea, *via* combined molecular modelling, NMR, and exciton splitting for diamide and hydroxyamide derivatives. Helv Chim Acta **77**: 1895
171. Hong TW, Jimenez DR, Molinski TF (1998) Agelastatins C and D, new pentacyclic bromopyrroles from the sponge *Cymbastela* sp., and potent arthropod toxicity of (−)-agelastatin A. J Nat Prod **61**: 158
172. (a) Anderson GT, Chase CE, Koh Y-H, Stien D, Weinreb SM, Shang MY (1998) Studies on total synthesis of the cytotoxic marine alakaloid agelastatin A. J Org Chem **63**: 7594; (b) Stien D, Anderson GT, Chase CE, Koh Y-H, Weinreb SM (1999) Total synthesis of the antitumor sponge alkaloid agelastatin A. J Am Chem Soc **121**: 9574
173. D'Ambrosio M, Guerriero A, Ripamonti M, Debitus C, Waikedre J, Pietra F (1996) The active centres of agelastatin A, a strongly cytotoxic alkaloid of the Coral Sea Anixellid sponge *Agelas dendromorpha*, as determined by comparative bioassays with semisynthetic derivatives. Helv Chim Acta **79**: 727
174. Olofson A, Yakushijin K, Horne DA (1997) Synthesis of mauritiamine. J Org Chem **62**: 7918
175. Keifer PA, Schwartz RE, Koker MES, Hughes Jr RG, Rittschof D, Rinehart KL (1991) Bioactive bromopyrole metabolites from the Caribbean sponge *Agelas conifera*. J Org Chem **56**: 2965
176. Shen X, Perry TL, Dunbar CD, Kelly-Borges M, Hamann MT (1998) Debromosceptrin, an alkaloid from the Caribbean sponge *Agelas conifera*. J Nat Prod **61**: 1302
177. Kobayashi J, Tsuda M, Ohizumi Y (1991) A potent actomyosin ATPase activator from the Okinawan marine sponge *Agelas* cf. *nemoechinata*. Experientia **47**: 301
178. Eder C, Proksch P, Wray V, van Soest RWM, Ferdinandus E, Pattisina LA, Sudarsono (1999) New bromopyrrole alkaloids from the indopacific sponge *Agelas nakamurai*. J Nat Prod **62**: 1295
179. Kawasaki I, Sakaguchi N, Fukushima N, Fujioka N, Nikaido F, Yamashita M, Ohta S (2002) Novel Diels-Alder-type dimerization of 5-ethenyl-2-phenylsulfanyl-1*H*-imidazoles and its application to biomimetic synthesis of 12,12′-dimethylageliferin. Tetrahedron Lett **43**: 4377
180. Kobayashi J, Tsuda M, Murayama T, Nakamura H, Ohizumi Y, Ishibashi M, Iwamura M, Ohta T, Nozoe S (1990) Ageliferins, potent actomyosin ATPase activators from the Okinawan marine sponge *Agelas* sp. Tetrahedron **46**: 5579
181. Williams DH, Faulkner DJ (1996) *N*-Methylated ageliferins from the sponge *Astrosclera willeyana* from Pohnpei. Tetrahedron **52**: 5381

182. Kobayashi J, Suzuki M, Tsuda M (1997) Konbu'acidin, a new bromopyrrole alkaloid with cdk4 inhibitory activity from *Hymenacidon* sponge. Tetrahedron **53**: 15681
183. (a) Kinnel RB, Gehrken H-P, Scheuer PJ (1993) Palau'amine. A cytotoxic immunosuppressive haxacyclic bisguanidine antibiotic from the sponge *Stylotella agminata*. J Am Chem Soc **115**: 3376; (b) Kinnel RB, Gehrken H-P, Swali R, Skoropowski G, Scheuer PJ (1998) Palau'amine and its congeners: A family of bioactive bisguanidines from the marine sponge *Stylotella aurantium*. J Org Chem **63**: 3281
184. Kato T, Shizuri Y, Izumida H, Yokoyama A, Endo M (1995) Styloguanidines, new chitinase inhibitors from the marine sponge *Stylotella aurantium*. Tetrahedron Lett **36**: 2133
185. Overman LE, Rogers BN, Tellew JE, Trenkle WC (1997) Stereocontrolled synthesis of the tetracyclic core of the bisguanidine alkaloids Palau'amine and styloguanidine. J Am Chem Soc **119**: 7159
186. Urban S, de Almeida Leone P, Carroll AR, Fechner GA, Smith J, Hooper JNA, Quinn RJ (1999) Axinellamines A–D. novel imidazoazolo-imidazole alkaloids from the Australian marine sponge *Axinella* sp. J Org Chem **64**: 731
187. Bascombe KC, Peter SR, Tinto WF, Bissada SM, Mclean S, Reynolds WF (1998) Axinellamines A and B, new pyrrole alkaloids of the marine sponge *Axinella* sp. Heterocycles **48**: 1461
188. Seki M, Mori K (2001) The absolute configuration of axinellamine A, a pyrrole alkaloid of the marine sponge *Axinella* sp., was determined as R by synthesizing its (S)-isomer. Eur J Org Chem 503
189. Miller SL, Tinto WF, Yang J-P, McLean S, Reynolds WF (1995) Axinellamide, a new alkaloid from the marine sponge *Axinella* sp. Tetrahedron Lett **36**: 5851
190. Cimino G, de Stefano S, Minale M (1975) Long alkyl chains, 3-substituted pyrrole-2-aldehyde (-2-carboxylic acid and methyl ester) from the marine sponge *Oscarella lobularis*. Experientia **31**: 1387
191. Muchowski JM, Naef R (1984) 3-Lithiopyrroles by halogen-metal interchange of 3-bromo-1-(triisopropylsilyl)pyrroles. Synthesis of verrucarin E and other 3-substituted pyrroles. Helv Chim Acta **67**: 1168
192. Stierle DB, Faulkner DJ (1980) Metabolites of the marine sponge *Laxosuberites* sp. J Org Chem **45**: 4980
193. Venkateswarlu Y, Rao MR, Farooq Biabani MA (1996) 5-Alkylpyrrole-2-carboxaldehydes from the sponges *Mycalecarmia monanchrorata* and *Mycale mytilorum*. Indian J Chem **B35**: 876
194. Compagnone R, Oliveri MC, Piña IC, Marques S, Rangel HR, Dagger F, Suàrez AI, Gòmez M (1999) 5-Alkylpyrrole-2-carboxaldehydes from the Caribbean sponges *Mycale microsigmatosa* and *Desmapsamma anchorata*. Nat Prod Lett **13**: 203
195. Venkatesham U, Rama Rao M, Venkateswarlu Y (2000) New 5-alkylpyrrole-2-carboxaldehyde derivatives from the sponge *Mycale tenuispiculata*. J Nat Prod **63**: 1318
196. Bowden BF, Clezy PS, Coll JC, Ravi BN, Tapiolas DM (1984) Studies of Australian soft corals. XXXIV. A new substituted pyrrole from a soft coral sponge association. Aust J Chem **37**: 227
197. Ortega MJ, Zubia E, Carballo JL, Salvà J (1997) New cytotoxic metabolites from the sponge *Mycale micracanthoxea*. Tetrahedron **53**: 331
198. Nabbs BK, Abell AD (1999) The synthesis and P388 cytotoxicity of mycalazol 11 and related 5-acyl-2-hydroxymethylpyrroles. Bioorg Med Chem Lett **9**: 505
199. (a) Cafieri F, De Napoli L, Fattorusso E, Santacrone C, Sica D (1977) Molliorin A: Unique scalarin-like pyrroloterpene from the sponge *Cacospongia mollior*.

Tetrahedron Lett 477; (b) Cafieri F, De Napoli L, Fattorusso E, Santacrone C (1977) Molliorin-B. a second scalarin-like pyrroloterpene from the sponge *Cacospongia mollior*. Experientia **33**: 994; (c) Cafieri F, De Napoli L, Iengo A, Santacrone C (1978) Molliorin-c. a further pyrroloterpene present in the sponge *Cacospongia mollior*. Ibid **34**: 300; (d) Cafieri F, De Napoli L, Iengo A, Santacrone C (1979) Minor pyrroloterpenoids from the marine sponge *Cacospongia mollior*. Ibid **35**: 157

200. Scheuer PJ (1990) Some marine ecological phenomena: Chemical basis and biomedical potential. Science **248**: 173 and references given therein
201. Urban S, Hickford SJH, Blunt JW, Munro MHG (2000) Bioactive Marine alkaloids. Curr Org Chem **4**: 765
202. Urban S, Butler MS, Capon RJ (1994) Lamellarins O and P: New aromatic metabolites from the Australian marine sponge *Dendrilla cactos*. Aust J Chem **47**: 1919
203. Urban S, Hobbs L, Hooper JNA, Capon: RJ (1995) Lamellarins Q and R: New aromatic metabolites from an Australian marine sponge *Dendrilla cactos*. Aust J Chem **48**: 1491
204. (a) Fürstner A, Weintritt H, Hupperts A (1995) A new, titanium-mediated approach to pyrroles: First synthesis of lukianol A and lamellarin O dimethyl ether. J Org Chem **60**: 6637; (b) Banwell MG, Flynn BL, Hamel E, Hockless DCR (1997) Convergent synthesis of the pyrrolic marine natural product lamellarin-O, lamellarin-Q, lukianol A and some more highly oxygenated congeners. J Chem Soc Chem Commun 207; (c) Boger DL, Boyce CW, Labroli MA, Sehon CA, Jin Q (1999) Total synthesis of ningalin A, lamellarin O, lukianol A, and permethyl storniamide A utilizing heterocyclic azadiene Diels-Alder reactions. J Am Chem Soc **121**: 54
205. Palermo JA, Rodriguez Brasco MF, Seldes AM (1996) Storniamides A–D: Alkaloids from a Patagonian sponge *Cliona* sp. Tetrahedron **52**: 2727
206. Ebel H, Terpin A, Steglich W (1998) A concise synthesis of storniamide A nonamethyl ether. Tetrahedron Lett **39**: 9165
207. Kashman Y, Koren-Goldschlager G, Garcia Gravalos MD, Schleyer M (1999) Halitulin, a new cytotoxic alkaloid from the marine sponge *Haliclona tulearensis*. Tetrahedron Lett **40**: 997
208. Frincke JM, Faulkner DJ (1982) Antimicrobial metabolites of the sponge *Reniera* sp. J Am Chem Soc **104**: 265
209. (a) Parker KA, Cohen ID, Padwa A, Dent W (1984) Cycloadditions of non-stabilized azomethine ylides and quinones. Synthesis of the *Reniera* isoindole. Tetrahedron Lett **25**: 4917; (b) Padwa A, Chen Y-Y, Dent W, Nimmesgern H (1985) Synthetic application of cyanoaminosilanes as azomethine ylide equivalents. J Org Chem **50**: 4006; (c) Schubert-Zsilavecz M, Schramm HW (1991) Synthese von 6-Methoxy-2,5-dimethyl-2*H*-isoindol-4,7-dion, einem Alkaloid aus *Reniera*-Arten. Liebigs Ann Chem 973
210. Aknin M, Miralles J, Kornprobst J-M, Faure R, Gaydou E-M, Boury-Esnault N, Kato Y, Clardy J (1990) Trikentramine, an unusual pyrrole derivative from the sponge *Trikentrion loeve* Carter. Tetrahedron Lett **31**: 2979
211. Loukaci A, Guyot M (1994) Trikendiol, an unusual red pigment from the sponge *Trikentrion loeve*, anti-HIV-1 metabolite. Tetrahedron Lett **35**: 6869
212. Treibs A, Jakob K, Dietl A (1967) Über isoindigoide Farbstoffe der Pyrrol-Reihe. Liebigs Ann Chem **702**: 112
213. Shin J, Rho J-R, Seo Y, Lee H-S, Cho KW, Sim CJ (2001) Sarcotragins A and B, new sesterterpenoid alkaloids from the sponge *Sarcotragus* sp. Tetrahedron Lett **42**: 3005

214. Pham AT, Carney JR, Yoshida WY, Scheuer PJ (1992) Haumanamide, a nitrogenous spongian derivative from a *Spongia* sp. Tetrahedron Lett **33**: 1147
215. For an excellent review see: Royles BJL (1995) Naturally occurring tetramic acids: Structure, isolation, and synthesis. Chem Rev **95**: 1981 and references given therein
216. Nolte MJ, Steyn PS, Wessels PL (1980) Structural investigations of 3-acylpyrrolidine-2,4-diones by nuclear magnetic resonance spectroscopy and X-ray crystallography. J Chem Soc Perkin Trans 1, 1057
217. Lacey RN (1954) Derivatives of acetoacetic acid. Part VII. α-Acetyltetramic acids. J Chem Soc 850
218. Harris SA, Fisher LV, Folkers K (1965) The synthesis of tenuazonic acid and congeneric tetramic acids. J Med Chem **8**: 478
219. Jones RCF, Begley MJ, Peterson GE, Sumaria S (1990) Acylation of pyrrolidine-2,4-diones: a synthesis of 3-acyltetramic acids. X-ray molecular structure of 3-[1-(difluoroboryloxy)ethylidene]-5-isopropyl-1-methyl-pyrrolidine-2,4-dione. J Chem Soc Perkin Trans 1, 1959
220. Rosen T (1989) The tetramic acids – an overview of their biological properties. Drugs Fut **14**: 153
221. Hofheinz W, Oberhänsli WE (1977) Dysidin, ein neuartiger, chlorhaltihger Naturstoff aus dem Schwamm *Dysidea herbacea*. Helv Chim Acta **60**: 660
222. Aoki S, Higuchi K, Ye Y, Satari R, Kobayashi M (2000) Melophlins A and B, novel tetramic acids reversing the phenotype of *ras*-transformed cells, from the marine sponge *Melophlus sarassinorum*. Tetrahedron **56**: 1833
223. Ohta S, Ohta E, Ikegami S (1997) Ancorinoside A: A novel tetramic acid glycoside from the marine sponge *Ancorina* sp. which specifically inhibits blastulation of starfish embryos. J Org Chem **62**: 6452
224. (a) Matsunaga S, Fusetani N, Kato Y, Hirota H (1991) Aurantosides A and B: Cytotoxic tetramic acid glycosides from the marine sponge *Theonella* sp. J Am Chem Soc **113**: 9690; (b) Schmidt EW, Harper MK, Faulkner DJ (1997) Mozamides A and B. cyclic peptides from a Theonellid sponge from Mozambique. J Nat Prod **60**: 779
225. Sata NU, Matsunaga S, Fusetani N, van Soest RWM (1999) Aurantosides D, E, and F: New antifungal tetramic acid glycosides from the sponge *Siliquariaspongia japonica*. J Nat Prod **62**: 969
226. Wolf D, Schmitz FJ, Qiu F, Kelly-Borges M (1999) Aurantoside C, a new tetramic acid glycoside from the sponge *Homophymia conferta*. J Nat Prod **62**: 170
227. Sata NU, Wada S, Matsunaga S, Watabe S, van Soest RWM, Fusetani N (1999) Rubrosides A-H. new bioactive tetramic acid glycosides from the marine sponge *Siliquariaspongia japonica*. J Org Chem **64**: 2331
228. Gunasekera SP, Gunasekera M, McCarthy P (1991) Discodemide: a new bioactive macrocyclic lactam from the marine sponge *Discoderma dissoluta*. J Org Chem **56**: 4830
229. Kanazawa S, Fusetani N, Matsunaga S (1993) Cylindramide: Cytotoxic tetramic acid lactam from the marine sponge *Halichondria cylindrata* Tanita & Hoshino. Tetrahedron Lett **34**: 1065
230. Andersen RJ, Faulkner DJ, Cun-heng He, Van Duyne GD, Clardy J (1985) Metabolites of the marine prosobranch mollusc *Lamellaria* sp. J Am Chem Soc **107**: 5492
231. Lindquist N, Fenical W, Van Duyne GD, Clardy J (1988) New alkaloids of the lamellarin class from the marine ascidian *Didemnun chartaceum* (Sluiter, 1909). J Org Chem **53**: 4570

232. Carroll AR, Bowden BF, Coll JC (1993) Studies of Australian ascidians. I. Six new lamellarin-class alkaloids from a colonial ascidian, *Didemnum* sp. Aust J Chem **46**: 489
233. Urban S, Capon RJ (1996) Lamellain-S: a new aromatic metabolite from an Australian tunicate, *Didemnum* sp. Aust J Chem **49**: 711
234. Davis RH, Carroll AR, Pierens GK, Quinn RJ (1999) New lamellarin alkaloids from the Australian Ascidian, *Didemnum chartaceum*. J Nat Prod **62**: 419
235. Reddy MVR, Faulkner DJ, Venkateswarlu Y, Rao MR (1997) New lamellarin alkaloids from an unidentified ascidian from the Arabian Sea. Tetrahedron **53**: 3457
236. Reddy MVR, Rao MR, Rhodes D, Hansen MST, Rubins K, Bushman FD, Venkateswarlu Y, Faulkner DJ (1999) Lamellarin alpha 20-sulfate, an inhibitor of HIV-1 virus in cell culture. J Med Chem **42**: 1901
237. (a) Köck M, Reif B, Fenical W, Griesinger C (1996) Differentiation of HMBC two- and three-bond correlations: A method to simplify the structure determination of natural products. Tetrahedron Lett **37**: 363; (b) Reif B, Köck M, Kerssebaum R, Kang H, Fenical W, Griesinger C (1996) ADEQUATE, a new set of experiments to determine the constitution of small molecules at natural abundance. J Magn Reson **A118**: 282
238. Ishibashi F, Miyazaki Y, Iwao M (1997) Total synthesis of lamellarin D and H. The first synthesis of lamellarin-class marine alkaloids. Tetrahedron **53**: 5951
239. (a) Heim A, Terpin A, Steglich W (1997) Biomimetic synthesis of lamellarin G trimethyl ether. Angew Chem **109**: 158; Angew Chem Int Ed Engl **36**: 155; (b) Ruchirawat S, Mutarapat T (2001) An efficient synthesis of lamellarin alkaloids: synthesis of lamellarin G trimethyl ether. Tetrahedron Lett **42**: 1205
240. Banwell MG, Flynn BL, Hockless DCR (1997) Convergent total synthesis of lamellarin K. J Chem Soc Chem Commun 2259
241. Peschko C, Winklhofer C, Steglich W (2000) Biomimetic total synthesis of lamellarin L by coupling of two different arylpyruvic acid units. Chem Eur J **6**: 1147
242. Yoshida WY, Lee KK, Carroll AR, Scheuer PJ (1992) A complex pyrrolo-oxazinone and its iodo derivative isolated from a tunicate. Helv Chim Acta **75**: 1721
243. (a) Gupton JT, Krumpe KE, Burnham BS, Webb TM, Shuford JS, Sikorski JA (1999) The application of vinylogous iminium salt derivatives to a regiocontrolled and efficient relay synthesis of lukianol A and related marine natural products. Tetrahedron **55**: 14515; (b) Liu J-H, Yang Q-C, Mak TCW, Wong HNC (2000) Highly regioselective synthesis of 2,3,4-trisubstituted 1*H*-pyrroles: A formal total synthesis of lukianol A. J Org Chem **65**: 3587; (c) Kim S, Son S, Kang H (2001) Efficient syntheses of 2-carbomethoxy-3,4-disubstituted pyrroles by the condensation of vinylogous amides with aminomalonate. Bull Korean Chem Soc **22**: 1403
244. Kang H, Fenical W (1997) Ningalins A–D: Novel aromatic alkaloids from western Australian ascidian of the genus *Didemnum*. J Org Chem **62**: 3254
245. Chan GW, Francis T, Thureen DR, Offen PH, Pierce NJ, Westley JW, Johnson RK, Faulkner DJ (1993) Purpurone, an inhibitor of ATP-citrate lyase: A novel alkaloid from the marine sponge *Iotrochota* sp. J Org Chem **58**: 2544
246. Boger DL, Soenen DR, Boyce CW, Hedrick MP, Jin Q (2000) Total synthesis of ningalin B utilizing a heterocyclic azadiene Diel-Alder reaction and discovery of a new class of potent multidrug resistant (MDR) reversal agents. J Org Chem **65**: 2479
247. (a) Peschko C, Steglich W (2000) First total synthesis of the marine alkaloids purpurone and ningalin. Tetrahedron Lett **41**: 9477; (b) Namsa-aid A, Ruchirawat S (2002) Efficient synthesis of ningalin C. Org Lett **4**: 2633
248. (a) Rudi A, Goldberg I, Stein Z, Frolow F, Benayahu Y, Schleyer M, Kashman Y (1994) Polycitone A and polycitrins A and B: New alkaloids from the marine ascidian

Polycitor sp. J Org Chem **59**: 999; (b) Rudi A, Evan T, Aknin M, Kashman Y (2000) Polycitone B and prepolycitrin A: Two novel alkaloids from the marine ascidian *Polycitor africanus*. J Nat Prod **63**: 832
249. Terpin A, Polborn K, Steglich W (1995) Biomimetic total synthesis of polycitrin A. Tetrahedron **51**: 9941
250. Beccalli EM, Clerici F, Marchesini A (2000) First total synthesis of the alkaloid polycitrin B Tetrahedron **56**: 2699
251. Kobayashi J, Cheng J, Kikuchi Y, Ishibashi M, Yamamura S, Ohizumi Y, Ohta T, Nozoe S (1990) Rigidin, a novel alkaloid with calmodulin antagonistic activity from the Okinavan marine tunicate *Eudistoma* cf. *rigida*. Tetrahedron Lett **31**: 4617
252. Edstrom ED, Wei Y (1993) Synthesis of a novel pyrrolo[2,3-d]pyrimidine alkaloid, rigidin. J Org Chem **58**: 403
253. (a) Sakamoto T, Kondo Y, Sato S, Yamanaka H (1994) Total synthesis of a marine alkaloid, rigidin. Tetrahedron Lett **35**: 2919; (b) Sakamoto T, Kondo Y, Sato S, Yamanaka H (1996) Condensed heteroaromatic ring systems. Part 24. Synthesis of rigidin, a pyrrolo [2,3-*d*]pyrimidine marine alkaloid. J Chem Soc Perkin Trans 1, 459
254. (a) Kobayashi J, Harbour GC, Gilmore J, Rinehart Jr. KL (1984) Eudistomins A, D, G, H, I, J, M, N, O, P, and Q, bromo-, hydroxy, pyrrolyl- and 1-pyrrolinyl-β-carbolines from the antiviral Caribbean tunicate *Eudistoma olivaceum*. J Am Chem Soc **106**: 1526; (b) Rinehart Jr. KL, Kobayashi J, Harbour GC, Gilmore J, Mascal M, Holt TG, Shield LS, Lafargue F (1987) Eudistomins A-Q, β-carbolines from the antiviral Caribbean tunicate *Eudistoma olivaceum*. J Am Chem Soc **109**: 3378
255. Carté B, Faulkner DJ (1983) Defensive metabolites from three nembrothid nudibranchs. J Org Chem **48**: 2314
256. Lindquist N, Fenical W (1991) New tambjamine class alkaloids from the marine ascidian *Atapozoa* sp. and its nudibranch predators. Origin of the tamjamines in *Atapozoa*. Experientia **47**: 504
257. Blackman AJ, Li C (1994) New tambjamine alkaloids from the marine bryozoan *Bugula dentata*. Aust J Chem **47**: 1625
258. Matsunaga S, Fusetani N, Hashimoto K (1986) Bioactive marine metabolites. VIII. Isolation of an antimicrobial blue pigment from the bryozoan *Bugula dentata*. Experientia **42**: 84
259. Kazlauskas R, Marwood JF, Murphy PT, Wells RJ (1982) A blue pigment from a compound ascidian. Aust J Chem **35**: 215
260. Carté B, Faulkner DJ (1986) Role of secondary metabolites in feeding associations between a predatory nudibranch, two grazing nudibranchs, and a bryozoan. J Chem Ecol **12**: 795
261. Melvin MS, Ferguson DC, Lindquist N, Manderville RA (1999) DNA binding by 4-methoxypyrrolic natural products. Preference for intercalation at AT sites by tambjamine E and prodigiosin. J Org Chem **64**: 6861
262. Pettit GR, Kamano Y, Dufresne C, Cerny RL, Herald CL, Schmidt JM (1989) Isolation and structure of the cytostatic linear depsipeptide dolastatin 15. J Org Chem **54**: 6005
263. Pettit GR, Herald DL, Singh SB, Thornton TJ, Mullaney JT (1991) Antineoplastic agents. 220. Synthesis of natural (−)-dolastatin 15. J Am Chem Soc **113**: 6692
264. Pettit GR, Thornton TJ, Mullaney JT, Boyd MR, Herald DL, Singh S-B, Flahive EJ (1994) The dolastatins 20. A convenient synthesis route to dolastatin 15. Tetrahedron **50**: 12097
265. Poncet J (1999) The dolastatins, a family of promising antineoplastic agents. Curr Pham Des 5: 139

266. Höfle G, Pohlan S, Uhlig G, Kabbe K, Schumacher D (1994) Keronopsins A and B. chemical defence substances of the marine ciliate *Pseudokeronopsis rubra* (protozoa): Identification by *ex vivo* HPLC. Angew Chem **106**: 1561; Angew Chem Int Ed Engl **33**: 1495
267. Cionga E (1935) Présence de la pyrryl-α-méthylcétone dans la valériane officinale stabilisée. Compt Rend **200**: 780
268. Szentpetéry GB, Nyomàrkay KM, Sàrkàny S, Horvtàh KB (1963) Untersuchung der nicht-flüchtigen Workstoffe der ungarischen Arzneibaldraine. Pharmazie **18**: 816
269. Sàndor P, Kovàch AGB, Horvàth KB, Szentpétery GB, Clauder O (1970) Pharmakologische Untersuchungen über die Wirkung von synthetischem α-Methylpyrrylketon auf das Zentralnervensystem und den Kreislauf. Arzneim-Forsch **20**: 29
270. Bricout J, Viani R, Müggler-Chavan F, Marion JP, Reymond D, Egli RH (1967) Sur la composition de l'arôme de thé noir II. Helv Chim Acta **50**: 1517
271. Naya Y, Kotake M (1968) Volatile constituents of hops. II. Nippon Kagaku Zasshi **89**: 1113 [Chem Abstr (1969) **70**: 80786]
272. Onishi I, Tomita H, Fukuzumi T (1956) Studies on the essential oils of tobacco leaves. Bull Agr Chem Soc Japan **20**: 61
273. Onishi I, Yamamoto K (1956) Studies on the essential oils of tobacco leaves. Part VI. Phenol fraction (II). Bull Agr Chem Soc Japan **20**: 70
274. Sannai A, Fujimori T, Katō K (1983) Neutral volatile components of "Kukoshi" (*Lycium chinense* M.). Agric Biol Chem **47**: 2397
275. Miyazawa M, Maruyama H, Kameoka H (1983) Essential oil constituents of "moutan radicis cortex" *Paeonia moutan* Sims (= *P. suffruticosa* Andrews). Agric Biol Chem **47**: 2925
276. Marion JP, Müggler-Chavan F, Viani R, Bricout J, Reymond D, Egli RH (1967) Sur la composition de l'arôme de cacao. Helv Chim Acta **50**: 1509
277. Van der Wal B, Sipma G, Kettenes DK, Semper ATJ (1968) Some constituents of roasted cocoa. Rec Trav Chim **87**: 238
278. Gianturco MA, Giammarino AS, Friedel P, Flanagan V (1964) The volatile constituents of coffee – IV Furanic and pyrrolic compounds. Tetrahedron **20**: 2951
279. Stoll M, Winter M, Gautschi F, Flament I, Willhalm B (1967) Recherches sur les arômes. 13e comm. Sur l'arôme de café I. Helv Chim Acta **50**: 628
280. Ito M, Shimura H, Watanabe N, Tamai M, Takahashi A, Tanaka Y, Arai I, Hanada K (1991) 2-Acetylpyrrole, a hepatoprotective compound from *Streptomyces* sp. A-5071. Agric Biol Chem **55**: 2117
281. Lloyd HA, Fales HM, Goldman ME, Jerina DM, Plowman T, Schultes RE (1985) Brunfelsamidine: A novel convulsant from the medicinal plant *Brunfelsia grandiflora*. Tetrahedron Lett **26**: 2623
282. Buschi CA, Pomilio AB (1987) Pyrrole-3-carboxamidine: A letal principle from *Nierenbergia hippomanica*. Phytochemistry **26**: 863
283. Blau F (1894) Zur Constitution des Nicotins. Chem Ber **27**: 2535
284. Cahours A, Etard A (1880) Recherches sur la nicotine. Bull Soc Chim Fr **34**: 449
285. Späht E, Kesztler F (1937) Über neue Basen des Tabaks (XIII. Mitteil. über Tabak-Alkaloide). Chem Ber **70**: 2450
286. Hutchinson CR, Hsia M-TS, Carver RA (1976) Biosynthetic studies $^{13}CO_2$ of secondary plant metabolites. *Nicotiana* alkaloids. 1. Initial experiments. J Am Chem Soc **98**: 6006 and references given therein
287. LaForge FB (1928) The preparation and properties of some new derivatives of pyridine J Am Chem Soc **50**: 2477

288. Richardson CH, Shepard HH (1930) The insecticidal action of some derivatives of pyridine and pyrrolidine and some aliphatic amines. J Agr Research **40**: 1007
289. El Sayed KA, Hamann MT, Abd El-Rahman HA, Zaghloul AM (1998) New pyrrole alkaloids from *Solanum sodomaeum*. J Nat Prod **61**: 848
290. Demole E, Demole C, Enggist P (1976) A chemical investigation of the volatile constituents of East Indian sandalwood oil (*Santalum album* L.). Helv Chim Acta **59**: 737
291. Gupta S, Jha A, Prasad AK, Rajwanshi VK, Jain SC, Olsen CE, Wengel J, Parmar VS (1999) A new amide, *N*-cinnamoylpyrrole and other constituents from *Piper argyrophyllum*. Indian J Chem Sect B: Org Chem Incl Med Chem **38**: 823
292. Likhitwitayawuid K, Ruangrungsi N, Lange GL, Decicco CP (1987) Structural elucidation and synthesis of new components isolated from *Piper sarmentosum* (Piperaceae). Tetrahedron **43**: 3689
293. Kapadia GJ, Highet RJ (1968) Peyote alkaloids IV. Structure of peyonine, novel β-phenetylpyrrole from *Lophophora williamsii*. J Pharm Sci **57**: 191
294. Jiang ZD, Gerwick WH (1991) Novel pyrroles from the Oregon red alga *Gracilariopsis lemaneiformis*. J Nat Prod **54**: 403
295. Jefford CW, Sienkiewicz K, Thornton SR (1994) A concise synthesis of two pyrroles of marine origin. Tetrahedron Lett **35**: 6271
296. Raffauf RF, Zennie TM, Onan KD, Le Quesne PW (1984) Funebrine, a structurally novel pyrrole alkaloid and other γ-hydroxyisoleucine-related metabolites of *Quararibea funebris* (Llave) Vischer (Bombacaceae). J Org Chem **49**: 2714
297. Zennie TM, Cassady JM, Raffauf RF (1986) Funebral. A new pyrrole lactone alkaloid from *Quararibea funebri*. J Nat Prod **49**: 695
298. Zennie TM, Cassady JM (1990) Funebradiol. a new pyrrole lactone alkaloid from *Quararibea funebri* flowers. J Nat Prod **53**: 1611
299. (a) Yu S-X, Le Quesne PW (1995) *Quararibea* metabolites. 3. Total synthesis of ($\pm$)-funebral, a rotationally restricted pyrrole alkaloid, using a novel Paal-Knorr reaction. Tetrahedron Lett **36**: 6205; (b) Dong Y, Pai NN, Ablaza SL, Yu S-X, Bolvig S, Forsyth DA, Le Quesne PW (1999) Quararibea Metabolites. 4. Total synthesis and conformational studies of ($\pm$)-Funebrine and ($\pm$)-Funebral. J Org Chem **64**: 2657
300. Ablaza SL, Pai NN, Le Quesne PW (1995) *Quararibea* Metabolites. 2. Efficient synthetic approaches to ($\pm$)-(2*S*,3*S*,4*R*)-γ-hydroxyisoleucine, the characteristic *Quararibea* amino acid. Nat Prod Lett **6**: 77
301. Lynn DG, Jaffe K, Cornwall M, Tramontano W (1987) Characterization of an endogeneous factor controlling the cell cycle of complex tissues. J Am Chem Soc **109**: 5858
302. Haidoune M, Mornet R, Laloue M (1990) Synthesis of 6-(3-methylpyrrol-1-yl)-9-β-ribofuranosyl purine, a novel metabolite of zeatin riboside. Tetrahedron Lett **31**: 1419
303. Banerji A, Majumder PL, Chatterjee A (1970) Occurrence of geissoschizine and other minor biogenetically related alkaloids in *Rhazya stricta*. Phytochemistry **9**: 1491
304. De Silva KT, Ratcliffe AH, Smith GF, Smith GN (1972) Rhazinilam, a neutral alkaloidal artefact from *Rhazya stricta* Decaisne. Tetrahedron Lett 913
305. Linde HA (1965) Die Alkaloide aus *Melodinus australis* (F. Mueller) Pierre (Apocynaceae). Helv Chim Acta **48**: 1822
306. Thoison O, Guénard D, Sévenet T, Kan-Fan C, Quirion J-C, Husson H-P, Deverre J-R, Chan K-C, Potier P (1987) Propriétés inhibitrices du rhazinilame sur la polymérisation de la tubuline en microtubules. C R Acad Sci Paris II, **304**: 157

307. Kam T-S, Tee Y-M, Subramanian G (1998) Rhazinal, a formylrhazinilam derivative from a Malayan *Kopsia*. Nat Prod Lett **12**: 307
308. Goh SH, Abdul R (1986) Ring-opened indole alkaloidal artefacts from *Leuconotis* species and the facile ring reclosure of leuconolam. Tetrahedron Lett 2501
309. Aimi N, Uchida N, Ohya N, Hosokawa H, Takayama H, Sakai S, Mendoza LA, Polz L, Stöckigt J (1991) Novel indole alkaloids from cell suspension cultures of *Aspidosperma quebracho blanco* Schlecht. Tetrahedron Lett **32**: 4949
310. (a) Ratcliffe AH, Smith GF, Smith GN (1973) The synthesis of rhazinilam. Tetrahedron Lett 5179; (b) Magnus P, Rainey T (2001) Concise synthesis of ($\pm$)-rhazinilam. Tetrahedron **57**: 8647
311. Abraham DJ, Rosenstein RD, Lyon RL, Fong HHS (1972) The structure elucidation of rhazinilam, a new class of alkaloids from the Apocinaceae. Tetrahedron Lett 909
312. David B, Sévenet T, Morgat M, Guénard D, Moisand A, Tollon Y, Thoison O, Wright M (1994) Rhazinilam mimics the cellular effects of taxol by different mechanisms of action. Cell Motil Cytoskel **28**: 317 [Chem Abstr (1994) **121**: 271 453]
313. Culvenor CCJ, Smith LW (1969) A quaternary *N*-dihydropyrrolizinomethyl derivative of heliotridine from *Heliotropium europaeum*. Tetrahedron Lett 3603
314. Borges del Castillo J, España de Aguirre AG, Bretòn JL, Gonzàlez AG, Trujillo J (1970) Loroquin, a new necine isolated from *Urechites karwinsky* Mueller. Tetrahedron Lett 1219
315. Robins DJ (1982) The pyrrolizidine alkaloids in: Herz W, Grisebach H, Kirby GW (eds) Progress in the chemistry of organic natural products vol. 41. Springer, Wien, p 116–203
316. Culvenor CCJ, Edgar JA, Smith LW, Tweeddale HJ (1969) Dihydropyrrolizine alkaloids of pyrroliizidine alkaloids. Tetrahedron Lett 3599 and references given therein
317. Bohlmann F, Gupta RK, Jakupovic J (1981) An acylpyrrole derivative and further constituents from Jamaican representatives of the tribe Senecioneae. Phytochemistry, **20**: 831
318. Bohlmann F, Zdero C, Jakupovic J, Grenz M, Castro V, King RM, Robinson H, Vincent: LPD (1986) Further pyrrolizidine alkaloids and furoeremophilanes from *Senecio* species. Phytochemistry **25**: 1151
319. Bohlmann F, Zdero C, Grenz M (1977) Natürlich vorkommende Terpen-Derivate. 78. Mitt. Weitere Inhaltsstoffe aus südafrikanischen *Senecio*-Arten. Chem Ber **110**: 474
320. Bohlmann F, Zdero C, Berger D, Suwita A, Mahanta P, Jeffrey C (1979) Neue Furanoeremophilane und weitere Inhaltsstoffe aus Südafrikanischen *Senecio*-Arten. Phytochemistry **18**: 79
321. Bohlmann F, Knoll K-H, Zdero C, Mahanta PK, Grenz M, Suwita A, Ehlers D, Ngo Le Van, Abraham W-R, Natu AA (1977) Terpen-Derivate aus *Senecio*-Arten. Phytochemistry **16**: 965
322. Bohlmann F, Knoll K-H (1978) Zwei neue Acylpyrrole aus *Kleinia kleinioides*. Phytochemistry **17**: 599
323. Bohlmann F, Zdero C, Snatzke G (1978) Zur Stereochemie der Acylpyrrole aus *Senecio*-Arten. Chem Ber **111**: 3009
324. Bohlmann F, Zdero C (1979) Neue C_{10}-Säureamide, Furanoeremophilane und andere Inhaltsstoffe aus bolivianischen *Senecio*-Arten. Phytochemistry **18**: 125
325. Close W, Nickisch K, Bohlmann F (1980) Synthese von 5,7a-Dihydroheliotridin-3-on, dem Grundkörper einer neuen Gruppe von Pyrrolizidin-Alkaloiden. Chem Ber **113**: 2694
326. (a) McNab H, Thornley C (1993) New chemistry of pyrrolizin-3-one. A concise route to 3,8-didehydroheliotridin-5-one. J Chem Soc Chem Commun 1570; (b) McNab H,

Thornley C (2000) Chemistry of pyrrolizinones. Part 1. Reactions of pyrrolizin-3-ones with electrophiles: synthesis of 3,8-didehydroheliotridin-5-one. J Chem Soc Perkin Trans 1 3584
327. Veen G, Greinwald R, Witte L, Wray V, Czygan F-C (1991) Alkaloids of *Virgilia divaricata* and *V. oroboides*. Phytochem **30**: 1891
328. White EP (1964) Alkaloids of the Leguminosae. Part XXVIII. Virgiline and O-(Pyrrol-2-ylcarbonyl)virgiline. J Chem Soc 5243
329. Manchanda AH, Nabney J, Young DW (1968) 2,3-Dehydro-O-(2-pyrrolylcarbonyl) virgiline, an alkaloid from *Readea membranacea* Gillespie (Rubiaceae). J Chem Soc (C) 615
330. Goosen A (1963) The alkaloids of the Leguminosae. Part I. The structure of calpurnine. J Chem Soc 3067
331. (a) Gerrans GC, Harley-Mason J (1963) The alkaloids of *Virgilia oroboides*. Chem and Ind (London) 1280, *ibid.* 1433; (b) Gerrans GC, Harley-Mason J(1964) The alkaloids of *Virgilia oroboides*. J Chem Soc 2202
332. Faugeras G, Paris R-R, Peltier M (1974) Alcaloïdes et polyphénols des Légumineuses. XXIX. – Sur le *Cadia ellisiana* Baker, Papilionacée de Madagascar: Alcaloïdes. Ann Pharm Fr **32**: 323
333. Faugeras G, Paris R-R, Debray M, Bourgeois J, Delabos C (1975) Alkaloïds and polyphenols from Leguminosae. Alkaloids of *Cadia ellisiana* bark. Isolation, identification and toxicity. Plant Med Phytother **9**: 37
334. Van Eijk JL, Radema MH (1977) Some alkaloids of Ethiopian *Calpurnea aurea* and *Cadia purpurea*. Planta Med **32**: 275
335. Lindner E, Kaiser J, Schacht U (1976) Hypotensive and antiarrhytmic effects of a new alkaloid, the 13-hydroxylupanine-2-pyrrolecarbonic acid ester from the Madagascar plant *Cadia ellisiana*. Arzneim-Forsch **26**: 1651
336. Asres K, Gibbons WA, Phillipson JD, Mascagni P (1986) Alkaloids of Ethiopian *Calpurnia aurea* subsp. *aurea*. Phytochemistry **25**: 1443
337. Mascagni P, Gibbons WA, Asres K, Phillipson JD, Niccolai N (1987) The determination of the molecular framework and conformation of the calpaurine alkaloid by 1D- and 2D-NMR methods. Tetrahedron **43**: 149
338. van Eijk JL, Radema MH (1976) New quinolizidine alkaloids from *Cadia purpurea*. Tetrahedron Lett 2053
339. Reinecke MG, Zhao Y-Y (1988) Phytochemical studies of the Chinese herb taizi-shen. *Pseudostellaria heterophylla*. J Nat Prod **51**: 1236
340. Rogers EF, Koniuszy FR, Shavel Jr J, Folkers K (1948) Plant insecticides. I. Ryanodine, a new alkaloid from *Ryania speciosa* Vahl. J Am Chem Soc **70**: 3086
341. Waterhouse AL, Holden I, Casida JE (1984) 9,21-Didehydroryanodine: a new principal toxic constituent of the botanical insecticide Ryania. J Chem Soc Chem Commun: 1265
342. Waterhouse AL, Holden I, Casida JE (1985) Ryanoid insecticides: Structural examination by fully coupled two-dimensional ^{1}H-^{13}C-shift correlation NMR spectroscopy. J Chem Soc Perkin Trans II 1011
343. Ruest L, Taylor DR, Deslongchamps P (1985) Investigation of the constituents of *Ryania speciosa*. Can J Chem **63**: 2840
344. Kelly RB, Whittingham DJ, Wiesner K (1951) The structure of ryanodine. I. Can J Chem **29**: 905
345. Babin DR, Forrest TP, Valenta Z, Wiesner K (1962) The complete structure and relative and absolute configuration of anhydroryanodine. Experientia, **18**: 549
346. Wiesner K (1963) The structure, stereochemistry, and absolute configuration of anhydroryanodine. Pure Appl Chem **7**: 285

347. Wiesner K, Valenta Z, Findlay JA (1967) The structure of ryanodine. Tetrahedron Lett 221
348. Srivastava SN, Przybylska M (1968) The molecular structure of ryanodol-p-bromo benzyl ether Can J Chem **46**: 795
349. Wiesner K (1972) The structure of ryanodine. Adv Org Chem **8**: 295
350. Belanger A, Berney DJF, Borschberg H-J, Brousseau R, Doutheau A, Durand R, Katayama H, Lapalme R, Leturc DM, Liao C-C, MacLachlan FN, Maffrand J-P, Marrazza F, Martino R, Moreau C, Saint-Laurent L, Saintonge R, Soucy P, Ruest L, Deslongchamps P (1979) Total synthesis of ryanodol. Can J Chem **57**: 3348
351. Deslongchamps P, Belanger A, Berney DJF, Borschberg H-J, Brousseau R, Doutheau A, Durand R, Katayama H, Lapalme R, Leturc DM, Liao C-C, MacLachlan FN, Maffrand J-P, Marrazza F, Martino R, Moreau C, Ruest L, Saint-Laurent L, Saintonge R, Soucy P (1990) The total synthesis of (+)-ryanodol. Part I. Can J Chem **68**: 115 Part II. *Ibid* 127; Part III. *Ibid* 153; Part IV. *Ibid* 186
352. Vierling W (1988) Ryanodine in mammalian heart ventricular muscle: indication for the induction of calcium leakage from the sarcoplasmic reticulum. Eur J Pharmacol **145**: 329
353. Procita L (1956) The action of ryanodine on mammalian skeletal muscle *in situ*. J Pharmacol Exp Ther **117**: 363
354. Jenden DJ, Fairhurst AS (1969) The pharmacology of ryanodine. Pharmacol Rev **21**: 1 and references given therein
355. Achenbach H, Hübner H, Vierling W, Brandt W, Reiter M (1995) Spiganthine, the cardioactive principle of *Spigelia anthelmia*. J Nat Prod **58**: 1092
356. Cheng D, Shao Y, Hartman R, Roder E, Zhao K (1994) Oligopeptides from *Aster tataricus*. Phytochem **36**: 945
357. Morita H, Nagashima S, Takeya K, Itokawa H (1995) Structure of a new peptide, astin J, from *Aster tataricus*. Chem Pharm Bull **43**: 271
358. Kosemura S, Ogawa T, Totsuka K (1993) Isolation and structure of asterin. a new halogenated cyclic pentapeptide from *Aster tataricus*. Tetrahedron Lett **34**: 1291
359. Morita H, Nagashima S, Takeya K, Itokawa H (1993) Astins A and B, antitumor cyclic pentapeptides from *Aster tataricus*. Chem Pharm Bull **41**: 992
360. Morita H, Nagashima S, Takeya K, Itokawa H, Iitaka Y (1995) Structures and conformation of antitumor cyclic pentapeptides, astins A, B and C, from *Aster tataricus*. Tetrahedron **51**: 1121 and references given therein
361. Jizba J, Samek Z, Novotný L (1977) A sesquiterpenic alkaloid, eremophilene lactam, from the rhizomes of *Petasites hybridus*. Collect Czech Chem Commun **42**: 2438
362. Wiedhopf RM, Trumbull ER, Cole JR (1973) Antitumor agents from *Jatropha macrorhiza* (Euphorbiaceae) I: Isolation and characterization of jatropham. J Pharm Sci **62**: 1206
363. Shimomura H, Sashida Y, Mimaki Y, Minegishi Y (1987) Jatropham glucoside from the bulbs of *Lilium hansonii*. Phytochemistry **26**: 582
364. Ori K, Mimaki Y, Mito K, Sashida Y, Nikaido T, Ohmoto T, Masuko A (1992) Jatropham derivatives and steroidal saponins from the bulbs of *Lilium hansonii*. Phytochemistry **31**: 2767
365. Haladovà M, Bučkovà A, Eisenreichovà E, Uhrìn D, Tomko J (1987) Jatropham in *Lilium candidum*. L. Chem Papers **41**: 835
366. Eisenreichovà E, Haladovà M, Bučkovà A, Ubik K, Uhrìn D (1991) Derivatives of pyrroline in *Lilium candidum* L. Chem Papers **45**: 709
367. Shimomura H, Sashida Y, Mimaki Y, Kudo Y, Maeda K (1988) New phenylpropanoid glycerol glucosides from the bulbs of *Lilium* species. Chem Pharm Bull **36**: 4841

368. Yakushijin K, Kozuka M, Ito Y, Suzuki R, Furukawa H (1980) Ring transformation of 2-furylcarbamates to 5-hydroxy-3-pyrrolin-2-ones, revised structure of jatropham. Heterocycles **14**: 1073
369. Yakushijin K, Suzuki R, Hattori R, Furukawa H (1981) Synthesis of (±)-jatrophan, an antitumor alkaloid from *Jatropha Macrorhiza*. Heterocycles **16**: 1157
370. Nagasaka T, Esumi S, Ozawa N, Kosugi Y, Hamaguchi F (1981) Synthesis of 5-hydroxy-3-methyl-3-pyrrolin-2-one [(±)-jatropham, an antitumor alkaloid] and its 4-methyl isomer. Heterocycles **16**: 1987
371. Fariña F, Martìn MV, Paredes MC, Ortega MC, Tito A (1984) Pseudoesters and derivatives. XXII. Synthesis of 5-hydroxy-3-pyrrolin-2-ones and 5-hydroxypyrrolidin-2-ones by ammonolysis of 5-methoxyfuran-2(5*H*)-ones and derivatives. Heterocycles **22**: 1733
372. Haladovà M, Eisenreichovà E, Bučkovà A, Tomko J, Uhrìn D, Ubik K (1991) Dimeric pyrroline alkaloids from *Lilium candidum* L. Collect Czech Chem Commun **56**: 436
373. Haladovà M, Eisenreichovà E, Bučkovà A, Tomko J, Uhrìn D (1988) New nitrogen-containing compounds in *Lilium candidum* L. Collect Czech Chem Commun **53**: 157
374. Eisenreichovà E, Haladovà M, Bučkovà A, Tomko J, Uhrìn D, Ubik K (1992) A pyrroline-pyrrolidine alkaloid from *Lilium candidum* bulbs. Phytochemistry **31**: 1084
375. Kuo R-Y, Chang F-R, Wu Y-C (2001) A new propentdyopent derivative, rollipyrrole, from *Rollinia mucosa* Baill. Tetrahedron Lett **42**: 7907
376. (a) Yamano K, Konno K, Shirahama H (1991) A new amino acid, L-3-(2-carboxy-4-pyrrolyl)alanine from the poisonous mushroom *Clytocybe acromelalga*. Chem Lett 1541; (b) Yamano K, Shirahama H (1992) New amino acids from the poisonous mushroom *Clytocybe acromelalga*. Tetrahedron **48**: 1457
377. Härri E, Loeffler W, Sigg HP, Stähelin H, Stoll C, Tamm C, Wiesinger D (1962) Über die Verrucarine und Roridine, eine Gruppe von cytostatisch hochwirksamen Antibiotica aus *Myrothecium*-Arten. Helv Chim Acta **45**: 839
378. Pfäffli P, Tamm C (1969) Revidierte Struktur von Verrucarin E. Eine Synthese des Antibioticums und verwandter β-Acetyl-Pyrrol-Derivate. Helv Chim Acta **52**: 1911
379. Groves JK, Cundasawmy NE, Anderson HJ (1973) Pyrrole chemistry. XV. Chemistry of some 3,4-disubstituted pyrroles. Can J Chem **5**: 1089
380. Gossauer A, Suhl K (1976) Totalsynthese des Verrucarins E sowie ihre Anwendung zur Herstellung eines ^{13}C-markierten Derivates desselben. Helv Chim Acta **59**: 1698
381. Sheldrick WS, Borkenstein A, Engel J (1978) The crystal structures of verrucarin E (4-acetyl-3-hydroxymethylpyrrole) and its 1/3 hydrate. Acta Crystallogr **B34**: 1248
382. Arndt RR, Holzapfel CW, Ferreira NP, Marsh JJ (1974) Structure and biogenesis of desoxyverrucarin E, a metabolite of *Eupenicillium hirayamae*. Phytochemistry **13**: 1865
383. Pfäffli P, Tamm C (1969) Über die Biosynthese des Antibiotikums Verrucarin E. Helv Chim Acta **52**: 1921
384. Chexal KK, Snipes C, Tamm C (1980) Biosynthesis of the antibiotic Verrucarin E. Use of [1-^{13}C]-, [2-^{13}C]-, [1,2-^{13}C]- and [2-^{13}C, 2-^{2}H$_3$]-acetates. Helv Chim Acta **63**: 761
385. Badar Y, Lockley WJS, Toube TP, Weedon BCL, Valadon LRG (1973) Natural and synthetic 2-pyrrolylpolyenes. J Chem Soc Perkin Trans 1, 1416
386. (a) Badar Y, Chopra AK, Dias HW, Hursthouse MB, Khokhar AR, Ito M, Toube TP, Weedon BCL (1977) Pyrrolylpolyenes. Part 2. Stereochemistry of wallemia A and synthesis of its (*E*)-isomer and other models. J Chem Soc Perkin Trans 1, 1372;

(b) Ito M, Tsukida K, Toube TP (1981) Pyrrolylpolyenes. Part 3. Synthesis of all-(E)-wallemia C and stereochemistry of natural wallemia C. J Chem Soc Perkin Trans 1, 3255
387. Ahmed FR, Buckingham MJ, Hawkes GE, Toube TP (1984) Pyrrolylpolyenes. Part 5. Revision of the structure of the principal pigment of *Wallemia sebi*. A nuclear Overhauser enhancement study. J Chem Res (S) 178
388. Ahmed FR, Toube TP (1984) Pyrrolylpolyenes. Part 6. Synthesis of wallemia A and wallemia E. J Chem Soc Perkin Trans 1, 1577
389. Yamagishi Y, Matsuoka M, Odagawa A, Kato S, Shindo K, Mochizuki J (1993) Rumbrin, a new cytoprotective substance produced by *Auxarthron umbrinum*. I. Taxonomy, production, isolation and biological activities. J Antibiotics **46**: 884
390. Yamagishi Y, Shindo K, Kawai H (1993) Rumbrin, a new cytoprotective substance produced by *Auxarthron umbrinum*. II. Physicochemical properties and structure determination. J Antibiotics **46**: 888
391. Numata A, Takahashi C, Ito Y, Minoura K, Yamada T, Matsuda C, Nomoto K (1996) Penochalasins, a novel class of cytotoxic cytochalasans from a *Penicillium* species separated from a marine alga: structure determination and solution conformation. J Chem Soc Perkin Trans 1, 239
392. Trigos A, Reyna S, Matamoros B (1995) Macrophominol, a diketopiperazine from cultures of *Macrophomina phaseolina*. Phytochemistry **40**: 1697
393. Rowan DD, Gaynor DL (1986) Isolation of feeding deterrents against Argentine stem weevil from riegrass infected with the endophyte *Acremonium loliae*. J Chem Ecol **12**: 647
394. Rowan DD, Hunt MB, Gaynor DL (1986) Peramine, a novel insect feeding deterrent from ryegrass infected with the endophyte *Acremonium loliae*. J Chem Soc Chem Commun 935
395. Dumas DJ (1988) Total synthesis of peramine (I). J Org Chem **53**: 4650
396. Brimble MA, Rowan DD (1990) Synthesis of the insect feeding deterrent peramine *via* Michael addition of a pyrrole anion to a nitroalkene. J Chem Soc Perkin Trans I, 311
397. Fröde R, Hinze C, Josten I, Schmidt B, Steffan B, Steglich W (1994) Isolation and synthesis of 3,4-bis(indol-3-yl)pyrrole-2,5-dicarboxylic acid derivatives from the slime mould *Lycogala epidendrum*. Tetrahedron Lett **35**: 1689
398. Hashimoto T, Yasuda A, Akazawa K, Takaoka S, Tori M, Asakawa Y (1994) Three novel dimethyl pyrroledicarboxylate, lycogarubins A-C, from the Myxomycetes *Lycogala epidendrum*. Tetrahedron Lett **35**: 2559
399. Nakanishi S, Matsuda Y, Iwahashi K, Kase H (1986) K-252b, c and d, potent inhibitors of protein kinase C from microbial origin. J Antibiot **39**: 1066
400. Yasuzawa T, Iida T, Yoshida M, Hirayama N, Takahashi M, Shirahata K, Sano H (1986) The structures of the novel protein kinase C inhibitors K-252a, b, c and d. J Antibiot **39**: 1072
401. Hoshino T, Koyima Y, Hayashi T, Uchiyama T, Kaneko K (1993) A new metabolite of tryptophan. chromopyrrolic acid, produced by *Chromobacterium violaceum*. Biosci Biotech Biochem **57**: 775
402. Buchanan MS, Hashimoto T, Asakawa Y (1996) Acylglycerols from the slime mould, *Lycogala epidendrum*. Phytochemistry **41**: 791
403. Shiomi K, Yang H, Xu Q, Arai N, Namiki M, Hayashi M, Inokoshi J, Takeshima H, Masuma, R, Komiyama K, Omura S (1995) Phenopyrrozin, a new radical scavenger produced by *Penicillium* sp. FO-2047. J Antibiot **48**: 1413

404. Rosset T, Sankhala RH, Stickings CE, Taylor MEU, Thomas R (1957) Studies in the biochemistry of microorganisms. 103. Metabolites of *Alternaria tenuis* Auct.: Culture filtrate products. Biochem J **67**: 390
405. Meronuck RA, Steele JA, Mirocha CJ, Christensen CM (1972) Tenuazonic acid, a toxin produced by *Alternaria alternata*. Appl Microbiol **23**: 613
406. Janardhanan KK, Husain A (1975) Isolation of tenuazonic acid, a phytotoxin from *Alternaria alternata* causing leaf blight of *Datura innoxia*. Indian J Exp Biol **13**: 321
407. Mikami Y, Nishijima Y, Iimura H, Suzuki A, Tamura S (1971) Chemical studies on brown-spot disease of tobacco plants. Part I. Tenuazonic acid as a vivotoxin of *Alternaria longipes*. Agr Biol Chem **35**: 611
408. Iwasaki S, Muro H, Nozoe S, Okuda S, Sato Z (1972) Isolation of 3,4-dihydro-3,4,8-trihydroxy-1(2*H*)-naphthalenone and tenuazonic acid from *Pyricularia oryzae* Cavarra. Tetrahedron Lett **1**: 13
409. (a) Umetsu N, Kaji J, Tamari K (1972) Investigation on the toxin production by several blast fungus strains and isolation of tenuazonic acid as a novel toxin. Agric Biol Chem **36**: 859; (b) Umetsu N, Kaji J, Tamari K (1973) Isolation of tenuazonic acid from blast-diseased rice plants. *Ibid* **37**: 451
410. Umetsu N, Kaji J, Aoyama K, Tamari K (1974) Toxins in blast-diseased rice plants. Agric Biol Chem **38**: 1867
411. Stickings CE (1959) Studies in the biochemistry of micro-organisms. 106. Metabolites of *Alternaria tenuis* Auct.: The structure of tenuazonic acid. Biochem J **72**: 332
412. Stickings CE, Townsend RJ (1961) Studies in the biochemistry of micro-organisms. 108. Metabolites of *Alternaria tenuis* Auct.: The biosynthesis of tenuazonic acid. Biochem J **78**: 412
413. Gatenbeck S, Sierankiewicz J (1973) On the biosynthesis of tenuazonic acid in *Alternaria tenuis*. Acta Chem Scand **27**: 1825
414. Holzapfel CW (1980) The biosynthesis of cyclopiazonic acid and related tetramic acids. In: Steyn PS (ed) The biosynthesis of mycotoxins. Academic Press, New York, pp 327–355
415. Kaczka EA, Gitterman CO, Dulaney EL, Smith MC, Hendlin D, Woodruff HB, Folkers K (1964) Discovery of inhibitory activity of tenuazonic acid for growth of human adenocarcinoma-1. Biochem Biophys Res Commun **14**: 54
416. Gitterman CO (1965) Antitumor, cytotoxic, and antibacterial activities of tenuazonic acid and congeneric tetramic acids. J Med Chem **8**: 483
417. Shigeura HT, Gordon CN (1963) The biological activity of tenuazonic acid. Biochemistry **2**: 1132
418. Battaner E, Vazquez D (1971) Inhibitors of protein biosynthesis by ribosomes of the 80-S type. Biochim Biophys Acta **254**: 316
419. Carrasco L, Vazquez D (1972) Survey of inhibitors in different steps of protein synthesis by mammalian ribosomes. J Antibiot **25**: 732
420. Muramatsu T, Umetsu N, Matsuda K, Tamari K (1974) Effect of tenuazonic acid on *in vitro* amino acid incorporation by rice embryo ribosomes. Agric Biol Chem **38**: 2049
421. Carrasco L, Vazquez D (1973) Differences in eukaryotic ribosomes detected by the selective action of an antibiotic. Biochem Biophys Acta **319**: 209
422. Lebrun MH, Nicolas L, Boutar M, Gaudemer F, Ranomenjanahary S, Gaudener A (1988) Relationships between the structure and the phytotoxicity of the fungal toxin tenuazonic acid Phytochemistry **27**: 77 and references given therein

423. Matsuda M, Kobayashi T, Nagao S, Ohta T, Nozoe S (1996) Laccarin, a new alkaloid from the mushroom *Laccaria vinaceoavellanea*. Heterocycles **43**: 685
424. Nowak A, Steffan B (1997) Physarorubinic acid, a polyenoyltetramic acid type plasmodial pigment from the slime mold *Physarum polycephalum* (Myxomycetes). Liebigs Ann Recl **1**: 1817
425. Dixon DJ, Ley SV, Longbottom DA (1999) Total synthesis of the plasmodial pigment physarorubinic acid, a polyenoyl tetramic acid. J Chem Soc Perkin Trans 1, 2231
426. Nowak A, Steffan B (1998) Polycephalin B and C: Unusual tetramic acids from plasmodia of the slime mold *Physarum polycephalum* (Myxomycetes). Angew Chem **110**: 3341; Angew Chem Int Ed **37**: 3139
427. Casser I, Steffan B, Steglich W (1987) Fungal pigments. Part 52. Chemistry of the plasmodial pigments of the slime mold *Fuligo septica* (myxomycetes). Angew Chem **99**: 597; Angew Chem Int Ed Engl **26**: 586
428. Ley SV, Smith SC, Woodward PR (1992) Further reactions of t-butyl 3-oxobutanthionate and t-butyl 4-diethylphosphono-3-oxobutanthionate: Carbonyl coupling reactions, amination, use in the preparation of 3-acyltetramic acids and application to the total synthesis of fuligorubin A. Tetahedron **48**: 1145
429. Steglich W (1989) Slime moulds (Myxomycetes) as a source of new biologically active metabolites. Pure Appl Chem **61**: 281
430. Howard BH, Raistrick H (1954) Studies in the biochemistry of micro-organisms. 92. The colouring matters of *Penicillium islandicum* Sopp. Part 4. Iridoskyrin, rubroskyrin and erythroskyrin. Biochem J **57**: 212
431. Shoji J, Shibata S (1964) The structure of erythroskyrine, a nitrogen-containing coloring matter of *Penicillium islandicum* Sopp. Chem and Ind (London) 419
432. Shoji J, Shibata S, Sankawa U, Taguchi H, Shibanuma Y (1965) Metabolic products of fungi. XXIV. The structure of erythroskyrine, a nitrogen-containing coloring matter of *Penicillium islandicum* Sopp. Chem Pharm Bull **13**: 1240
433. Ueno Y, Kato Y, Enomoto M (1975) Erythroskyrine, the third mycotoxin from *Penicillium islandicum* Scopp. Jpn J Exp Med **45**: 525
434. Beutler JA, Hilton BD, Clark P, Tempesta MS, Corley DG (1988) Absolute and relative configuration of erythroskyrin. J Nat Prod **51**: 562
435. Dixon DJ, Ley SV, Gracza T, Szolcsanyi P (1999) Total synthesis of the polyenetramic acid mycotoxin erythroskyrine. J Chem Soc Perkin Trans 1, 839
436. Shibata S, Sankawa U, Taguchi H, Yamasaki K (1966) Biosynthesis of natural products. III. Biosynthesis of erythoskyrine, a coloring matter of *Penicillium islandicum* Sopp. Chem Pharm Bull **14**: 474
437. (a) Ono M, Sakuda S, Suzuki A, Isogai A (1997) Aflastatin A, a novel inhibitor of aflatoxin production by aflatoxigenic fungi. J Antibiotics **50**: 111; (b) Sakuda S, Ono M, Furihata K, Nakayama J, Suzuki A, Isogai A (1996) Aflastatin A, a novel inhibitor of aflatoxin production of *Aspergillus parasiticus*, from *Streptomyces*. J Am Chem Soc **118**: 7855
438. Ono M, Sakuda S, Ikeda H, Furihata K, Nakayama J, Suzuki A, Isogai A (1998) Structures and biosynthesis of aflastatines: Novel inhibitors of aflatoxin production by *Aspergillus parasiticus*. J Antibiot **51**: 1019
439. Longbottom DA, Morrison AJ, Dixon DJ, Ley SV (2002) Total synthesis of polycephalin C and determination of the absolute configurations at the 3″,4″ ring junction. Angew Chem **114**: 2910; Angew Chem Int Ed Engl **41**: 2786
440. Vesonder RF, Tjarks LW, Rohwedder WK, Burmeister HR, Laugal JA (1979) Equisetin, an antibiotic from *Fusarium equiseti* NRRL 5537, identified as a derivative of *N*-methyl-2,4-pyrrolidone. J Antibiot **32**: 759

441. Phillips NJ, Goodwin JT, Fraiman A, Cole RJ, Lynn DG (1989) Characterization of the *Fusarium* toxin equisetin: The use of phenylboronates in structure assignment. J Am Chem Soc **111**: 8223
442. Turos E, Audia JE, Danishefsky S (1989) Total synthesis of the *Fusarium* toxin equisetin: Proof of the stereohemical relationship of the tetramate and terpenoid sectors. J Am Chem Soc **111**: 8231
443. Singh MP, Zaccardi J, Greenstein M (1998) LL-49F233α, a novel antibiotic produced by an unknown fungus: Biological and mechanistic activities. J Antibiot **51**: 1109
444. Sugie Y, Dekker KA, Inagaki T, Kim Y-J, Sakakibara T, Sakemi S, Sugiura A, Brennan L, Duignan J, Sutcliffe JA, Kojima Y (2002) A novel antibiotic CJ-17,572 from a fungus, *Pezicula* sp. J Antibiot **55**: 19
445. Holzapfel CW (1968) The isolation and structure of cyclopiazonic acid, a toxic metabolite of *Penicillium cyclopium* Westling. Tetrahedron **24**: 2101
446. Kozikowski AP, Greco MN, Springer JP (1984) Total synthesis of the unique mycotoxin α-cyclopiazonic acid (α-CA): An unusual dimethylzinc-mediated replacement of a phenylthio substituent by a methyl group and a contrathermodynamic Raney nickel desulfurization reaction. J Am Chem Soc **106**: 6873
447. Muratake H, Natsume M (1985) Total synthesis of (±)-α-cyclopiazonic acid. Heterocycles **23**: 1111
448. van Rooyen PH (1992) Structure of α-cyclopiazonic acid. Acta Crystallogr **C48**: 551
449. McGrath RM, Steyn PS, Ferreira NP (1973) α-Acetyl-γ-(β-indolyl)methyltetramic acid. A biosynthetic intermediate of cyclopiazonic acid and of bis-secodehydrocyclopiazonic acid. J C S Chem Commun 812
450. Holzapfel CW, Hutchison RD, Wilkins DC (1970) The isolation and structure of two new indole derivatives from *Penicillium cyclopium* Westling. Tetrahedron **26**: 5239
451. Ohmomo S, Sugita M, Abe M (1973) Isolation of cyclopiazonic acid, cyclopiazonic acid imine and bissecodehydrocyclopiazonic acid from the cultures of *Aspergillus versicolor* (Vuill.) Tiraboshi. J Agric Chem Soc Jpn **47**: 57
452. (a) Holzapfel CW, Gildenhuys PJ (1977) The synthesis of DL-β-cyclopiazonic acid. S Afr J Chem **30**: 125; (b) Holzapfel CW, Kruger FWH (1992) The synthesis of optically pure β-cyclopiazonic acid, an indolic fungal metabolite. Aust J Chem **45**: 99
453. McGrath RM, Steyn PS, Ferreira NP, Neethling DC (1976) Biosynthesis of cyclopiazonic acids in *Penicillium cyclopium*: The isolation of dimethylallylpyrophosphate: *cyclo*-acetoacetyltryptophanyl dimethylallyltransferase. Bioorg Chem **4**: 11
454. de Jesus AE, Steyn PS, Vleggaar R, Kirby GW, Varley MJ, Ferreira NP (1981) Biosynthesis of α-cyclopiazonic acid. Steric course of proton removal during the cyclisation of β-cyclopiazonic acid in *Penicillium griseofulvum*. J Chem Soc Perkin Trans 1, 3292
455. Steyn PS, Vleggaar R, Ferreira NP, Kirby GW, Verley MJ (1975) Steric course of proton removal during the cyclization of β-cyclopiazonic acid in *Penicillium cyclopium*. J Chem Soc Chem Commun 465
456. Chalmers AA, Gorst-Allman CP, Steyn PS (1982) Biosynthesis of α-cyclopiazonc acid: Stereochemical aspects of D-ring formation. J Chem Soc Chem Commun 1367
457. Schabort JC, Wilkens DC, Holzapfel CW, Potgieter DJJ, Neitz AW (1971) β-Cyclopiazonate oxydocyclase from *Penicillium cyclopium*. I. Assay methods, isolation and purification. Biochim Biophys Acta **250**: 311
458. Schabort JC, Potgieter DJJ (1971) β-Cyclopiazonate oxydocyclase from *Penicillium cyclopium*. II. Studies on electron acceptors, inhibitors, enzyme kinetics, amino acid composition, flavin prosthetic group and other properties. Biochim Biophys Acta **250**: 329

459. Omura S, Tanaka H, Shimoi K, Liu JR (1994) Pyrrole propionic acid amide (cystamidine A) manufacture with *Streptomyces*, and its use as calpain inhibitor. Eur Pat Appl EP 569,122. Chem Abstr (1994) **120**: 105 131
460. Bihovsky R, Pendrak I (1996) Synthesis of cystamidin A (pyrrole-3-propanamide), a reported calpain inhibitor. Bioorg Med Chem Lett **6**: 1541
461. Keller-Schierlein W, Müller A, Hagmann L, Schneider U, Zähner H (1985) Stoffwechselprodukte von Mikroorganismen. 232. Mitt. (*E*)-3-(1*H*-Pyrrol-3-yl)-2-propensäure und (*E*)-3-(1*H*-Pyrrol-3-yl)-2-propensäureamid aus *Streptomyces parvulus*, Stamm Tü 2480. Helv Chim Acta **68**: 559
462. Kato S, Shindo K, Kawai H, Odagawa A, Matsuoka M, Mochizuki J (1993) Pyrrolostatin, a novel lipid peroxidation inhibitor from *Streptomyces chrestomyceticus*. J Antibiotics **46**: 892
463. Fumoto Y, Eguchi T, Uno H, Ono N (1999) Synthesis of pyrrolostatin and its analogues. J Org Chem **64**: 6518
464. Andersen RJ, Wolfe MS, Faulkner DJ (1974) Autotoxic antibiotic production by a marine Chromobacterium. Mar Biol (Berlin) **27**: 281
465. Shomura T, Amano S, Yoshida J, Ezaki N, Itō T, Niida T (1983) *Actinosporangium vitaminophilum* sp. nov. Int J Syst Bacteriol **33**: 557
466. Ezaki N, Shomura T, Koyama M, Niwa T, Kojima M, Inouye S, Itō T, Niida T (1981) New chlorinated nitropyrrole antibiotics pyrrolomycin A and B (SF-2080 A and B). J Antibiot **34**: 1363
467. Koyama M, Kodama Y, Tsuruoka T, Ezaki N, Niwa T, Inouye S (1981) Structure and synthesis of pyrrolomycin A, a chlorinated nitropyrrole antibiotic. J Antibiot **34**: 1569
468. Umezawa K, Mimura S, Matsushima T, Muramatsu S, Sawa T, Takeuchi T (1982) Enhancement of haemolysis and cellular arachidonic acid release by pyrrolomycins. Biochem Biophys Res Commun **105**: 82
469. Kaneda M, Nakamura S, Ezaki N, Iitaka Y (1981) Structure of pyrrolomycin B, a chlorinated nitropyrrole antibiotic. J Antibiot **34**: 1366
470. Yano K, Oono J, Mogi K, Asaoka T, Nakashima T (1987) Pyrroxamycin, a new antibiotic. Taxonomy, fermentation, isolation, structure determination and biological properties. J Antibiotics **40**: 961
471. Nakamura H, Shiomi K, Iinuma H, Naganawa H, Obata T, Takeuchi T, Umezawa H, Takeuchi Y, Iitaka Y (1987) Isolation and characterization of a new antibiotic, dioxapyrrolomycin, related to pyrrolomycins. J Antibiotics **40**: 899
472. Rengaraju S, Narayanan S, Ganju PL, Amin MA, Iyengar MRS, Itoh J, Takeuchi Y, Fujita K, Miyadoh S, Shomura T, Sezaki M, Kojima M (1985) A new antibiotic Al-R2081 related to pyrrolomycin B. Sci Reports of Meiji Seika Kenkyu Nenpo (Japan), **24**: 48 [Chem Abstr (1987) **106**: 152653]
473. Ono Y, Yano K, Asaoka T, Mogi K, Nakajima T (S. S. Pharm. Co. Ltd.) (1987) Antibiotic SS 46506A. Japan Pat 61 152 295. Chem Abstr (1987) **106**: 31379
474. Carter GT, Nietsche JA, Goodman JJ, Torrey MJ, Dunne TS, Borders DB, Testa RT (1987) LL-F42248α, a novel chlorinated pyrrole antibioc. J Antibiotics **40**: 233
475. Masuda K, Suzuki K, Ishida-Okawara A, Mizuno S, Hotta K, Miyadoh S, Hara O, Koyama M (1990) Pyrrolomycin group antibiotics inhibit substance P-induced release of myeloperoxidase from human polymorphonuclear leukocytes. J Antibiotics **44**: 533
476. Carter GT, Nietsche JA, Goodman JJ, Torrey MJ, Dunne TS, Siegel MM, Borders DB (1989) Direct biochemical nitration in the biosynthesis of dioxapyrrolomycin. A unique mechanism for the introduction of nitro groups in microbial products. J Chem Soc Chem Commun 1271

477. Ezaki N, Koyama M, Shomura T, Tsuruoka T, Inouye S (1983) Pyrrolomycins C, D and E, new members of pyrrolomycins. J Antibiot **36**: 1263
478. Koyama M, Ezaki N, Tsuruoka T, Inouye S (1983) Structural studies of pyrrolomycins C, D and E. J Antibiot **36**: 1483
479. Ezaki N, Koyama M, Kodama Y, Shomura T, Tashiro K, Tsuruoka T, Inouye S, Sakai S (1983) Pyrrolomycins F_1, F_{2a}, F_{2b} and F_2, new metabolites produced by the addition of bromine to the fermentation. J Antibiot **36**: 1431
480. Takeda R (1958) Structure of a new anibiotic, pyolutorin. J Am Chem Soc **80**: 4749
481. Takeda R (1959) *Pseudomonas* pigments. III. Derivatives of pyoluteorin. Bull Agr Chem Soc Japan **23**: 126
482. Takeda R (1959) *Pseudomonas pigments*. IV. The structure of pyoluteorin. Bull Agr Chem Soc Japan **23**: 165
483. Birch AJ, Hodge P, Rickards RW, Takeda R, Watson TR (1964) The structure of pyoluteorin. J Chem Soc 2641
484. Elix JA, Sargent MV (1967) The synthesis of some substituted pyoluteorins. J Chem Soc (C) 1718
485. Birchhall GR, Hughes CG, Rees AH (1970) Newer syntheses of the pyoluteorin antibiotics. Tetrahedron Lett 4879
486. Bailey K, Rees AH (1970) Pyoluteorin, a synthesis. Can J Chem **48**: 2257
487. Davis D, Hodge P (1970) A synthetic route to pyoluteorin. Tetrahedron Lett 1673
488. Bayley DM, Johnson RE (1970) The synthesis of pyoluteorin. Tetrahedron Lett 3555
489. Durham DG, Hughes CG, Rees AH (1972) The chlorination of pyrroles. Part III. Can J Chem **50**: 3223
490. Cue Jr BW, Dirlam JP, Czuba LJ, Windisch WW (1981) A practical synthesis of pyoluteorin. J Heterocycl Chem **18**: 191
491. Bayley DM, Johnson RE, Salvador UJ (1973) Pyrrole antibacterial agents. 1. Compounds related to pyoluteorin. J Med Chem **16**: 1298
492. Utsumi Y, Furusaki A, Tomiie Y (1970) The crystal and molecular structure of O,O',N-Trimethylpyoluteorin, $C_{14}\,H_{13}NCl_2$. Bull Chem Soc Japan **43**: 2640
493. Ohmori T, Hagiwara S, Ueda A, Minoda Y, Yamada K (1978) Production of pyoluteorin and its derivatives from n-paraffin by *Pseudomonas aeruginosa* S10B2. Agric Biol Chem **43**: 2031
494. Cuppels DA, Howell CR, Stipanovic RD, Stoessi A, Stothers JB (1986) Biosynthesis of pyoluteorin: A mixed polyketide tricarboxylic acid cycle origin demonstratted by $[1,2\text{-}^{13}C_2]$-acetate incorporation. Z Naturforsch **41c**: 532
495. Burkholder PR, Pfister RM, Leitz FM (1966) Production of a pyrrole antibiotic by a marine bacterium. Appl Microbiol **14**: 649
496. Laatsch H, Pudleiner H (1989) Marine Bakterien, 1. Synthese von Pentabrompseudilin, einem cytotoxischen Phenylpyrrol aus *Alteromonas luteo-violaceus*. Liebigs Ann Chem 863
497. Lovell FM (1966) The structure of a bromine-rich marine antibiotic. J Am Chem Soc **88**: 4510
498. Hanessian S, Kaltenbronn JS (1966) Synthesis of a bromine-rich antibiotic. J Am Chem Soc **88**: 4509
499. ApSimon JW, Durham DG, Rees AH (1973) A new synthetic route to the marine antibiotic of *Pseudomonas bromoutilis*. Chem Ind (London) 275
500. Pudleiner H, Laatsch H (1990) Marine Bakterien, II. Synthese cyclischer und sterisch gehinderter Pseudiline. Liebigs Ann Chem 423
501. Hanefeld U, Laatsch H (1991) Synthesis of isopentabromopseudilin. Liebigs Ann Chem 865

502. Renneberg B, Kellner M, Laatsch H (1993) Synthese halogenierter Benzyl- und Benzoylpyrole. Liebigs Ann Chem 847
503. Laatsch H, Pudleiner H, Pelizaeus B, Van Pée K-H (1994) Enzymatische Bromierung von Pseudilinen und verwandten Heteroarylphenolen mit der Chlorperoxidase aus *Streptomyces aureofaciens* Tü 24. Liebigs Ann Chem 65
504. Nogami T, Shigihara Y, Matsuda N, Takahashi Y, Naganawa H, Nakamura H, Hamada M, Muraoka Y, Kakita T, Iitaka Y, Takeuchi T (1990) Neopyrrolomycin, a new chlorinated phenylpyrrole antibiotic. J Antibiotics **43**: 1192
505. Tatsuta K, Itoh M (1993) Total synthesis of chlorinated phenylpyrrole antibiotics, (+)- and (−)-neopyrrolomycins. Tetrahedron Lett **34**: 8443
506. Tatsuta K, Itoh M (1994) Total synthesis of (+)- and (−)-neopyrrolomycins, chlorinated phenylpyrrole antibiotics. Bull Chem Soc Japan **67**: 1449
507. Tatsuta K, Itoh M (1994) Synthesis and antibacterial activities of less chlorinated analogs of neopyrrolomycin. J Antibiot **47**: 602
508. Imanaka H, Kousaka M, Tamura G, Arima K (1965) Studies on pyrrolnitrin, a new antibiotic. II. Taxonomic studies on pyrrolnitrin-producing strain. J Antibiot (Tokyo) **A18**: 205
509. Arima K, Imanaka H, Kousaka M, Fukuta A, Tamura G (1964) Pyrrolnitrin, a new antibiotic substance, produced by *Pseudomonas*. Agr Biol Chem (Tokyo) **28**: 575
510. Arima K, Imanaka H, Kousaka M, Fukuda A, Tamura G (1965) Studies on pyrrolnitrin, a new antibiotic. I. Isolation and properties of pyrrolnitrin. J Antibiot (Tokyo) **18A**: 201
511. Arima K, Tamura G, Imanaka H, Kosaka M, Fukuda A (Fujisawa Pharmaceutical Co., Ltd.) (1966) Pyrrolnitrin, a new antibiotic substance Japan 7838 ('66) Chem Abstr (1966) **65**: 13 662b
512. Arima K, Tamura G, Imanaka H, Kosaka M, Fukuda A (Fujisawa Pharmaceutical Co., Ltd.) Pyrrolnitrin. Japan 21.750 ('67). Chem Abstr (1968) **68**: 38 175
513. Lively DH, Gorman M, Haney ME, Mabe JA (1966) Metabolism of tryptophans by *Pseudomonas aureofaciens*. I. Biosynthesis of pyrrolnitrin. Antimicrob Agents Chemother 462
514. Elander RP, Mabe JA, Hamill, RL, Gorman M (1968) Metabolism of tryptophans by *Pseudomonas aureofaciens*. VI. Production of pyrrolnitrin by selected *Pseudomonas* species Appl Microbiol **16**: 753
515. Elander RP, Mabe JA, Hamill RL, Gorman M (1971) Biosynthesis of pyrrolnitrins by analogue-resistant mutants of *Pseudomonas fluorescens*. Folia Microbiol (Prag) **16**: 156 [Chem Abstr (1971) **75**. 72 810]
516. Roitman JN, Mahoney NE, Janisiewicz WJ, Benson M (1990) A new chlorinated phenylpyrrole antibiotic produced by the antifungal bacterium *Pseudomonas cepacia*. J Agric Food Chem **38**: 538
517. Roitman JN, Mahoney NE, Janisiewicz WJ (1990) Production and composition of phenylpyrrole metabolites produced by *Pseudomonas cepacia*. Appl Microbiol Biotechnol **34**: 381
518. Fujisawa Pharmaceutical Co., Ltd (1966) Antibiotic pyrrolnitrin. Neth Appl 6.503.853 (1965) Chem Abstr (1966) **65**: 18 434
519. Imanaka H, Kousaka M, Tamura G, Arima K (1965) Studies on pyrrolnitrin, a new antibiotic. III. Structure of pyrrolnitrin. J Antibiot (Tokyo) **18A**: 207
520. Nakano H, Umio S, Kariyone K, Tanaka K, Kishimoto T, Noguchi H, Ueda I, Nakamura H, Morimoto Y (1966) Total synthesis of pyrrolnitrin, a new antibiotic. Tetrahedron Lett 737

521. Umio S, Kariyone K, Tanaka K et al. (1969) Total synthesis of pyrrolnitrin. Chem Pharm Bull (Tokyo) **17**: I. Synthesis of 3-arylpyrrole derivatives by Knorr's condensation. 559; II. Synthesis of ethyl 3-aryl-5-methyl-2-pyrrolecarboxylate. (1) 567; III. Synthesis of ethyl 3-aryl-5-methyl-2-pyrrolecarboxylate. (2). 576; IV. Synthesis of ethyl 3-aryl-5-methyl-2-pyrrolecarboxylate. (3). 582; VI. Synthesis of nitrochloro-2-aminoacetophenones and 1-aryl-1,3-butanediones. 596; VII. Synthesis of 2,5-dimethyl-3-arylpyrrole. 605; VIII. A new method of pyrrole ring closure using aminoacetal. 611; IX. Synthesis of dialkyl 3-aryl-2,5-pyrroledicarboxylates and alkyl β-aryl-5-methyl-2-pyrrolecarboxylates. 616
522. Umio S, Kariyone K, Tanaka K, Ueda I, Morimoto Y (1969) Total synthesis of pyrrolnitrin. V. Synthesis of pyrrolnitrin and its analogues. (1). Chem Pharm Bull **17**: 588
523. Tanaka K, Kariyone K, Umio S (1969) Total synthesis of pyrrolnitrin. X. Synthesis of pyrrolnitrin. (2). Chem Pharm Bull **17**: 622
524. Gosteli J (1972) Eine neue Synthese des Antibioticums Pyrrolnitrin. Helv Chim Acta **55**: 451
525. Morimoto Y, Hashimoto M, Hattori K (1968) The crystal structure of pyrrolnitrin. Tetrahedron Lett: 209
526. Gordee RS, Matthews TR (1969) Systemic antifungal activity of pyrrolnitrin. Appl Microbiol **17**: 690
527. Nishida M, Matsubara T, Watanabe N (1965) Pyrrolnitrin. a new anifungal antibiotic. Microbiological and toxicological observations. J Antibiot **A18**: 211
528. Murphy PJ, Williams TL (1972) Biological inactivation of pyrrolnitrin. Identification and synthesis of pyrrolnitrin metabolites. J Med Chem **15**: 137
529. Gordee RS, Matthews TR (1967) Evaluation of the *in vitro* and *in vivo* antifungal activity of pyrrolnitrin. Antimicrob Agents Chemother 378
530. Nose M, Arima K (1969) On the mode of action of a new antifungal antibiotic, pyrrolnitrin. J Antibiot (Tokyo), **22**: 135
531. Lambowitz AM, Slayman CW (1972) Effect of pyrrolnitrin on electron transport and oxidative phosphorylation in mitochondria isolated from *Neurospora crassa*. J Bacteriol **112**: 1020
532. Tripathi RK, Gottlieb D (1969) Mechanism of action of the antifungal antibiotic pyrrolnitrin. J Bacteriol **100**: 310
533. Wong DT, Airall JM (1970) The mode of action of antifungal agents: Effect of pyrrolnitrin on mitochondrial electron transport. J Antibiot **23**: 55
534. Wong DT, Horng J-S, Gordee RS (1971) Respiratory chain of a pathogenic fungus, *Microsporum gypseum*: Effect of the antifungal agent pyrrolnitrin. J Bacteriol **106**: 168
535. Umio S, Kariyone K, Tanaka K, Kishimoto T, Nakamura H, Nishida M (1970) Structure-activity studies of pyrrolnitrin analogues. Chem Pharm Bull (Tokyo) **18**: 1414
536. Massa S, DiSanto R, Costi R, Mai A, Artico M, Retico A, Apuzzo G, Artico M, Simonetti G (1993) Antifungal agents. 4. Synthesis of *neo*-isopyrrolnitrin and other 3-aryl-4-nitro-1*H*-pyrroles via TosMIC. Med Chem Res **3**: 192
537. Gorman M, Lively DH (1967) Pyrrolnitrin: A new mode of tryptophan metabolism. Antibiotics **2**: 433
538. Hamill RL, Elander R, Mabe J, Gorman M (1967) Metabolism of tryptophan by *Pseudomonas aureofaciens*. V. Conversion of tryptophan to pyrrolnitrin. Antimicrob Agents Chemother 388
539. Hamill RL, Elander RP, Mabe JA, Gorman M (1970) Metabolism of tryptophan by *Pseudomonas aureofaciens*. III. Production of substituted pyrrolnitrins from tryptophan analogs. Appl Microbiol **19**: 721

540. Martin LL, Chang C-J, Floss HG, Mabe JA, Hagaman EW, Wenkert E (1972) A ^{13}C-nuclear magnetic resonance study on the biosynthesis of pyrrolnitrin from tryptophan by *Pseudomonas*. J Am Chem Soc **94**: 8942
541. Chang CJ, Floss HG, Hook DJ, Mabe JA, Manni PE, Martin LL, Schröder K, Shieh TL (1981) The biosynthesis of the antibiotic pyrrolnitrin by *Pseudomonas aureofaciens*. J Antibiot **34**: 555
542. Hohaus K, Altmann A, Burd W, Fischer I, Hammer PE, Hill DS, Ligon JM, van Pée K-H (1997) NADH-dependent halogenases are more likely to be involved in metabolite biosynthesis than haloperoxidases. Angew Chem **109**: 2102; Angew Chem Int Ed Engl **36**: 2012
543. Ajisaka M, Kariyone K, Jomon K, Yazawa H, Arima K (1969) Isolation of the bromo analogues of pyrrolnitrin from *Pseudomonas pyrrolnitrica*. Agr Biol Chem (Tokyo) **33**: 294
544. Ajisaka M, Kariyone K, Jomon K, Yazaw H (Fujisawa Pharm. Co.) (1972) Fermentative production of antibiotic bromonitrins A, B and C. Japan Pat 72 14 916. Chem Abstr (1972) **77**: 60049
545. van Pee K-H, Salcher O, Fischer P, Bokel M, Lingens F (1983) The biosynthesis of brominated pyrrolnitrin derivatives by *Pseudomonas aureofaciens*. J Antibiot **36**: 1735
546. Hashimoto M, Hattori K (1968) A new metabolite from *Pseudomonas pyrrolnitrica*. Chem Pharm Bull (Tokyo) **16**: 1144
547. Hashimoto M, Hattori K (1966) Isopyrrolnitrin: A metabolite from *Pseudomonas*. Bull Chem Soc Japan **39**: 410
548. Hashimoto M, Hattori K (1966) Oxypyrrolnitrin: A metabolite of *Pseudomonas*. Chem Pharm Bull (Tokyo) **14**: 1314
549. Shindo K, Yamagishi Y, Kawai H (1993) Thiazohalostatin, a new cytoprotective substance produced by *Actinomadura*. II. Physico-chemical properties and structure determination. J Antibiotics **46**: 1638
550. Kawamura N, Sawa R, Takahashi Y, Issiki K, Sawa T, Kinoshita N, Naganawa H, Hamada M, Takeuchi T (1995) Pyralomycins, new antibiotics from *Actinomadura spiralis*. J Antibiot **48**: 435
551. Kawamura N, Kinoshita N, Sawa R, Takahashi Y, Sawa T, Naganawa H, Hamada M, Takeuchi T (1996) Pyralomycins, novel antibiotics from *Microtetraspora spiralis*. I. Taxonomy and production. J Antibiot **49**: 706
552. Kawamura N, Sawa R, Takahashi Y, Isshiki K, Sawa T, Naganawa H, Takeuchi T (1996) Pyralomycins, novel antibiotics from *Microtetraspora spiralis*. II. Structure determination. J Antibiot **49**: 651
553. Kawamura N, Nakamura H, Sawa R, Takahashi Y, Sawa T, Naganawa H, Takeuchi T (1997) Pyralomycins, novel antibiotics from *Microtetraspora spiralis*. IV. Absolute configuration. J Antibiot **50**: 147
554. Tatsuta K, Takahashi M, Tanaka N (1999) The first total synthesis of pyralomycin 2c. Tetrahedon Lett **40**: 1929
555. Hayakawa Y, Kawakami K, Seto H, Furihata K (1992) Structure of a new antibiotic, roseophilin. Tetrahedron Lett **33**: 2701
556. (a) Fürstner A, Weintritt H (1997) Total synthesis of the potent antitumor agent roseophilin: A concise approach to the macrotricyclic core. J Am Chem Soc **119**: 2944; (b) Fürstner A, Weintritt H (1998) Total synthesis of roseophilin. J Am Chem Soc **120**: 2817
557. Mochizuki T, Itoh E, Shibata N, Nakatani S, Katoh T, Terashima S (1998) Studies toward the total synthesis of antibiotic roseophilin: A novel synthesis of the macrotricyclic part. Tetrahedron Lett **39**: 6911

558. Fürstner A, Gastner T, Weintritt H (1999) A second generation synthesis of roseophilin and chromophore analogues: J Org Chem **64**: 2361
559. (a) Robertson J, Hatley RJD (1999) Formal synthesis of roseophilin. Chem Commun 1455; (b) Robertson, J Hatley RJD, Watkin D (2000) Preparation of the tricyclic ketopyrrole core of roseophilin by radical macrocyclization and Paal-Knorr condensation. J Chem Soc Perkin Trans 1, 3389
560. Harrington PE, Tius MA (2001) Synthesis and absolute stereochemistry of roseophilin. J Am Chem Soc **123**: 8509
561. Ninet L, Bénazet F, Charpentié Y, Dubost M, Florent J, Mancy D, Preud'Homme J, Threlfall TL, Vuillemin B, Wright DE, Abraham A, Cartier M, de Chezelles N, Godard C, Theilleux J (1972) La chlorobiocine (18.631 R.P.), nouvel antibiotique chloré produit par plusieurs espèces de streptomyces. C R Acad Sci Paris, **275**: 455
562. Dolak L (1973) Structure of RP 18,631. J Antibiot **26**: 121
563. Kominek LA, Sebek OK (1974) Biosynthesis of novobiocin and related coumarin antibiotics. Dev Ind Microbiol 60
564. Kawaguchi H, Okanishi M, Miyaki T (1965) Coumermycin and its salts. US Pat. No. 3 201 386. Chem Abstr (1965) **63**: 14011
565. Kawaguchi H, Tsukiura H, Okanishi M, Miyaki T, Ohmori T, Fujisawa K, Koshiyama H (1965) Studies on coumermycin. A new antibiotic. I. Production, isolation and characterization of coumermycin A_1. J Antibiot (Tokyo) **A18**: 1
566. Kawaguchi H, Naito T, Tsukiura H (1965) Studies on coumermycin. A new antibiotic. II. Structure of coumermycin A_1. J Antibiot (Tokyo) **A18**: 11
567. Kawaguchi H, Miyaki T, Tsukiura H (1965) Studies on coumermycin. A new antibiotic. III. Structure of coumermycin A_2. J Antibiot (Tokyo) **A18**: 220
568. Berger J, Schocher AJ, Batcho AD, Pecherer B, Keller O, Maricq J, Karr AE, Vaterlaus BP, Furlenmeier A, Speigelberg H (1965) Production, isolation and synthesis of the coumermycins (sugordomycins), a new Streptomycete antibiotic complex. Antimicrob Agents Chemother 778
569. Hoffmann-LaRoche & Co., AG (1966) Preparation of antibiotics. Belg. Pat. 665 237. Chem Abstr (1966) **65**: 11304
570. Karr AE (Hoffmann-LaRoche & Co., AG) (1970) Preparation of pure coumermycin A_1. Ger. Offen. 1.905.328 (1969). Chem Abstr (1970) **72**: 11385
571. Wick AE, Blount JF, Leimgruber W (1976) Stability determining factors in anomeric coumerosides. Tetrahedron **32**: 2057
572. Whitaker WD (Hoffmann-LaRoche & Co AG) (1968) 3–(Coumarinylaminocarbonyl)- 2,4 - bis[4 - hydroxy - 8-methyl-7-[5,5-dimethyl-4-O-methyl-3-O-(5-methyl-2-pyrroyl)-α-L-lyxopyranosyloxy]-3-methylpyrrole. Brit. Pat. 1 114 468. Chem Abstr (1968) **69**: 67671
573. Furlenmeier A, Schocher AJ, Spiegelberg H, Vaterlaus BP, Batcho AD, Berger J, Keller O, Pecherer B (Hoffmann-La Roche & Co, AG) (1969) Sugordomycin antibiotics. Swiss Pat. 467,799. Chem Abstr (1969) **71**: 39352
574. Laurin P, Ferroud D, Klich M, Dupuis-Hamelin C, Mauvais P, Lassaigne P, Bonnefoy A, Musicki B (1999) Synthesis and *in vitro* evaluation of novel highly potent coumarin inhibitors of gyrase B. Bioorg Med Chem Lett **9**: 2079
575. Umezawa H, Hamada M, Takita T, Naganawa H (Microbiochemical Research Foundation) (1971) Coumermycin A_1, Jap. Pat. 71 15,675. Chem Abstr (1971) **75**: 87081
576. Duma RJ, Warner JF (1969) *In vitro* activity of coumermycin A_1 against *Mycobacterium tuberculosis* var. *hominis*. Appl Microbiol **18**: 404
577. Fedorko J, Katz S, Allnoch H (1969) *In vitro* activity of coumermycin A_1. Appl Microbiol **18**: 869

578. Michaeli D, Meyers BR, Weinstein L (1969) *In vitro* studies of the activity of coumermycin-A1 against Staphylococci resistant to methicillin and cephalothin. J Infec Dis **120**: 488
579. Devine LF, Hagerman CR (1970) Spectra of susceptibility of *Neisseria meningitidis* to antimicrobial agents *in vitro*. Appl Microbiol **19**: 329
580. Grunberg E, Cleeland R, Titsworth E (1966) Further observations on chemotherapeutic activity of coumermycin A1. I. Activity against *Neisseria meningitidis* type A and Meningopneumonitis. Antimicrob Agents Chemother 397
581. Michaeli D, Meyers BR, Weinstein L (1969) Microbiological and pharmacological study of a new antibiotic, coumermycin A_1. Antimicrob Agents Chemother 463
582. Gellert M, O'dea MH, Itoh T, Tomizawa J (1976) Novobiocin and coumermycin inhibit DNA supercoiling catalyzed by DNA gyrase. Proc Natl Acad Sci USA **73**: 4474
583. Ferroud D, Collard J, Klich M, Dupuis-Hamelin C, Mauvais P, Lassaigne P, Bonnefoy A, Musicki B (1999) Synthesis and biological evaluation of coumarincarboxylic acids as inhibitors of gyrase B. L-Rhamnose as an effective substitute for L-noviose. Bioorg Med Chem Lett **9**: 2881
584. Ràdl S (1990) Structure-ativity relationships in DNA gyrase inhibitors. Pharmac Ther **48**: 1 and refeences given therein
585. Kaplan SA (1970) Pharmacokinetic profile of coumermycin A_1. J Pharm Sci **59**: 309
586. Newmark HL, Berger J (1970) Coumermycin A_1-Biopharmaceutical studies I. J Pharm Sci **59**: 1246
587. Newmark HL, Berger J, Carstensen JT (1970) Coumermycin A_1 – Biopharmaceutical studies II. J Pharm Sci **59**: 1249
588. Price KE, Chisholm DR, Godfrey JC, Misiek M, Gourevitch A (1970) Semisynthetic coumermycins: Structure-activity relationships. Appl Microbiol **19**: 14
589. Keil JG, Godfrey JC, Cron MJ, Hooper IR, Nettleton DE, Price KE, Schmitz H (1971) Structure-activity relations in chemically modified coumermycin. Pure Appl Chem **28**: 571
590. Scannell J, Kong YL (1969) Biosynthesis of coumermycin A_1: Incorporation of L-proline in the pyrrole groups. Antimicrob Agents Chemother 139
591. Celmer WD, Cullen WP, Moppett CE, Jefferson MT, Huang LH, Shibakawa R, Tone J (Pfizer Inc., New York) (1979) Antibiotics produced by new species of Nocardia. US Pat. 4 148 883. Chem Abstr (1979) **91**: 54595
592. Celmer WD, Chmurny GN, Moppett CE, Ware RS, Watts PS, Whipple EB (1980) Structure of natural antibiotic CP-47,444. J Am Chem Soc **102**: 4203
593. Whaley HA, Chidester CG, Mizsak SA, Wnuk RJ (1980) Nodusmicin: The structure of a new antibiotic. Tetrahedron Lett **21**: 3659
594. Whaley HA, Coats JH (Upjohn Co.) (1983) Antibiotic composition of matter. US 4,351,769, 1982. Chem Abstr (1983) **98**: 15 465
595. Cane DE, Yang C-C (1985) Nargenicin biosynthesis: Late stage oxidations and absolute configuration. J Antibiot **38**: 423
596. Magerlein BJ, Reid RJ (1982) Synthesis of 18-deoxynargenicin A_1 (antibiotic 367c) from nargenicin A_1. J Antibiot **35**: 254
597. Plata DJ, Kallmerten J (1988) Total synthesis of (+)-18-deoxynargenicin A_1. J Am Chem Soc **110**: 4041
598. Cane DE, Yang C-C (1984) Biosynthetic origin of the carbon skeleton and oxygen atoms of nargenicin A_1. J Am Chem Soc **106**: 784
599. Snyder WC, Rinehart Jr KL (1984) Biosynthesis of nargenicin and nodusmicin. J Am Chem Soc **106**: 787

600. Cane DE, Tan W, Ott WR (1993) Nergenicin biosynthesis. Incorporation of polyketide chain, elongation, intermediates, and support for a proposed intramolecular Diels-Alder Cyclization. J Am Chem Soc **115**: 527
601. Takiguchi Y, Mishima H, Okuda M, Terao M (1980) Milbemycins, a new family of macrolide antibiotics: Fermentation, isolation and physico-chemical properties. J Antibiot **33**: 1120
602. Ding W, Williams DR, Northcote P, Siegel MM, Tsao R, Ashcroft J, Morton GO, Alluri M, Abbanat D, Maiese WM, Ellestad GA (1994) Pyrroindomycins, novel antibiotics produced by *Streptomyces rugosporus* sp. LL-42D005. J Antibiotics **47**: 1250
603. Singh MP, Petersen PJ, Jacobus NV, Mroczenski-Wildey MJ, Maiese WM, Greenstein M, Steinberg DA (1994) Pyrroindomycins, novel antibiotics produced by *Streptomyces rugosporus* LL-42D005: II. Biological activities. J Antibiotics **47**: 1258
604. Schönewolf M, Grabley S, Hütter K, Machinek R, Wink J, Zeeck A, Rohr J (1991) Glycerinopyrin, a novel metabolite from *Streptomyces violaceus*. Liebigs Ann Chem 77
605. Schönewolf M, Rohr J (1991) Biogenesis of the carbon framework in glycerinopyrin: A new type biosynthesis pathway of pyrrole. Angew Chem **103**: 211; Angew Chem Int Ed Engl **30**: 183
606. Kikuchi M, Kumagai K, Ishida N, Ito Y, Yamaguchi T, Furumai T, Okuda T (1965) Isolation, purification, and properties of kikumycins A and B. J Antibiot **A18**: 243
607. Takaishi T, Sugawara Y, Suzuki M (1972) Structure of kikumycins A and B. Tetrahedron Lett 1873
608. Takaishi T, Suzuki M, Tatematsu A (1974) Mass spectrometry of antibiotic kikumycin A, B and related compounds. Org Mass Spectrom **9**: 635
609. Lee M, Lown JW (1987) Synthesis of (4S)- and (4R)-methyl 2 amino-1-pyrroline-5-carboxylates and their application to the preparation of (4S)-(+)- and (4R)-(−)-dihydrokikumycin B. J Org Chem **52**: 5717
610. Finlay AC, Hochstein FA, Sobin BA, Murphy FX (1951) Netropsin. a new anibiotic produced by a Streptomyces. J Am Chem Soc **73**: 341
611. Watanabe K (1956) Sinanomycin (Netropsin) produced by *Streptomyces* sp. No. 7618. J Antibiotics (Tokyo) **A9**: 102
612. Weiss MJ, Webb JS, Smith Jr JM (1957) The structure of antibiotic T-1384. Synthesis of the degradation fragments. J Am Chem Soc **79**: 1266
613. Thrum H (1959) Eine neue, von einer Spezies der *Streptomycetes-reticuli*-Gruppe gebildete Antibiotikakombination. Naturwiss **46**: 87
614. van Tamelen EE, White DM, Kogon IC, Powell ADG (1956) Structural studies on the antibiotic netropsin. J Am Chem Soc **78**: 2157
615. Waller CW, Wolf CF, Stein WJ, Hutschings BL (1957) The structure of antibiotic T-1384. J Am Chem Soc **79**: 1265
616. Julia M, Joseph N (1956) Premières études sur la structure chimique d'un nouvel antibiotique, la congocidine. Compt Rend **243**: 961
617. Nakamura S, Yonehara H, Umezawa S (1964) On the structure of netropsin. J Antibiot **A17**: 220
618. Cosar C, Ninet L, Pinnert-Sindico S, Preud'Homme J (1952) Activité trypanocide d'un antibiotique produit par un streptomyces. Compt Rend **234**: 1498
619. (a) Julia M, Préau-Joseph N (1963) Structure et synthèse de la congocidine. Compt Rend. **257**: 1115; (b) Julia M, Préau-Joseph N (1967) Amidines et guanidines apparentées à la congocidine. I. Structure de la congocidine. Bull Soc Chim Fr 4348

620. Lown JW, Krowicki K (1985) Efficient total synthesis of the oligopeptide antibiotics netropsin and distamycin. J Org Chem **50**: 3774
621. Wildfeuer ME (1964) The biosynthesis of netropsin. Diss. Abstr. University Ann Arbor, Mich. **24**: 3090. [Chem Abstr (1964) **60**: 14854]
622. Arcamone F, Orezzi PG, Barbieri W, Nicolella V, Penco S (1967) Distamicina A – Nota I. Isolamento e struttura dell'agente antivirale distamicina A. Gazz chim ital **97**: 1097
623. Arcamone F, Penco S, Orezzi P, Nicolella V, Pirelli A (1964) Structure and synthesis of Distamycin A. Nature **203**: 1064
624. Penco S, Redaelli S, Arcamone F (1967) Distamicina A – Nota II. Sintesi totale. Gazz chim ital **97**: 1110
625. Grokhovskii SL, Zhuze AL, Gottikh BP (1975) Ligands possessing affinity for definite pairs of DNA bases. I. Synthesis of distamycin A and its analogs with different numbers of *N*-methyl and *N*-propylpyrrole residues in the molecule. Bioorg Khim **1**: 1616
626. Bialer M, Yagen B, Mechoulam R (1978) A total synthesis of distamycin A (V), an antiviral antibiotic. Tetrahedron **34**: 2389
627. Grehn L, Ragnarsson U (1981) Novel efficient total synthesis of antiviral antibiotic distamycin A. J Org Chem **46**: 3492
628. Arcamone F, Penco S, Delle Monache F (1969) Distamicina A – Nota III. Sintesi di analoghi contenenti modificazioni nelle catene laterali. Gazz chim ital **99**: 620
629. Arcamone F, Nicolella V, Penco S, Redaelli S (1969) Distamicina A – Nota III. Sintesi di analoghi con diverso numero di resti dell'acido 1-metil-4-amminopirrolo-2-carbossilico. Gazz chim ital **99**: 632
630. Bialer M, Yagen B, Mechoulam R, Becker Y (1979) Structure-activity relationships of pyrrole amidine antiviral antibiotics. 1. Modifications of the alkylamidine side chain. J Med Chem **22**: 1296
631. Lown JW, Krowicki K, Bhat UG, Skorobogaty A, Ward B, Dabrowiak JC (1986) Molecular recognition between oligopeptides and nucleic acids: Novel imidazole-containing oligopeptides related to netropsin that exhibit altered DNA sequence specificity. Biochemistry **25**: 7408
632. Cozzi P, Beria I, Caldarelli M, Capolongo L, Geroni C, Mongelli N (2000) Cytotoxic α-bromoacrylic derivatives of distamycin analogues modified at the amidino moiety. Bioorg Med Chem Lett **10**: 1273
633. Cozzi P, Beria I, Caldarelli M, Geroni C, Mongelli N, Pennella G (2000) Cytotoxic Halogenoacrylic derivatives of distamycin A. Bioorg Med Chem Lett **10**: 1269
634. Cozzi P, Beria I, Caldarelli M, Capolongo L, Geroni C, Mazzini S, Ragg E (2000) Phenyl sulfur mustard derivatives of distamycin A. Bioorg Med Chem Lett **10**: 1653
635. Chandra P (1974) Molecular approaches for designing antiviral and antitumor compounds. Top Curr Chem **52**: 99 and references given therein
636. Hahn FE (1975) Distamycin A and netropsin in Corcoran JW, Hahn FE (eds) Antibiotics. Mechanism of Action of Antimicrobial and Antitumor Agents, vol 3. Springer Verlag, New York, pp 79–100 and references given therein
637. Zimmer C (1975) Effects of the antibiotics netropsin and distamycin A on the structure and function of nucleic acids: Progr Nucleic Acid Res Mol Biol **15**: 285 and references given therein
638. Wartell RM, Larson JE, Wells RE (1974) Netropsin. Specific probe for A-T regions of duplex deoxyribonucleic acid. J Biol Chem **249**: 6719
639. Kopka ML, Yoon C, Goodsell D, Pjura P, Dickerson RE (1985) The molecular origin of DNA-drug specifity in netropsin and distamycin. Proc Natl Acad Sci USA **82**: 1376

640. Baker BF, Dervan PB (1985) Sequence-specific cleavage of double-helical DNA by *N*-bromoacetyl-distamycin. J Am Chem Soc **107**: 8266
641. Zimmer C, Wähnert U (1986) Nonintercalating DNA-binding ligands: Specifity of the interaction and their use as tools in biophysical, biochemical and biological investigations of the genetic material. Progr Biophys Mol Biol **47**: 31
642. Patel DJ, Shapiro L (1986) Sequence-dependent recognition of DNA duplexes. Netropsin complexation to the AATT site of the d(G-G-A-A-T-T-C-C) duplex in aqueous solution. J Biol Chem **261**: 1230
643. Klevit RE, Wemmer DE, Reid BR (1986) ^{1}H NMR studies on the inteaction between distamycin A and a symmetrical DNA dodecamer. Biochemistry **25**: 3296
644. Lown JW, Kowicki K, Bath UG, Skorobogaty A, Ward B, Dabrowiak JC (1986) Molecular recognition between oligopeptides and nucleic acids: Novel imidazole-containing oligopeptides related to netropsin that exhibit altered DNA sequence specifity. Biochemistry **25**: 7408
645. Lee M, Krowicki K, Hartley JA, Pon RT, Lown JW (1988) Molecular regognition between oligopeptides and nucleic acids: Influence of van der Waals contacts in determining the 3'-terminus of DNA sequences read by monocationic lexitropsins. J Am Chem Soc **110**: 3641
646. Probst GW, Hoehn MM, Woods BL (1965) Anthelvencins, new antibotics with anthelmintic properties. Antimicrob Agents Chemother 789
647. Lee M, Shea RG, Hartley JA, Kissinger K, Pon RT, Vesnaver G, Breslauer KJ, Dabrowiak JC, Lown JW (1989) Molecular recognition between oligopeptides and nucleic acids: Sequence-specific binding of the naturally occurring antibiotic (4*S*)-(+)-anthelvencin A and its (4*R*)-(−)-enantiomer to deoxyribonucleic acids deduced from ^{1}H NMR, footprinting, and thermodynamic data. J Am Chem Soc **111**: 345
648. Lam KS, Hesler GA, Gustavson DR, Berry RL, Tomita K, MacBeth JL, Forenza S (1996) Pyrrolosporin A, a new antitumor antibiotic from *Micromonospora* sp. C39217-R109-7 I. Taxonomy of the producing organism, fermentation and biological activity. J Antibiotics **49**: 860
649. Schröder DR, Colson KL, Klohr SE, Lee MS, Matson JA, Brinen LS, Clardy J (1996) Pyrrolosporin A, a new antitumor antibiotic from *Micromonospora* sp. C39217-R109-7. II. Isolation, physico-chemical properties, spectroscopic study and X-ray analysis. J Antibiot **49**: 865
650. Deschamps M, Floc HF, Jung G, Margraff R (Rhone-Poulenc Sante) (1988) Immunosuppressive compounds, their preparation by cultivating *Streptomyces* (CBS 162.86) and pharmaceutical compositions containing them. Eur Pat Appl EP 246,975. Chem Abstr (1988) **109**: 21631
651. Andres N, Wolf H, Zähner H, Rössner E, Zeeck A, König WA, Sinnwell V (1989) Stoffwechselprodukte von Mikroorganismen. 253. Mitt.: Hormaomycin, ein neues Peptidlacton mit morphogener Aktivität auf Streptomyceten. Helv Chim Acta **72**: 426
652. Rössner E, Zeeck A, König WA (1990) Elucidation of the structure of hormaomycin. Angew Chem **102**: 84; Angew Chem Int Ed Engl **29**: 64
653. Kavai Y, Furihata K, Seto H, Otake N (1985) The structure of a new antibiotic, chromoxymycin. Tetrahedron Lett **26**: 3273
654. Séquin U, Furukawa M (1978) The structure of the antibiotic hedamycin. III ^{13}C-NMR spectra of hedamycin and kidamycin. Tetrahedon **34**: 3623
655. Reed PW, Lardy HA (1972) A-23187: a divalent cation ionophore. J Biol Chem **247**: 6970
656. Chaney MO, Demarco PV, Jones ND, Occolowitz JL (1974) The structure of A23187, a divalent cation ionophore. J Am Chem Soc **96**: 1932

657. Cresp TM, Probert CL, Sondheimer F (1978) An approach to the synthesis of ionophores related to A23187. Tetrahedron Lett 3955
658. Grieco PA, Kanai K, Williams E (1979) Studies directed the total synthesis of calcimycin (A-23187). Heterocycles **12**: 1623
659. Evans DA, Sacks CE, Kleschick WA, Taber TR (1979) Polyether antibiotics synthesis. Total synthesis and absolute configuration of the ionophore A-23187. J Am Chem Soc **101**: 6789
660. Grieco PA, Williams E, Tanaka H, Gilman SJ (1980) Elaboration of the C(3)-C(12) carbon fragment of calcimycin (A-23187). Formal synthesis of calcimycin. J Org Chem **45**: 3537
661. Hanessian S, Tyler PC, Chapleur Y (1981) Reaction of lithium dimethylcuprate with conformationally biased β-acyloxy enol esters – Regio and stereocontrolled access to functionalized six-carbon chiral synthons. Tetrahedron Lett **22**: 4583
662. Martinez GR, Grieco PA, Williams E, Kanai K, Srinivasan CV (1982) Stereocontrolled total synthesis of antibiotic A-23187 (calcimycin). J Am Chem Soc **104**: 1436
663. Hoppe D (1982) Total Synthese von Calcimycin. Nach Chem Tech Lab **30**: 581 and references given therein
664. Nakahara Y, Fujita A, Ogawa T (1984) Total Synthesis of Antibiotic A23187 (Calcimycin) from D-Glucose. J Carbohydr Chem **3**: 487
665. Nakahara Y, Fujita A, Beppu K, Ogawa T (1986) Total synthesis of antibiotic A23187 (calcimycin) from D-glucose. Tetrahedron **42**: 6465
666. Negri DP, Kishi Y (1987) A total synthesis of polyether antibiotic (−)-A23187 (calcimycin). Tetrahedron Lett **28**: 1063
667. Boeckman Jr RK, Charette AB, Asberom T, Johnston BH (1987) A convergent general synthetic protocol for construction of spirocyclic ketal ionophores: An application to the total synthesis of (−)-A23187 (calcimycin). J Am Chem Soc **109**: 7553
668. Ziegler FE, Cain DM (1989) Formal synthesis of (−)-calcimycin (A-23187) *via* the 3-methyl-γ-butyrolactone approach. J Org Chem **54**: 3347
669. Boeckman Jr RK, Charette AB, Asberom T, Johnston BH (1991) The chemistry of cyclic vinyl ethers. 6. Total synthesis of polyether ionophore antibiotics of the calcimyin (A-23187) class. J Am Chem Soc **113**: 5337
670. Pfeiffer DR, Taylor RW, Lardy HA (1978) Ionphore A23187; Cation binding and transport properties. Ann N Y Acad Sci **307**: 402 and references given therein
671. Tissier C, Juillard J, Dupin M, Jeminet G (1979) Mode d'action de la calcimycine (A. 23187). I. Equilibre avec les ions alcalins et alcalino-terreux en milieu homogène. J Chim Phys **76**: 611
672. Thomas TP, Pfeiffer DR, Taylor RW (1987) Formation and dissociation kinetics of nickel(II) with ionophore A23187 in 80% methanol-water. J Am Chem Soc **109**: 6670
673. Pfeiffer DR, Reed PW, Lardy HA (1974) Ultraviolet and fluorescent spectral properties of the divalent cation chromophore A23187 and its metal ion complexes. Biochemistry **19**: 4007
674. Tissier C, Juillard J, Boyd DW, Albrecht-Gary AM (1985) Mode of action of calcimycin (A23187) – II – A study of its interactions with alkali and alkali-earth cations in methanol-water mixtures. J Chim Phys **82**: 899
675. Chapman CJ, Puri AK, Taylor RW, Pfeiffer DR (1987) Equilibria between ionophore A23187 and divalent cations: Stability of 1:1 complexes in solutions of 80% methanol/water. Biochemistry **26**: 5009
676. Nelson A (1991) Electrochemical studies of Antibiotic 23187 (A23187). Mediated permeability to divalent heavy-metal ions in phospholipid monolayers adsorbed on mercury electrodes. J Chem Soc Faraday Trans: Phys Chem **87**: 1851

677. Pfeiffer DR, Deber CM (1979) Isosteric metal complexes of ionophore A23187. A basis for cation selectivity. FEBS Lett **105**: 360
678. Smith GD, Duax WL (1976) Crystal and molecular structure of the calcium ion complex of A23187. J Am Chem Soc **98**: 1578
679. Alléaume M, Barrans Y (1985) Structure cristalline du complexe de magnésium de la calcimycine (A. 23187). Can J Chem **63**: 3482
680. Kolber MA, Haynes DH (1981) Fluorescence study of the divalent cation-transport mechanisn of ionophore A23187 in phospholipid membranes. Biophys J **36**: 369
681. Yaginuma S, Awata M, Muto N, Kinoshita K, Mizuno K (1987) A novel polyether antibiotic, AC7230 (3-hydroxycezomycin or its stereoisomer). J Antibiot **40**: 239
682. David L, Kergomard A (1982) Production by controlled biosynthesis of a novel ionophore antibiotic, cezomycin (Demethylamino A23187). J Antibiot **35**: 1409
683. Westley JW, Liu C-M, Blount JF, Sello LH, Troupe N, Miller PA (1983) Isolation and characterization of a novel polyether antibiotic of the pyrolether class. Antibiotic X-14885A. J Antibiotics **36**: 1275
684. Liu C-M, Chin M, Prosser B La T, Palleroni NJ, Westley JW, Miller PA (1983) X-14885A. a novel divalent cation ionophore produced by a *Streptomyces* culture: Discovery, fermentation, biological as well as ionophore properties and taxonomy of the producing culture. J Antibiot **36**: 1118
685. Albrecht-Gary AM, Blanc-Parasote S, Boyd DW, Dauphin G, Jeminet G, Juillard J, Prudhomme M, Tissier C (1989) X-14547 A: An ionophore closely related to calcimycin (A-23187). NMR, thermodynamic, and kinetic studies of cation selectivity. J Am Chem Soc **111**: 8598
686. Celmer WD, Cullen WP, Maeda H, Tone J (1986) Polyether antibiotic from *Streptomyces*. U S Pat 4,547, 523. Chem Abstr (1986) **104**: 49 844
687. Diez-Martin D, Kotecha NR, Ley SV, Mantegani S, Menéndez JC, Organ HM, White AD, Banks BJ (1992) Total synthesis of the ionophore antibiotic CP-61,405 (routiennocin). Tetrahedron **48**: 7899
688. Edwards MP, Ley SV, Lister SG, Palmer BD, Williams DJ (1984) Total synthesis of ionophore antibiotic X-145474 (Indamomycin). J Org Chem **49**: 3503
689. Westley JW, Evans Jr. RH, Liu C-M, Hermann T, Blount JF (1978) Structure of antibiotic X-14547 A, a carboxylic acid ionophore produced by *Streptomyces antibioticus*, NRRL 8167. J Am Chem Soc **100**: 6784
690. Liu C-M, Hermann TE, Liu M, Bull DN, Palleroni NJ, Prosser B La T, Westley JW, Miller PA (1979) X-14547A, a new ionophorous antibiotic produced by *Streptomyces antibioticus* NRRL 8167. J Antibiot **32**: 95
691. Westley JW, Evans Jr. RH, Sello LH, Troupe N, Liu C-M, Blount JF (1979) Isolation and characterization of antibiotic X-14547A. A novel monocarboxylic acid ionophore produced by *Streptomyces antibioticus* NRRL 8167. J Antibiot **32**: 100
692. (a) Nicolaou KC, Papahatjis DP, Claremon DA, Dolle III, RE (1981) Total synthesis of ionophore antibiotic X-14547A. 1. Enantioselective synthesis of the tetrahydropyran and tetrahydroindan building blocks. J Am Chem Soc **103**: 6967; (b) Nicolaou KC, Claremon DA, Papahatjis DP, Magolda RL (1981) Total synthesis of ionophore antibiotic X-14547 A. Part 2. Coupling of the tetrahydropyran and tetrahydroindan systems and construction of the butadienyl and ketopyrrole moieties. *Ibid* **103**: 6969
693. Roush WR, Peseckis SM, Walts AE (1984) Synthesis of antibiotic X-14547 A. J Org Chem **49**: 3429
694. Nicolaou KC, Papahatjis DP, Claremon DA, Magolda RL, Dolle RE (1985) Total synthesis of ionophore antibiotic X-14547A. J Org Chem **50**: 1440

695. Boeckman Jr. RK, Enholm EJ, Demko DM, Charette AB (1986) An efficient enantioselective total synthesis of (−)-X-14547 A (indanomycin). J Org Chem **51**: 4743
696. Burke SD, Piscopio AD, Kort ME, Matulenko ME, Parker MA, Armistead DM, Shankaran K (1994) Total synthesis of ionophore antibiotic X-14547A (Indanomycin). J Org Chem **59**: 332
697. Beloeil JC, Desluc MA, Lallemand JY, Dauphin G, Jeminet G (1984) Application of the homonuclear and heteronuclear two-dimensional chemical shift correlation NMR spectroscopy to the complete assignment of ^{1}H- and ^{13}C-NMR spectra of ionophore antibiotic X.14547 A. J Org Chem **49**: 1797
698. Edwards MP, Ley SV (1984) Novel rearrangements of the ionophore antibiotic X-14547 A (indanomycin) and related derivatives induced by lithium tetrafluoroborate. J Chem Soc Perkin Trans I, 1761
699. Grandjean J, Laszlo P (1984) Synergistic transport of Pr^{3+} across lipid bilayers in the presence of two chemically distinct ionophores. J Am Chem Soc **106**: 1472
700. Westley JW (1977) Polyether antibiotics: Versatile carboxylic acid ionophores produced by *Streptomyces*. Adv Appl Microbiol **22**: 177
701. Klika KD, Haansuu JP, Ovcharenko VV, Haahtela KK, Vuorela PM, Pihlaja K (2001) Frankiamide, a highly unusual macrocycle containing the imide and orthoamide functionalities from the symbiotic Actinomycete *Frankia*. J Org Chem **88**: 4065
702. Pouteau-Thouvenot M, Choussy M, Gaudemer A, Barbier M (1970) Sur la structure chimique de l'anhydro-pro-ferrorosamine B. Bull Soc Chim Biol **52**: 51 [Chem Abstr (1970) **73**: 55934]
703. (a) Helbling AM, Viscontini M (1976) Naturstoffe aus Mikroorganismen. 8. Mitt. Optimale Gewinnung von Proferrorosamin A aus *Pseudomonas roseus fluorescens* J. C. Marchal 1973. Helv Chim Acta **59**: 2278; (b) Naturstoffe aus Mikroorganismen. 9. Mitt. Über die Biogenese von Proferrorosamin A. *Ibid* **59**: 2284
704. (a) Anthony C, Zatman LJ (1964) The microbial oxidation of methanol. 2. The methanol-oxidizing enzyme of *Pseudomonas* sp M27. Biochem J **92**: 614; (b) Anthony C, Zatman LJ (1967) The prosthetic group of the alcohol dehydrogenase of *Pseudomonas* sp. M27: A new oxidoreductase prosthetic group. *Ibid* **104**: 960
705. Salisbury SA, Forrest HS, Cruse WBT, Kennard O (1979) A novel coenzyme from bacterial alcohol dehydrogenases. Nature **280**: 843
706. Duine JA, Frank Jr. J, Van Zeeland JK (1979) Glucose dehydrogenase from *Acinetobacter calcoaceticus*. FEBS Lett **108**: 443
707. Corey EJ, Tramontano AJ (1981) Total synthesis of the quinoid alcohol dehydrogenase coenzyme (1) of methylotropic bacteria. J Am Chem Soc **103**: 5599
708. Gainor JA, Weinreb SM (1982) Synthesis of the bacterial coenzyme methoxatin. J Org Chem **47**: 2833
709. (a) Hendrickson JB, de Vries JG (1982) A convergent total synthesis of methoxatin. J Org Chem **47**: 1148; (b) Hendrickson JB, de Vries JG (1985) *ibid* **50**: 1688
710. MacKenzie AR, Moody CJ, Rees CW (1986) Synthesis of the bacterial conzyme methoxatin. Tetrahedron **42**: 3259
711. Martin P, Steiner E, Auer K, Winkler T (1993) Zur Herstellung von PQQ in kg-Mengen. Helv Chim Acta **76**: 1667
712. Anthony C (1986) Bacterial oxidation of methane and methanol. Adv Microbiol Physiol **27**: 113 and references given therein
713. Duine JA, Frank Jzn J, Jongejan JA (1987) Enzymology of quinoproteins. Adv Enzymol **59**: 169

714. Anthony C (1993) Methanol dehydrogenase Gram-negative bacteria. In: Davidson VL (ed) Principles and Applications of Quinoproteins. Dekker, New York, p 17
715. Anthony C, Ghosh M, Blake CCF (1994) The structure and function of methanol dehydrogenase and related quinoproteins containing pyrrolo-quinoline quinone. Biochem J **304**: 665
716. Duine JA, Jongejan JA (1989) Pyrroloquinoline quinone: A novel cofactor. Vitam Horm **45**: 223
717. Duine JA (1991) Quinoproteins: enzymes containing the quinoid cofactor pyrroloquinoline quinone, topaquinone or tryptophan-tryptophan quinone. Eur J Biochem **200**: 271 and references given therein
718. Matsushita K, Adachi O (1993) Bacterial quinoproteins glucose dehydrogenase and alcohol dehydrogenase. In: Davidson VL (ed) Principles and Applications of Quinoproteins. Dekker, New York, p 47
719. Anthony C (1996) Quinoprotein-catalyted reactions. Biochem J **320**: 697
720. Matsushita K, Toyama H, Adachi O (1994) Respiratory chains and bioenergetics of acetic acid bacteria. Adv Microbial Physiol **36**: 247
721. Duine JA, Frank Jr J (1980) (a) Studies on methanol dehydrogenase from *Hyphomicrobium* X. Biochem J **187**: 213–219; (b) The prosthetic group of methanol dehydrogenase. *Ibid* **187**: 221
722. (a) Duine JA, Frank Jzn J, Verwiel PEJ (1980) Structure and activity of the prosthetic group of methanol dehydrogenase. Eur J Biochem **108**: 187; (b) Duine JA, Frank Jzn J, Verwiel PEJ (1981) Characterization of the second prosthetic group in methanol dehydrogenase from *Hyphomicrobium* X. *Ibid* **118**: 395
723. Dekker RH, Duine JA, Frank J, Verwiel PEJ, Westerling J (1982) Covalent addition of H_2O, enzyme substrates and activators to pyrrolo-quinoline quinone, the coenzyme of quinoproteins. Eur J Biochem **125**: 69
724. Frank J, Dijkstra M, Balny C, Verwiel PEJ, Duine JA (1989) Methanol dehydrogenase: Mechanism of action. In: Jongejan JA, Duine JA (eds) PQQ and Quinoproteins. Kluwer, Dordrecht, p 13
725. Ohshiro Y, Itoh S (1993) The chemistry of PQQ and related compounds. In: Davidson VL (ed) Principles and Applications of Quinoproteins. Dekker, New York, p 309
726. Duine JA, Frank Jzn J, De Ruiter LGJ (1979) Isolation of a methanol dehydrogenase with a functional coupling to cytochrome c. J Gen Microbiol **115**: 523
727. Dijkstra M, Frank Jzn J, Duine JA (1989) Studies on electron transfer from methanol dehydrogenase to cytochrome c_L, both purified from *Hyphomicrobium* X. Biochem J **257**: 87
728. Frank Jzn J, Dijkstra M, Duine JA, Balny C (1988) Kinetic and spectral studies on the redox forms of methanol dehydrogenase from *Hyphomicrobium* X. Eur J Chem **174**: 331
729. Frank Jr J, van Krimpen SH, Verwiel PEJ, Jongejan JA, Mulder AC, Duine JA (1989) On the mechanism of inhibition of methanol dehydrogenase by cyclopropane-derived inhibitors. Eur J Biochem **184**: 187
730. Goodwin MG, Anthony C (1996) Characterization of a novel methanol dehydrogenase containing a Ba^{2+} ion at the active site. Biochem J **318**: 673
731. Goodwin MG, Avezoux A, Dales SL, Anthony C (1996) Reconstitution of the quinoprotein methanol dehydrogenase from inactive Ca^{2+}-free enzyme with Ca^{2+}, Sr^{2+} or Ba^{2+}. Biochem J **319**: 839
732. Houk DR, Hanners JL, Unkefer CJ (1988) Biosynthesis of pyrroloquinoline quinone. 1. Identification of biosynthetic precursors using ^{13}C-labelling and NMR spectroscopy. J Am Chem Soc **110**: 6920

733. Houk DR, Hanners JL, Unkefer CJ (1991) Biosynthesis of pyrroloquinoline quinone. 2. Biosynthetic assembly from glutamate and tyrosine J Am Chem Soc **113**: 3162
734. Unkefer CJ, Houk DR, Britt BM, Sosnick, TR, Hanners JL (1995) Biogenesis of pyrroloquinoline quinone from ^{13}C-labeled tyrosine. Meth Enzymol **258**: 227
735. Velterop JS, Sellink E, Meulenberg JJM, David S, Bulder I, Postma PW (1995) Synthesis of pyrroloquinoline quinone *in vivo* and *in vitro* and detection of an intermediate in the biosynthetic pathway. J Bacteriol **177**: 5088
736. Ikegami S, Isomura H, Tsuchimori N, Osano YT, Hayase T, Yugami T, Ohkishi H, Matsuzaki T (1990) Structure of pyrrolosine: A new inhibitor of RNA synthesis, from the actinomycete *Streptomyces albus*. J Am Chem Soc **112**: 9668
737. Box SJ, Corbett DF (1981) Structure elucidation and synthesis of (2*S*)-4-oxo-1-azabicyclo[3.3.0]octa-5,7-diene-2-carboxylic acid, a new metabolite isolated from *Streptomyces olivaceus*. Tetrahedron Lett **22**: 3293
738. (a) Kojiri K, Nakajima S, Suzuki H, Okura A, Suda H (1993) A new antitumor substance, BE-18591, produced by a streptomycete. I. Fermentaion, isolation, physico-chemical and biological properties. J Antibiotcs **46**: 1799; (b) Nakajima S, Kojiri K, Suda H (1993) A new antitumor substance, BE-18591, produced by a streptomycete. II. Structure determination. J Antibiotcs **46**: 1894
739. Momen AZMR, Mizuoka T, Hoshino T (1998) Studies on the biosynthesis of violacein. Part 9. Green pigments possessing tetraindole and dipyrromethene moieties, chomoviridans and desoxy-chromoviridans, produced by a cell-free extract of *Chromobacterium violaceum* and their biosynthetic origins. J Chem Soc Perkin Trans 1, 3087
740. Variyar PS, Chander R, Venkatachalam SR, Bongirwar DR (2002) A new red pigment from an alkalophilic Micrococcus species. Indian J Chem Sect B: Org Chem Incl Med Chem **41**: 232
741. Gaughran ERL (1969) From superstition to science: The history of a bacterium. Trans N Y Acad Sci **II31**: 3
742. Kraft E (1902) PhD. doctoral thesis University of Würzburg (Germany)
743. Williams RP, Hearn WR (1967) Prodigiosin. In: Gottlieb D, Shaw PD (eds) Antibiotics, vol 2. Springer, Berlin, p 410. Addenda: *ibid*, p 449
744. Feofilova EP (1968) Formation of prodigiosine and prodigiosine-like pigments by microorganisms. Usp Microbiol **5**: 147
745. Wrede F, Hettche O (1929) Über das Prodigiosin, den roten Farbstoff des *Bacillus prodigiosus*. (I. Mitteil.) Chem Ber **62**: 2678
746. Lewis SM, Corpe WA (1964) Prodigiosin-producing bacteria from marine sources. Appl Microbiol **12**: 13
747. Gandhi NM, Nazareth PV, Divekar PV, Kohl H, de Souza NJ (1973) Magnesidin, a novel magnesium-containing antibiotic. J Antibiot **26**: 797
748. D'Aoust JY, Gerber NN (1974) Isolation and purification of prodigiosin from *Vibrio psychoerythreus*. J Bacteriol **118**: 756
749. Gerber NN (1983) Cycloprodigiosin from *Beneckea gazogenes*. Tetrahedron Lett **24**: 2797
750. Sveshnikova MA, Timyk OE, Borisova VN, Fedorova GB (1983) *Streptomyces variegatus* sp nov, a new actinomycetous species producing an antibiotic from the α-hydroxyketopentaene group. Antibiotiki (Moscow) **28**: 723 [Chem Abstr (1983) **99**: 209 437]
751. Wrede F, Rothhaas A (1933) Über das Prodigiosin, den roten Farbstoff des *Bacillus prodigiosus*. IV. Hoppe-Seyler's Z Physiol Chem **219**: 267

752. Wrede F, Rothhaas A (1934) Über das Prodigiosin, den roten Farbstoff des *Bacillus prodigiosus*. VI. Hoppe-Seyler's Z Physiol Chem **226**: 95
753. Treibs A, Zimmer-Galler R (1960) Zur Konstitution des Prodigiosins Hoppe-Seyler's Z Physiol Chem **318**: 12
754. Wasserman HH, McKeon JE, Smith L, Forgione P (1960) Prodigiosin. Structure and partial synthesis. J Am Chem Soc **82**: 506
755. Wasserman HH, McKeon JE, Smith L, Forgione P (1966) Studies on prodigiosin and the bipyrrole precursor. Tetrahedron Suppl **8** Part II: 647 and references given therein
756. Rapoport H, Holden KG (1962) The synthesis of prodigiosin. J Am Chem Soc **84**: 635
757. Hearn WR, Elson MK, Williams RH, Medina-Castro J (1970) Prodigiosene [5-(2-pyrryl)-2,2'-dipyrrylmethene] and some substituted prodigiosenes. J Org Chem **35**: 142
758. Boger DL, Patel M (1988) Total synthesis of prodigiosin, prodigiosene, and desmethoxyprodigiosin: Diels-Alder reactions of heterocyclicazadienes and development of an effective palladium(II)-promoted 2,2'-bipyrrole coupling procedure. J Org Chem **53**: 1405
759. Wasserman HH, Lombardo LJ (1989) The chemistry of vicinal tricarbonyls: A total synthesis of prodigiosin. Tetrahedron Lett **30**: 1725
760. Castro AJ, Corwin AH, Waxham FJ, Beilby AL (1959) Products from *Serratia marcescens*. J Org Chem **24**: 455
761. Jackson AH, Kenner GW, Budzikiewicz H, Djerassi C, Wilson JM (1967) Pyrroles and related compounds – X. Mass spectra of linear di-, tri- and tetrapyrrolic compounds. Tetrahedron **23**: 603
762. Gerber NN (1969) Prodigiosin-like pigments from *Actinomadura* (*Nocardia*) *pelletieri* and *Actinomadura madurae*. Appl Microbiol **18**: 1
763. Morgan EN, Tanner EM (1955) Prodigiosin. J Chem Soc 3305
764. Hearn WR, Medina-Castro J, Elson MK (1968) Colour change of prodigiosin. Nature **220**: 170
765. Dimitrov DP, Boyanoff TA, Todorov TA (1970) A study of the association of prodigiosin, isolated from *Serratia marcescens*. Z Naturforsch **25b**: 46
766. Cruz-Camarillo R, Sànchez-Zùñiga AA (1968) Complex protein-prodigiosin in *Serratia marcescens*. Nature **218**: 567
767. Tsang JC, Kallvy DM (1971) Association of prodigiosin with outer cell wall components. Trans Ill State Acad Sci **64**: 22
768. Yoshida S (1962) A study of a water-soluble complex of prodigiosin produced by a strain of *Serratia marcescens*. Can J Biochem Physiol **40**: 1019
769. Roth MM (1967) The photosensitizing ability of prodigiosin. Photochem Photobiol **6**: 923
770. Boryu SI (1957) Mechanism of antibiotic action of *Bacterium prodigiosum* (*Serratia marcescens*). Mikrobiologiya **26**: 464 [Chem Abstr (1958) **52**: 7432]
771. Castro AJ, Gale GR, Means GE, Tertzakian G (1967) Antimicrobial properties of pyrrole derivatives. J Med Chem **10**: 29
772. Barbagallo C, Maugeri TL, Pavone P, Terrasi TC (1979) Attività citologica della prodigiosina sul test *Allium cepa* L. Riv Biol Norm Patol **5**: 25
773. Castro AJ (1967) Antimalarial activity of prodigiosin. Nature **213**: 903
774. Gerber NN (1975) A new prodiginine (prodigiosin-like) pigment from *Streptomyces*, antimalarial activity of several prodiginines. J Antibiot **28**: 194
775. Wasserman HH, Friedland DJ, Morrison DA (1968) A novel dipyrrolyldipyrromethene prodigiosin analog from *Serratia marcescens*. Tetrahedron Lett 641

776. Hubbard R, Rimington C (1950) The biosynthesis of prodigiosin, the tripyrrylmethene pigment from *Bacillus prodigiosus* (*Serratia marcescens*). Biochem J **46**: 220
777. Santer UV, Vogel HJ (1956) Prodigiosin synthesis in *Serratia marcescens*: Isolation of a pyrrole-containing precursor. Biochem Biophys Acta **1–9**: 578
778. Wasserman HH, McKeon JE, Santer UV(1960) Studies related to the biosynthesis of prodigiosin in *Serratia marcescens*. Biochem Biophys Res Commun **3**: 146
779. Tanaka WK, de Medina LB, Hearn WR (1972) Labeling patterns in prodigiosin biosynthesis. Biochem Biophys Res Commun **46**: 731
780. Marks GS, Bogorad L (1960) Studies on the biosynthesis of prodigiosin in *Serratia marcescens*. Proc Natl Acad Sci U S **46**: 25
781. Shrimpton DM, Marks GS, Bogorad L (1963) Studies on the biosynthesis of prodigiosin in *Serratia marcescens*. Biochim Biophys Acta **71**: 408
782. Williams RP, Gott CL, Qadri SMH (1971) Induction of pigmentation in nonproliferating cells of *Serratia marcescens* by addition of single amino acids. J Bacteriol **106**: 444
783. Goldschmid MC, Williams RP (1968) Thiamine-induced formation of the monopyrrole moiety of prodigiosin. J Bacteriol **96**: 609
784. Hussain Qadri SM, Williams RP (1971) Incorporation of methionine into prodigiosin. Biochim Biophys Acta **230**: 181
785. Cushley RJ, Anderson DR, Lipsky SR, Sykes RJ, Wasserman HH (1971) Carbon-13 Fourier transform nuclear magnetic resonance spectroscopy. II. The pattern of biosynthetic incorporation of [1-^{13}C]- and [2-^{13}C]-acetate into prodigiosin. J Am Chem Soc **93**: 6284
786. Cushley RJ, Sykes RJ, Shaw C-K, Wasserman HH (1973), Carbon-13 Fourier transform nuclear magnetic resonance. IX. Complete assignments of some prodigiosins. Bioincorporation of label. Can J Chem **53**: 148
787. Wasserman HH, Sykes RJ, Peverada P, Shaw CK, Cushley RJ, Lipsky SR (1973) Biosynthesis of prodigiosin. Incorporation patterns of ^{13}C-labeled alanine, proline, glycine, and serine elucidated by Fourier transform nuclear magnetic resonance. J Am Chem Soc **95**: 6874
788. Lim DV, Hussain Qadri SM, Nichols C, Williams RP (1977) Biosynthesis of prodigiosin by non-proliferating wild-type *Serratia marcescens* and mutants deficient in catabolism of alanine, histidinem and proline. J Bacteriol **129**: 124
789. Hearn WR, Worthington RE, Burgus RC, Williams RP (1964) Norprodigiosin: Occurrence in a mutant of *Serratia marcescens*. Biochem Biophys Res Commun **17**: 517
790. Deol BS, Alden JR, Still JL, Robertson AV, Winkler J (1974) Isolation and structure confirmation of norprodigiosin from *Serratia marcescens* mutant. Aust J Chem **27**: 2657
791. Gerber NN, Gauthier MJ (1979) New prodigiosin-like pigment from *Alteromonas rubra*. Appl Environ Microbiol **37**: 1176
792. Laatsch H, Thomson RH (1983) A revised structure for cycloprodigiosin. Tetrahedron Lett **24**: 2701
793. Wasserman HH, Fukuyama JM (1984) The synthesis of (±)-cycloprodigiosin. Tetrahedron Lett **25**: 1387
794. Brockmann H, Pini H (1947) Actinorhodin, ein roter Farbstoff aus Actinomyceten. Naturwissenschaften **34**: 190
795. Dietzel E (1949) Über prodigiosin-ähnliche Farbstoffe bei Actonomyceten. Hoppe-Seyler's Z Physiol Chem **284**: 262

796. Gerber NN (1975) Prodigiosin-like pigments. CRC Crit Rev Microbiol **3**: 469 and refeences given therein
797. Gerber NN, Lechevalier MP (1976) Prodiginine (prodigiosin-like) pigments from *Streptomyces* and other aerobic Actinomycetes. Can J Microbiol **22**: 658 and references given therein
798. Harashima K, Tsuchida N, Nagatsu J (1966) Prodigiosin-25 C. A new prodigiosin-like pigment. Agr Biol Chem (Tokyo) **30**: 309
799. Harashima K, Tsuchida N, Tanaka T, Nagatsu J (1967) Prodigiosin 25C. Isolation and the chemical structure. Agr Biol Chem (Tokyo) **31**: 481
800. Wasserman HH, Rodgers, Jr GC, Keith DD (1966) The structure and synthesis of undecylprodigiosin. A prodigiosin analogue from *Streptomyces*. J Chem Soc Chem Commun 825
801. Tsao S-W, Rudd BAM, He X-G, Chang C, Floss HG (1985) Identification of a red pigment from *Streptomyces coelicolor* A3(2) as a mixture of prodigiosin derivatives. J Antibiot **38**: 128
802. Gerber NN (1971) Prodigiosin-like pigments from *Actinomadura (Nocardia) pelletieri*. J Antibiot (Tokyo) **24A**: 636
803. Gerber NN, Stahly DP (1975) Prodiginine (prodigiosin-like) pigments from *Streptoverticillium rubrireticuli*, an organism that causes pink staining of polyvinyl chloride. Appl Microbiol **30**: 807
804. Wasserman HH, Rodgers GC, Keith DD (1976) Undecylprodigiosin. Tetrahedron **32**: 1851
805. D'Alessio R, Rossi A (1996) Short synthesis of undecylprodigiosine. A new route to 2,2'-bipyrroylpyrromethene systems. Synlett 513
806. Thirumalachar MJ, Bringi NV, Deshmukh PV, Rahalkar PW, Indira R, Gopalkrishan KS (1964) Streptorubrin A and B. New antibiotics with cytostatic properties. Hind Antibiot Bull **7**: 18
807. Wasserman HH, Rodgers GC, Keith DD (1969) Metacycloprodigiosin, a tripyrrole pigment from *Streptomyces longisporus ruber*. J Am Chem Soc **91**: 1263
808. Wasserman HH, Keith DD, Rodgers GC (1976) The structure of metacycloprodigiosin. Tetrahedron **32**: 1855
809. (a) Wasserman HH, Keith DD, Nadelson J (1969) The synthesis of metacycloprodigiosin. J Am Chem Soc **91**: 1264; (b) Wasserman HH, Keith DD, Nadelson J (1976) The synthesis of metacycloprodigiosin. Tetrahedron **32**: 1867
810. Fürstner A, Szillat H, Gabor B, Mynott R (1998) Platinum- and acid-catalyzed enyne methatesis reactions: Mechanistic studies and applications to the synthesis of streptorubin B and metacycloprodigiosin. J Am Chem Soc **120**: 8305
811. Korenyako AI, Gavrilova OA (1962) Preparation of vitamycin.Vestn Akad Nauk SSSR **32**: 80 [Chem Abstr (1962) **57**: 17202]
812. Khokhlova YuM, Sergeeva LN, Vul'fson NS, Zaretskii VI, Zaikin VG, Sheichenko VI, Khokhlov AS (1968) Difference of vitamycin A from other natural analogs. Khim Prirod Soedin **4**: 307
813. Laatsch H, Kellner M, Weyland H (1991) Butyl-*meta*-cycloheptylprodigiosine – A revison of the structure of the former *ortho*-isomer. J Antibiot **44**: 187
814. Gerber NN (1970) A novel, cyclic, tetrapyrrole pigment from *Actinomadura (Nocardia) madurae*. Tetrahedron Lett 809
815. Gerber NN (1973) Minor prodiginine pigments from *Actinomadura madurae* and *Actinomadura pelletieri*. J Heterocycl Chem **10**: 925
816. Gerber NN, McInnes AG, Smith DG, Walter JA, Wright JLC, Vining LC (1978) Biosynthesis of prodiginines. ^{13}C resonance assigments and enrichment patterns

in nonyl-, cyclononyl-, methylcyclodecyl-, and butylcycloheptylprodiginine produced by actinomycete cultures supplemented with ^{13}C-labeled acetate and 15N-labeled nitrate. Can J Chem **56**: 1155
817. Wasserman HH, Shaw CK, Sykes RJ (1974) Biosynthesis of prodigiosin. III. Carbon-13 Fourier transform NMR. Biosynthesis of metacycloprodigiosin and undecylprodigiosin. Tetrahedron Lett 2787
818. Magae J, Miller MW, Nagai K, Shearer GM (1996) Effect of metacycloprodigiosin, an inhibitor of killer T cells, on murine skin and heart transplantations. J Antibiot **49**: 86 and references given therein
819. Grote R, Zeeck A, Stüpfel J, Zähner H (1990) Metabolic products of microorganisms, 256. Pyrrolams, new pyrrolizidones produced by *Streptomyces olivaceus*. Liebigs Ann Chem 525
820. Aoyagi Y, Manabe T, Ohta A, Kurihara T, Pang G-L, Yuhara T (1996) First total synthesis of pyrrolam A. Tetrahedron **52**: 869
821. Kuroda Y, Okuhara M, Goto T, Yamashita M, Iguchi E, Kohsaka M, Aoki H, Imanaka H (1980) FR-900148, a new antibiotic. I. Taxonomy, fermentation, isolation and characterization. J Antibiot **33**: 259
822. Kuroda Y, Okuhara M, Goto T, Okamoto M, Yamashita M, Kohsaka M, Aoki H, Imanaka H (1980) FR-900148, a new antibiotic. II. Structure determination of FR-900148. J Antibiot **33**: 267
823. Chaiet L, Monaghan RL, Zimmerman SB, Sheldon B, Fernandez MIM (Merck and Co, Inc) (1985) (*R*)-(*Z*)-4-amino-3-chloro-2-pentenedioic acid, antibacterial agent. Eur Pat Appl EP 137,498. Chem Abstr (1985) **103**: 21204
824. Nagle DG, Paul VJ, Roberts MA (1996) Ypaoamide, a new broadly acting feeding deterrent from the marine cyanobacterium *Lyngbya majuscula*. Tetrahedron Lett **37**: 6263
825. Moore RE, Entzeroth M (1988) Majusculamide D and deoxymajusculamide D, two cytotoxins from *Lyngbya majuscula*. Phytochemistry **27**: 3101
826. (a) Koehn FE, Longley RE, Reed JK (1992) Microcolins A and B, new immunosuppressive peptides from the blue-green alga *Lyngbya majuscula*. J Nat Prod **55**: 613; (b) Koehn FE, MacConnel OJ, Longley RE, Sennett SH, Reed JK (1994) Analogs of the marine immunosuppressant microcolin A: Preparation and biological activity. J Med Chem **37**: 3181
827. Decicco CP, Grover P (1996) Total asymmetric synthesis of the potent immunosuppressive marine natural product microcolin A. J Org Chem **61**: 3534
828. Cardellina II JH, Marner F-J, Moore RE (1979) Malyngamide A, a novel chlorinated metabolite of the marine cyanophyte *Lyngbya majuscula*. J Am Chem Soc **101**: 240
829. Kan Y, Sakamoto B, Fujita T, Nagai H (2000) New malyngamides from the Hawaiian Cyanobacterium *Lyngbya majuscula*. J Nat Prod **63**: 1599
830. Milligan KE, Marquez B, Williamson RT, Davies-Coleman M, Gerwick WH (2000) Two new malyngamides from a Madagascan *Lyngbya majuscula*. J Nat Prod **63**: 965
831. Simmons CJ, Marner F-J, Cardellina II JH, Moore RE, Seff K (1979) Pukeleimide C, a novel pyrrolic compound from the marine cyanophyte *Lyngbya majuscula*. Tetrahedron Lett **22**: 2003
832. Cardellina II JH, Moore RE (1979) The structures of pukeleimides A, B, D, E, F, and G. Tetrahedron Lett 2007
833. James GD, Mills SD, Pattenden G (1993) Total synthesis of pukeleimide A, a 5-ylidenepyrrol-2(5*H*)-one from blue geen algae. J Chem Soc Perkin Trans 1 2581
834. Yamaguchi H, Nakayama Y, Takeda K, Tawara K, Maeda K, Takeuchi T, Umezawa H (1957) A new antibiotic, althiomycin. J Antibiot **A10**: 195

835. Sensi P, Ballota R, Gallo GG (1959) Matamycin, a new antibiotic. II. Isolation and characterization. Antibiot Chemother **9**: 76
836. Kunze B, Reichenbach H, Augustiniak H, Höfle G (1983) Isolation and identification of althiomycin from *Cystobacter fuscus* (Myxobacteriales). J Antibiot **35**: 635
837. Cram DJ, Theander O, Jager H, Stanfield MK (1963) Mold metabolites. IX. Contribution to the elucidation of the structure of althiomycin. J Am Chem Soc **85**: 1430
838. Bycroft BW, Pinchin R (1975) Structure of althiomycin, a highly modified peptide antibiotic. J Chem Soc Chem Commun 121
839. (a) Nakamura H, Iitaka Y, Sakakibara H, Umezawa H (1974) The molecular and crystal structure determination of bisanhydroalthiomycin by the X-ray diffraction method. J Antibiot **27**: 894; (b) Sakakibara H, Naganawa H, Ohno M, Maeda K, Umezawa H (1974) The structure of althiomycin. J Antibiot **27**: 897
840. Kirst H, Szymanski EF, Dorman DE, Occolowitz JL, Jones ND, Chaney MO, Hamill RL, Hoehn MM (1975) Structure of althiomycin. J Antibiot **28**: 286
841. (a) Inami K, Shiba T (1984) Total synthesis of antibiotic althiomycin. Tetrahedron Lett: **25**: 2009; (b) Inami K, Shiba T (1985) Total synthesis of antibiotic althiomycin. Bull Chem Soc Jpn **58**: 352
842. Toogood PL, Hollenbeck JJ, Lam HM, Li L (1998) A formal synthesis of althiomycin. Bioorg Med Chem Lett **6**: 1543
843. (a) Fujimoto H, Kinoshita T, Suzuki H, Umezawa H (1970) Studies on the mode of action of althiomycin. J Antibiot **23**: 271; (b) Pestka S, Brot N (1971) Studies on the formation of transfer ribonucleic acid-ribosome complexes. XV. Effects of antibiotics on steps of bacterial protein synthesis: some new ribosomal inhibitors of translocation. J Biol Chem **246**: 7715; (c) Burns DJ, Cundliffe E (1973) Bacterial-protein synthesis. A novel system for studying antibiotic action *in vivo*. Eur J Biochem **37**: 570
844. Batelaan JG, Barnick JWFK, van der Baan JL, Bickelhaupt F (1972) Structure of the antibiotic K 16: I. The dipeptide side chain. Tetrahedron Lett 3103; II. Chromophore and total structure. *Ibid* 3107
845. Van der Baan JL, Barnick JWFK, Bickelhaupt F (1978) The total synthesis of the antibiotic malonomicin (K16). Tetrahedron **34**: 223
846. Shigemori H, Bae M-A, Yazawa K, Sasaki T, Kobayashi J (1992) Alteramide A. a new tetracyclic alkaloid from a bacterium *Alteromonas* sp. associated with the marine sponge *Halichondria okadai*. J Org Chem **57**: 4317
847. Bae M-A, Yamada K, Ijuin Y, Tsuji T, Yazawea K, Tomono Y, Uemura D (1996) Aburatubolactam A, a novel inhibitor of superoxide anion generation from a marine microorganism. Heterocycl Comm **2**: 315
848. (a) Aizawa S, Akutsu H, Satomi T, Nagatsu T, Taguchi R, Seino A (1979) Capsimyicin, a new antibiotic. I. Production, isolation and properties. J Antibiot **32**: 193; (b) Seto H, Yonehara H, Aizawa S, Akutsu H, Clardy J, Arnold E, Tanabe M, Urano S (1979) Structural studies of capsimycin and biosynthetic studies of ikarugamycin. Koen Yoshishu-Tennen Yuki Kagobutsu Toronkai, 22nd. p 394 [Chem Abstr (1980) **92**: 211459]
849. Jomon K, Kuroda Y, Ajisaka M, Sasaki H (1972) A new antibiotic, ikarugamycin. J Antibiot **25**: 271
850. (a) Ito S, Hirata Y (1972) Ikarugamycin: I. Chromophore and partial structure. Tetrahedron Lett 1181; II. Structure of ikarugamycin. *Ibid* 1185; III. Stereochemistry of ikarugamycin. *Ibid* 2557; (b) Ito S, Hirata Y (1977) Structure of ikarugamycin, an acyltetramic acid antibiotic possesing a unique *as*-hydrindacene skeleton. Bull Chem Soc Jpn **50**: 1813

851. Boeckman Jr RK, Napier JJ, Thomas EW, Sato RI (1983) Stereocontrol in the intramolecular Diels-Alder Reaction. 5. Preparation of a tetracyclic intermediate for ikarugamycin. J Org Chem **48**: 4152
852. (a) Kurth MJ, Burns DH, O'Brien MJ (1984) Ikaguramycin: Total snthesis of the decahydro-*as*-indacene portion. J Org Chem **49**: 731; (b) Whitesell JK, Minton MA (1987) A novel synthesis of ikaguramycin: The carbocyclic portion. J Am Chem Soc **109**: 6403
853. Paquette LA, Romine JL, Lin H-S, Wright J (1990) Total synthesis of (+)-ikarugamycin. 1. Stereocontrolled construction of the decahydro-*as*-indacene subunit. J Am Chem Soc **112**: 9284
854. Roush WR, Wada CK (1994) Application of η^4-diene iron tricarbonyl complexes in acyclic stereocontrol: Asymmetric synthesis of the *as*-indacene unit of ikaguramycin (a formal total synthesis). J Am Chem Soc **116**: 2151
855. Boeckman Jr RK, Weidner CH, Perni RB, Napier JJ (1989) An enantioselective and highly convergent synthesis of (+)-ikarugamycin. J Am Chem Soc **111**: 8036
856. Paquette LA, Macdonald D, Anderson L, Wright J (1989) A triply convergent enantioselective total synthesis of (+)-ikarugamycin. J Am Chem Soc **111**: 8037
857. Paquette LA, Macdonald D, Anderson L (1990) Total synthesis of (+)-ikarugamycin. 2. Elaboration of the macrocyclic lactam and tetramic acid substructures and complete assembly of tha antibiotic. J Am Chem Soc **112**: 9292
858. Schwarz O, Schmalz H-G (2000) Total synthesis of Ikaguramycin. Org Synth Highlights IV: 207
859. (a) Hayakawa Y, Kanamaru N, Morisaki N, Seto H, Furihata K (1991) Structure of lydicamycin, a new antibiotic of a novel skeletal type. Tetrahedron Lett **32**: 213; (b) Hayakawa Y, Kanamaru N, Shimazu H, Seto H (1991) Lydicamycin, a new antibiotic of a novel skeletal type. I. Taxonomy, fermentation, isolation and biological activity. J Antibiot **44**: 282
860. Hayakawa Y, Kanamaru N, Morisaki N, Furihata K, Seto H (1991) Lydicamycin, a new antibiotic of a novel skeletal type. II. Physico-chemical properties and structure elucidation. J Antibiot **44**: 288
861. Kunze B, Schabacher K, Zähner H, Zeeck A (1972) Stoffwechselprodukte von Mikroorganismen. 111. Mitteilung. Lipomycine. I. Isolierung, Charakterisierung und erste Untersuchungen zur Konstitution und Wirkungsweise. Arch Microbiol **86**: 147
862. Gyimesi J, Ott I, Horvàth I, Koczka I, Magyar K (1971) Antibiotics produced by *Streptomyces*. VIII. A new polyenic antibiotic, oleficin, exhibiting antibacterial activity. J Antibiot **24**: 277
863. Schabacher K, Zeeck A (1973) Lipomycins. II. Structure of α- and β-lipomycin. Tetrahedron Lett 2691
864. Zeeck A (1975) Lipomycine. III. Isolierung und Zuordnung der Methyl-2,6-didesoxy-D-*ribo*hexoside. Justus Liebigs Ann Chem 2079
865. Gyimesi J, Méhesfalvi-Vajna Z, Horvàth G (1978) Reinvestigation of structure of the polyenic antibiotic oleficin. J Antibiot **31**: 626
866. Horvàth G, Gyimesi J, Mehesfalvi-Vajna Z (1973) The structure of oleficin, a new antibiotic exhibiting antibacterial activity. Tetrahedron Lett 3643
867. Barashkova NP, Shenin YuD, Myasnikova LG (1976) *Actinomyces pneumonicus* var. nov. – a producer of new altamycin antibiotics. Antibiotiki (Moscow) **21**: 487 [Chem Abstr (1976) **85**: 106442]
868. Shenin YuD (1986) Physicochemical properties and structure of altamycin A, a nonmacrolide polyenic antibiotic. Antibiot Med Biotekhnol **31**: 835 [Chem Abstr **106** (1987): 66971]

869. Reusser F (1979) Tirandamycin. In: Hahn FE (ed) Antibiotics. Mechanism of action of antibacterial agents, vol 5/1. Springer, New York, p 361
870. Rinehart Jr KL, Beck JR, Epstein WW, Spicer LD (1963) Streptolydigin. I. Streptolic acid. J Am Chem Soc **85**: 4035
871. Eble TE, Large CM, DeVries WH, Crum GF, Shell JW (1956) Streptolydigin: A new antimicrobial antibiotic. II. Isolation and characterization. Antibiot Ann 1955–1956, 893
872. Rinehart Jr KL, Borders DB (1963) Streptolydigin. II. Ydiginic acid. J Am Chem Soc **85**: 4037
873. Rinehart Jr KL, Beck JR, Borders DB, Kinstle TH, Krauss D (1963) Streptolydigin. III. Chromophore and Structure. J Am Chem Soc **85**: 4038
874. Lee VJ, Branfman AR, Herrin TR, Rinehart Jr KL (1978) Synthesis of 3-dienoyl tetramic acids related to streptolydigin and tirandamycin. J Am Chem Soc **100**: 4225
875. Lee VJ, Rinehart Jr KL (1980) ^{13}C NMR spectra of streptolydigin, tirandamycin and related degradation products. J Antibiot **33**: 408
876. Karwowski JP, Jackson M, Theriault RJ, Barlow GJ, Coen L, Hensey DM, Humphrey PE (1992) Tirandalydigin, a novel tetramic acid of the tirandamycin-streptolydigin type. I. Taxonomy of the producing organism, fermentation and biological activity. J Antibiot **45**: 1125
877. Brill GM, McAlpine JB, Whittern D (1988) Tirandalydigin, a novel tetramic acid of the tirandamycin-streptolydigin type. II. Isolation and structure characterization. J Antibiot **41**: 36
878. MacKellar FA, Grostic MF, Olson EC, Wnuk RJ, Branfman AR, Rinehart Jr KL (1971) Tirandamycin. I. Structure assignment. J Am Chem Soc **93**: 4943
879. Meyer CE (1971) Tirandymycin, a new antibiotic isolation and characterization. J Antibiot **24**: 558
880. Hagenmaier H, Jaschke KH, Santo L, Scheer M, Zaehner H (1976) Stoffwechselprodukte von Mikroorganismen. 158. Mitteilung. Tyrandamycin B. Arch Microbiol **109**: 65
881. Duchamp DJ, Branfman AP, Button AC, Rinehart Jr KL (1973) X-ray structure of tirandamycic acid p-bromophenacyl ester. Complete stereochemical assignments of tirandamycin and streptolydigin. J Am Chem Soc **95**: 4077
882. Pearce CJ, Rinehart Jr KL (1983) The use of doubly-labeled ^{13}C-acetate in the study of streptolydigin biosynthesis. J Antibiot **36**: 1536
883. Brazhnikova MG, Konstantinova NV, Potatova NP, Tolstykh IV (1977) Physicochemical characteristics of nocamycin, a new antitumor antibiotic. Antibiotiki (Moscow) **22**: 486 [Chem Abstr (1977) **87**: 100613]
884. Horvàth G, Brazhnikova MG, Konstantinova NV, Tolstykh IV, Potatova NP (1979) The structure of nocamycin, a new antitumor antibiotic. J Antibiot **32**: 555
885. Brazhnikova MG, Konstantinova NV, Potatova NP, Tolstykh IV, Rubasheva LM, Rozynov BV, Horvàth G (1981) Structure of the antitumor antibiotics nocarmycin I and II. Bioorg Khim **7**: 298 [Chem Abstr (1981) **95**: 62092]
886. Nakagawa S, Naito T, Kawaguchi H (1979) Structures of Bu-2313 A and B, new antianaerobic antibiotics and syntheses of their analogs. Heterocycles **13**: 477
887. Tsukiura H, Tomita K, Hanada M, Kobaru S, Tsunakawa M, Fujisawa K, Kawaguchi H (1980) Bu-2313, a new antibiotic complex active against anaerobes. I. Production, isolation and properties of Bu-2313 A and B. J Antibiot **33**: 157
888. Toda S, Nakagawa S, Naito T, Kawaguchi H (1980) Bu-2313, a new antibiotic complex active against anaerobes. III. Semisynthesis of Bu-2313 A and B, and their analogs. J Antibiot **33**: 173

889. Tsunakawa M, Toda S, Okita T, Hanada M, Nakagawa S, Tsukiura H, Naito T, Kawaguchi H (1980) Bu-2313, a new antibiotic complex active against anaerobes. II. Structure determination of Bu-2313 A and B. J Antibiot **33**: 166
890. Meyers E, Cooper R, Dean L, Johnson JH, Slusarchyk DS, Trejo WH, Singh PD (1985) Catacandins, novel anticandidal antibiotics of bacterial origin. J Antibiot **38**: 1642

Author Index

Numbers printed in *italics* refer to Reference numbers

Abbanat, D. *602*
Abbaspour Tehrani, K. *44*(h)
Abd El-Rahman, H.A. *289*
Abdul, R. *308*
Abe, M. *451*
Abell, A.D. *198*
Ablaza, S.L. *299*(b), *300*
Abraham, A. *561*
Abraham, D.J. *311*
Abraham, W.-R. *321*
Achenbach, H. *355*
Adachi, O. *718*, *720*
Agarwal, V.K. *136*
Ahmed, F.R. *387*, *388*
Ahond, A. *105*, *110*(a), *111*, *118*, *139*(a)
Aimi, N. *309*
Airall, J.M. *533*
Aizawa, S. *848*(a), *848*(b)
Ajisaka, M. *543*, *544*, *849*
Akazawa, K. *398*
Aknin, M. *210*, *248*(b)
Akutsu, H. *848*(a), *848*(b)
Al Mourabit, A. *111*, *118*, *139*(a)
Albizati, K.F. *145*
Albrecht-Gary, A.M. *674*, *685*
Albuquerque, E.X. *39*
Alden, J.R. *790*
Alléaume, M. *679*
Allen, T.M. *77*
Allnoch, H. *577*
Alluri, M. *602*
Altmann, A. *542*
Amade, P. *89*
Amano, S. *465*
Amin, M.A. *472*
Andersen, R.J. *230*, *464*
Anderson, D.R. *785*
Anderson, G.T. *172*(a), *172*(b)
Anderson, H.J. *379*, *70*(a)
Anderson, L. *856*, *857*
Anderson, T. *2*

Andrade, P. *146*
Andres, N. *651*
Annoura, H. *154*
Anthony, C. *704*(a), *704*(b), *712*, *714*, *715*, *719*, *730*, *731*
Antipin, M.Yu. *168*
Aoki, H. *821*, *822*
Aoki, S. *222*
Aoyagi, Y. *820*
Aoyama, K. *410*
ApSimon, J.W. *499*
Apuzzo, G. *536*
Arai, I. *280*
Arai, N. *403*
Arcamone, F. *622*–*624*, *628*, *629*
Arihara, S. *85*, *88*
Arima, K. *508*–*512*, *519*, *530*, *543*
Armistead, D.M. *696*
Arndt, R.R. *382*
Arnold, E. *848*(b)
Artico, M. *536*
Asakawa, Y. *398*, *402*
Asaoka, T. *470*, *473*
Asberom, T. *667*, *669*
Ashcroft, J. *602*
Asres, K. *336*, *337*
Assmann, M. *107*(a), *122*
Attygalle, A.B. *52*, *53*
Audia, J.E. *442*
Auer, K. *711*
Augustiniak, H. *836*
Avezoux, A. *731*
Awata, M. *681*

Babin, D.R. *345*
Badar, Y. *385*, *386*(a)
Bae, M.-A. *846*, *847*
Bailey, K. *486*
Baker, B.F. *640*
Ballota, R. *835*
Balny, C. *724*, *728*

Banerji, A. *303*
Banks, B.J. *687*
Banwell, M.G. *140*, *204*(b), *240*
Barashkova, N.P. *867*
Barbagallo, C. *772*
Barbier, M. *702*
Barbieri, W. *622*
Barlow, G.J. *876*
Barnick, J.W.F.K. *844*, *845*
Barrans, Y. *679*
Barrios Sosa, A.C. *132*, *139*(b), *144*(b)
Barrow, R.A. *67*
Barton, D.H.R. *44*(d)
Bascombe, K.C. *187*
Batcho, A.D. *568*, *573*
Batelaan, J.G. *844*
Bates, R.B. *148*
Bath, U.G. *644*
Batlle, A.M. del C. *18*
Battaner, E. *418*
Battersby, A.R. *3*, *6*, *16*, *17*
Baumann, H. *45*, *46*, *49*, *50*
Bayley, D.M. *488*, *491*
Bayne, W. *23*(a), *23*(b)
Beccalli, E.M. *250*
Beck, J.R. *870*, *873*
Beck, R. *23*(c)
Becker, Y. *630*
Beehler, B.M. *34*
Begley, M.J. *219*
Behrens, C. *98*
Beilby, A.L. *760*
Belanger, A. *350*, *351*
Bell, C.A. *2*, *2*(a), *2*(b)
Bell, T.W. *64*(a), *64*(b)
Beloeil, J.C. *697*
Benayahu, Y. *248*(a), *93*
Bénazet, F. *561*
Benjamin, L.E. *108*
Benson, M. *516*
Beppu, K. *665*
Berger, C.R.A. *29*(a)
Berger, D. *320*
Berger, J. *568*, *573*, *586*, *587*
Beria, I. *632–634*
Berner, H. *37*
Berner-Fenz, L. *37*
Berney, D.J.F. *350*, *351*
Bernthsen, A. *2*
Berrée, F. *110*(c)

Berry, R.L. *648*
Beutler, J.A. *89*, *434*
Bezold, G. *19*(a)
Bhat, U.G. *631*
Bialer, M. *626*, *630*
Bickelhaupt, F. *844*, *845*
Bihovsky, R. *460*
Birch, A.J. *483*
Birchhall, G.R. *485*
Bissada, S.M. *187*
Blackman, A.J. *257*
Blake, C.C.F. *715*
Blanc-Parasote, S. *685*
Blank, D.H. *27*
Blankespoor, C.L. *52*
Blau, F. *68*, *283*
Blount, J.F. *571*, *683*, *689*, *691*
Blunt, J.W. *201*
Blyth, P.C. *76*
Blythe, T.A. *7*
Bobal, P. *14*(a)
Boeckman, Jr. R.K. *667*, *669*, *695*, *851*, *855*
Boger, D.L. *123*, *204*(c), *246*, *758*
Bogorad, L. *10*, *780*, *781*
Bohlmann, F. *317–325*
Bokel, M. *545*
Bolvig, S. *299*(a)
Bongirwar, D.R. *740*
Bonnefoy, A. *574*, *583*
Boppré, M. *60*, *62*(a), *62*(b), *64*(a)
Borders, D.B. *474*, *476*, *872*, *873*
Bordner, J. *43*
Borges del Castillo, J. *314*
Boriack, C.J. *60*, *62*(a)
Borisova, V.N. *750*
Borkenstein, A. *381*
Borremans, D. *44*(h)
Borschberg, H.-J. *350*, *351*
Boryu, S.I. *770*
Bourauel, T. *61*
Bourgeois, J. *333*
Boury-Esnault, N. *210*
Boutar, M. *422*
Bouthillier, L.P. *28*
Bowden, B.F. *196*, *232*
Box, S.J. *737*
Boyanoff, T.A. *765*
Boyce, C.W. *204*(c), *246*
Boyd, D.W. *674*, *685*

Boyd, M.R. *167, 264*
Braeckman, J.-C. *51, 120*
Brandt, W. *355*
Branfman, A.R. *874, 878, 881*
Bray, A.M. *140*
Brazhnikova, M.G. *883–885*
Breitmaier, E. *44*(c)
Brennan, L. *444*
Breslauer, K.J. *647*
Bretòn, J.L. *314*
Bricout, J. *270, 276*
Brill, G.M. *877*
Brimble, M.A. *396*
Brinen, L.S. *649*
Bringi, N.V. *806*
Bringmann, G. *142*
Britt, B.M. *734*
Brockman, P.E. *11*(b)
Brockmann, H. *794*
Brot, N. *843*(b)
Brousseau, R. *350, 351*
Brower, L.P. *54*
Brownlee, R.G. *40*(a), *40*(b)
Bruck, M. *148*
Bučkovà, A. *365, 366, 372–374*
Buchanan, M.S. *402*
Büchi, G. *164*
Buckingham, M.J. *387*
Budzikiewicz, H. *761*
Bulder, I. *735*
Bull, D.N. *690*
Burd, W. *542*
Burgus, R.C. *789*
Burke, S.D. *696*
Burkholder, P.R. *162, 163*(a), *495*
Burmeister, H.R. *440*
Burnham, B.S. *243*(a)
Burns, D.H. *852*(a)
Burns, D.J. *843*(c)
Buschi, C.A. *282*
Bushman, F.D. *236*
Butler, M.S. *202*
Button, A.C. *881*
Buyer, J.S. *150*
Bycroft, B.W. *133, 838*

Cafieri, F. *79, 99–101, 123, 137, 199*
Cahours, A. *284*
Cain, D.M. *668*
Caldarelli, M. *632–634*

Camou, F. *148*
Cane, D.E. *595, 598, 600*
Capolongo, L. *632, 634*
Capon, R.J. *67, 84, 202, 203, 233*
Carballo, J.L. *197*
Carboni, B. *110*(c)
Cardellina II, J.H. *828, 831, 832*
Carney, J.R. *214*
Carnuccio, R. *79, 101*
Carrasco, L. *419, 421*
Carroll, A.R. *58, 186, 232, 234, 242*
Carroll, B. *41*
Carroll, R. *41*
Carstensen, J.T. *587*
Carté, B. *255, 260*
Carter, G.T. *474, 476*
Cartier, M. *561*
Carver, R.A. *286*
Casida, J.E. *341, 342*
Cassady, J.M. *297, 298*
Casser, I. *427*
Castro, A.J. *760, 771, 773*
Castro, V. *318*
Celmer, W.D. *591, 592, 686*
Cerny, R.L. *262*
Chaiet, L. *823*
Chalmers, A.A. *456*
Chan, G.W. *245*
Chan, K.-C. *306*
Chander, R. *740*
Chandra, P. *635*
Chaney, M.O. *656, 840*
Chang, C. *801*
Chang, C.-J. *540, 541*
Chang, F.-R. *375*
Chapleur, Y. *661*
Chapman, C.J. *675*
Chapuis, J.C. *167*
Charette, A.B. *667, 669, 695*
Charpentié, Y. *561*
Chase, C.E. *172*(a), *172*(b)
Chatterjee, A. *303*
Chen, Y.-Y. *209*(b)
Cheng, D. *356*
Cheng, J. *251*
Chevolot, L. *76*
Chexal, K.K. *384*
Chiasera, G. *170*
Chib, J.S. *92*
Chidester, C.G. *593*

Chiles, S.A. *121*
Chin, M. *684*
Chisholm, D.R. *588*
Chistoffersen, M.W. *98*
Chmurny, G.N. *592*
Cho, K.W. *213*
Chopra, A.K. *386*(a)
Choussy, M. *702*
Christensen, C.M. *405*
Cimino, G. *45, 102, 104, 190*
Cionga, E. *267*
Clardy, J. *109, 210, 230, 231, 649, 848*(b)
Claremon, D.A. *692*(a), *692*(b), *694*
Clark, P. *434*
Clark, W.D. *93*
Clauder, O. *269*
Cleeland, R. *580*
Clerici, F. *250*
Clewlow, P.J. *148*
Clezy, P.S. *196*
Close, W. *325*
Coats, J.H. *594*
Coen, L. *876*
Cohen, I.D. *209*(a)
Cole, J.R. *362*
Cole, R.J. *441*
Coll, J.C. *196, 232*
Collard, J. *583*
Collas, M. *44*(f)
Colson, K.L. *649*
Compagnone, R. *194*
Conner, W.E. *63*
Conova, S. *97*(b)
Cook, J.H. *44*(g)
Cookson, G.H. *3, 6*
Cooper, R. *890*
Corbett, D.F. *737*
Corbett, T. *93*
Corey, E.J. *707*
Corley, D.G. *434*
Cornforth, J. *44*(e)
Cornwall, M. *301*
Corpe, W.A. *9, 746*
Corwin, A.H. *760*
Cosar, C. *618*
Costi, R. *536*
Cozzi, P. *632–634*
Cram, D.J. *837*
Cranick, S. *73*
Cresp, T.M. *657*

Crews, P. *93, 128*
Cron, M.J. *589*
Cross, S.S. *121*
Crum, G.F. *871*
Cruse, W.B.T. *705*
Cruz-Camarillo, R. *766*
Cue, Jr. B.W. *490*
Cullen, W.P. *591, 686*
Culvenor, C.C.J. *59, 313, 316*
Cundasawmy, N.E. *379*
Cundliffe, E. *843*(c)
Cun-heng, He *230*
Cuppels, D.A. *494*
Cushley, R.J. *785–787*
Czuba, L.J. *490*
Czygan, F.-C. *327*

D'Alessio, R. *805*
D'Ambrosio, M. *169, 170, 173*
D'Aoust, J.Y. *748*
Dabrowiak, J.C. *631, 644, 647*
Dagger, F. *194*
Dales, S.L. *731*
Daloze, D. *46, 51, 120*
Daly, J. *30, 31*
Daly, J.W. *33, 34, 38, 39*
Daninos, S. *118*
Daninos-Zeghal, S. *111*
Danishefsky, S. *442*
Dauphin, G. *685, 697*
David, B. *312*
David, L. *682*
David, S. *735*
Davies-Coleman, M. *830*
Davis, D. *487*
Davis, R.H. *234*
de Almeida Leone, P. *186*
de Chezelles, N. *561*
De Guzman, F. *149*(a), *149*(b)
de Jesus, A.E. *454*
De Kimpe, N. *44*(h)
de Medina, L.B. *779*
de Nanteuil, G. *105, 110*
De Napoli, L. *199*
De Rosa, S. *104*
De Ruiter, L.G.J. *726*
De Silva, K.T. *304*
de Souza, N.J. *747*
De Stefano, S. *102, 104, 190*
de Vries, J.G. *709*(a), *709*(b)

Dean, L. *890*
Deber, C.M. *677*
Debitus, C. *169, 173*
Debray, M. *333*
Decicco, C.P. *292, 827*
Declercq, J.-P. *51*
Dekker, K.A. *444*
Dekker, R.H. *723*
Delabos, C. *333*
Della Lucia, T.M.C. *42*
Delle Monache, F. *628*
Demarco, P.V. *656*
Demko, D.M. *695*
Demole, C. *290*
Demole, E. *290*
Dent, W. *209*(a), *209*(b)
Deol, B.S. *790*
Dervan, P.B. *640*
Deschamps, M. *650*
Deshmukh, P.V. *806*
Deslongchamps, P. *343, 350, 351*
Desluc, M.A. *697*
Deverre, J.-R. *306*
Devine, L.F. *579*
DeVries, W.H. *871*
Dhar, K.L. *136*
Dias, H.W. *386*(a)
Dickerson, R.E. *639*
Diehl, E.W. *64*(a)
Dietl, A. *212*
Dietzel, E. *795*
Diez-Martin, D. *687*
Dijkstra, M. *724, 727, 728*
Dimitrov, D.P. *765*
Ding, W. *602*
Dirlam, J.P. *490*
DiSanto, R. *536*
Divekar, P.V. *747*
Dixon, D.J. *425, 435, 439*
Djerassi, C. *761*
Do Nascimento, R.R. *42*
Dolak, L. *562*
Dolle, R.E. *692*(a), *694*
Dong, Y. *7, 299*(a)
Dorman, D.E. *840*
Doubek, D.L. *148, 167*
Doutheau, A. *350, 351*
Duax, W.L. *678*
Dubost, M. *561*
Duchamp, D.J. *881*

Dufresne, C. *262*
Duignan, J. *444*
Duine, J.A. *706, 713, 716, 717, 721*(a), *722*(a), *722*(b), *723, 724, 726–729*
Dulaney, E.L. *415*
Duma, R.J. *576*
Dumas, D.J. *395*
Dumbacher, J.P. *34*
Dunbar, C.D. *176*
Dunne, T.S. *474, 476*
Dupin, M. *671*
Dupuis-Hamelin, C. *574, 583*
Durand, R. *350, 351*
Durham, D.G. *489, 499*

Ebel, H. *206*
Eble, T.E. *871*
Eder, C. *142, 178*
Edgar, J.A. *59, 316*
Edstrom, E.D. *252*
Edwards, M.P. *688, 698*
Egli, R.H. *270, 276*
Eguchi, T. *463*
Ehlers, D. *321*
Eisenreichovà, E. *365, 366, 372–374*
Eisner, T. *52–54, 56, 58, 60, 63*
El Sayed, K.A. *289*
Elander, R.P. *514, 515, 538, 539*
Elix, J.A. *484*
Ellestad, G.A. *602*
Elliot, J.E. *25*
Elson, M.K. *757, 764*
Emmerling, A. *1*
Emrich, R. *66*
Endo, M. *184*
Engel, J. *381*
Enggist, P. *290*
Enholm, E.J. *695*
Enomoto, M. *433*
Entzeroth, M. *825*
Epstein, W.W. *870*
Escalona de Motta, G. *159*(b)
España de Aguirre, A.G. *314*
Esumi, S. *370*
Etard, A. *284*
Evan, T. *248*(b)
Evans, D.A. *659*
Evans, Jr. R.H. *689, 691*
Ezaki, N. *465–467, 469, 477–479*

Fairhurst, A.S. *354*
Fales, H.M. *281*
Fariña, F. *371*
Farooq Biabani, M.A. *193*
Fathi-Afshar, R. *77*
Fattorusso, E. *71, 79, 99–101, 123, 129, 137, 199*(a), *199*(b)
Faugeras, G. *332, 333*
Faulkner, D.J. *45, 84, 109, 145, 156, 181, 192, 208, 224*(b), *230, 235, 236, 245, 255, 260, 464*
Faure, R. *210*
Fechner, G.A. *186*
Fedoreev, S.A. *74, 151, 165, 168*
Fedorko, J. *577*
Fedorova, G.B. *750*
Fenical, W. *65, 231, 237*(a), *237*(b), *244, 256*
Feofilova, E.P. *122, 744*
Ferdinandus, E. *142, 178*
Ferguson, D.C. *261*
Fernandez, M.I.M. *823*
Ferramola, A.M. *18*
Ferreira, N.P. *382, 449, 453–455*
Ferroud, D. *574, 583*
Findlay, J.A. *347*
Finlay, A.C. *610*
Fischer, H. *2*
Fischer, I. *542*
Fischer, P. *545*
Fisher, L.V. *218*
Flahive, E.J. *264*
Flament, I. *279*
Flanagan, V. *278*
Floc, H.F. *650*
Florent, J. *561*
Floss, H.G. *540, 541, 801*
Flynn, B.L. *204*(b), *240*
Foley, L.H. *164*
Folkers, K. *340, 415*
Fong, H.H.S. *311*
Forenza, S. *71, 648*
Forgione, P. *754, 755*
Forrest, H.S. *705*
Forrest, T.P. *345*
Forsyth, D.A. *299*(a)
Fraiman, A. *441*
Francis, T. *245*
Francke, S. *46*
Francke, W. *45–47, 50*

Frank, Jr. J. *706, 713, 721*(a), *722*(a), *772*(b), *723, 724, 726–729*
Fresneda, P.M. *125*
Friedel, P. *278*
Friedland, D.J. *775*
Frincke, J.M. *208*
Fröde, R. *397*
Frolow, F. *248*(a)
Frydman, B. *12*(b), *13*(b)
Frydman, R.B. *12*(b), *13*(b)
Fryer, R.I. *108*
Fu, X. *78, 149*(a), *149*(b)
Fujimori, T. *274*
Fujimoto, H. *843*(a)
Fujioka, N. *179*
Fujisawa, K. *565, 887*
Fujita, A. *664, 665*
Fujita, K. *472*
Fujita, T. *829*
Fukuda, A. *510–512*
Fukushima, N. *179*
Fukuta, A. *509*
Fukuyama, J.M. *793*
Fukuzumi, T. *272*
Fumoto, Y. *463*
Furihata, K. *437*(b), *438, 555, 653, 859*(a), *860*
Furlenmeier, A. *568, 573*
Fürstner, A. *204*(a), *556*(a), *556*(b), *558, 810*
Furukawa, H. *368, 369*
Furukawa, M. *654*
Furumai, T. *606*
Furusaki, A. *492*
Fusetani, N. *90, 91, 96, 224*(a), *225, 227, 229, 258*

Gabor, B. *810*
Gainor, J.A. *708*
Gale, G.R. *771*
Gallo, G.G. *835*
Gandhi, N.M. *747*
Ganem, B. *97*(a), *97*(b)
Ganju, P.L. *472*
Garcia Gravalos, M.D. *207*
Garcia, E.E. *108*
Garraffo, H.M. *34*
Gastner, T. *558*
Gatenbeck, S. *413*
Gaudemer, A. *702*

Gaudemer, F. *422*
Gaudener, A. *422*
Gaughran, E.R.L. *741*
Gauthier, M.J. *791*
Gautschi, F. *279*
Gavrilova, O.A. *811*
Gaydou, E.-M. *210*
Gaynor, D.L. *393, 394*
Gehrken, H.-P. *183*(a), *183*(b)
Gellert, M. *582*
Gerber, N.N. *748, 749, 762, 774, 791, 796, 797, 802, 803, 814–816*
Geroni, C. *632–634*
Gerrans, G.C. *77, 331*(a), *331*(b)
Gerwick, W.H. *294, 830*
Ghosh, M. *715*
Giammarino, A.S. *278*
Gianturco, M.A. *278*
Gibbons, W.A. *336, 337*
Gilardi, R.D. *36*
Gildenhuys, P.J. *452*(a)
Gilman, S.J. *660*
Gilmore, J. *254*(a), *254*(b)
Girard-Le Bleis, P. *110*(c)
Gitterman, C.O. *87, 415, 416*
Godard, C. *561*
Godfrey, J.C. *588, 589*
Goh, S.H. *308*
Goldberg, I. *248*(a), *95*
Goldman, M.E. *281*
Goldschmid, M.C. *783*
Gomez, M. *194*
Gonzàlez, A.G. *314*
Goodman, J.J. *474, 476*
Goodsell, D. *639*
Goodwin, J.T. *441*
Goodwin, M.G. *730, 731*
Goosen, A. *77, 330*
Gopalkrishan, K.S. *806*
Gordee, R.S. *526, 529, 534*
Gordon, C.N. *417*
Gorman, M. *100, 513–515, 537, 538, 539*
Gorst-Allman, C.P. *456*
Gossauer, A. *13*(a), *380*
Gössinger, E. *37*
Gosteli, J. *524*
Goto, T. *821, 822*
Gott, C.L. *782*
Gottikh, B.P. *625*

Gottlieb, D. *532*
Gourevitch, A. *588*
Grabley, S. *604*
Gracza, T. *435*
Graf, W. *37*
Gram, L. *98*
Grandjean, J. *699*
Granick, S. *10*
Gray, C.H. *11*(b)
Greco, M.N. *446*
Green, S. *95*
Greenstein, M. *443, 603*
Grehn, L. *627*
Greinwald, R. *327*
Grenz, M. *318, 319, 321*
Gribble, G.W. *26, 27*
Grieco, P.A. *94, 658, 660, 662*
Griesinger, C. *237*(a), *237*(b)
Grinstein, M. *18*
Grokhovskii, S.L. *625*
Grostic, M.F. *878*
Groszek, G. *153*
Grote, R. *819*
Grover, P. *827*
Groves, J.K. *379*
Grunberg, E. *580*
Guénard, D. *306, 312*
Guella, G. *89*
Guerrero, A. *63*
Guerriero, A. *169, 170, 173*
Guilhem, J. *105*
Gunasekera, M. *228*
Gunasekera, S.P. *68, 73, 228*
Gupta, R. *148*
Gupta, R.K. *317*
Gupta, S. *291*
Gupton, J.T. *243*(a)
Gurne, D. *20*
Gustavson, D.R. *648*
Guyot, M. *211*
Gyimesi, J. *862, 865, 866*

Haahtela, K.K. *701*
Haansuu, J.P. *701*
Hagaman, E.W. *540*
Hagenmaier, H. *880*
Hagerman, C.R. *579*
Hagiwara, S. *493*
Hagmann, L. *461*
Hahn, F.E. *636*

Haidoune, M. *302*
Haladovà, M. *365, 366, 372–374*
Hamada, M. *504, 550, 551, 575*
Hamaguchi, F. *370*
Hamann, M.T. *176, 289*
Hamel, E. *204*(b)
Hamill, R.L. *514, 515, 538, 539, 840*
Hammer, P.E. *542*
Hanada, K. *280*
Hanada, M. *887, 889*
Hanefeld, U. *501*
Hanessian, S. *498, 661*
Haney, M.E. *513*
Hanners, J.L. *732–734*
Hansen, K. *64*(a)
Hansen, M.S.T. *236*
Hara, O. *475*
Harashima, K. *798, 799*
Harbour, G.C. *254*(a), *254*(b)
Harley-Mason, J. *77, 331*(a), *331*(b)
Harper, M.K. *224*(b)
Harrington, P.E. *560*
Härri, E. *377*
Harris, S.A. *218*
Hartley, J.A. *645, 647*
Hartman, R. *356*
Hashimoto, K. *258*
Hashimoto, M. *525, 546–548*
Hashimoto, T. *398, 402*
Hatley, R.J.D. *559*(a), *559*(b)
Hattori, K. *525, 546–548*
Hattori, R. *369*
Hawkes, G.E. *387*
Hayakawa, Y. *555, 859*(a), *859*(b), *860*
Hayase, T. *736*
Hayashi, M. *403*
Hayashi, T. *401*
Haynes, D.H. *680*
He, X.-G. *801*
Hearn, W.R. *122, 123, 743, 757, 764, 779, 789*
Hedrick, M.P. *246*
Heim, A. *239*(a)
Helbling, A.M. *703*(a)
Hendlin, D. *415*
Hendrickson, J.B. *709*(a), *709*(b)
Hensey, D.M. *876*
Herald, C.L. *148, 262*
Herald, D.L. *167, 263, 264*
Hermann, T. *689, 690*

Herrin, T.R. *874*
Hesler, G.A. *648*
Hettche, O. *745*
Hickford, S.J.H. *201*
Highet, R.J. *293*
Higuchi, K. *222*
Hill, D.S. *542*
Hilton, B.D. *434*
Hinze, C. *397*
Hirata, Y. *114, 116, 141, 159*(a), *850*(a), *850*(b)
Hirayama, N. *400*
Hirota, H. *224*(a), *90, 91, 96*
Hobbs, L. *203*
Hochgürtel, M. *107*(b), *113, 117, 119*(a)
Hochstein, F.A. *610*
Hockless, D.C.R. *204*(b), *240*
Hodge, P. *483, 487, 70*(b)
Hoehn, M.M. *646, 840*
Hoffmann, H. *107*(b), *119*(a), *124, 127*
Hofheinz, W. *221*
Höfle, G. *266, 836*
Hohaus, K. *542*
Holden, I. *341, 342*
Holden, K.G. *123, 756*
Hollenbeck, J.J. *842*
Holt, T.G. *254*(b)
Holzapfel, C.W. *382, 414, 445, 450, 452*(a), *452*(b), *457*
Hong, T.W. *171*
Hook, D.J. *541*
Hooper, I.R. *589*
Hooper, J.N.A. *186, 203*
Hoppe, D. *663*
Horne, D.A. *112, 132, 139*(b), *143, 144*(a), *144*(b), *147, 166, 174*
Horng, J.-S. *534*
Horsley, S.B. *64*(a)
Horvàth, G. *865, 866, 884, 885*
Horvàth, I. *862*
Horvàth, K.B. *268, 269*
Hoshino, T. *401, 739*
Hosokawa, H. *309*
Hossain, M.B. *149*(a), *149*(b), *78*
Hotta, K. *475*
Houck, M.A. *52, 53*
Houk, D.R. *732–734*
Howard, B.H. *89, 430*
Howell, C.R. *494*
Hsia, M.-T.S. *286*

Huang, L.H. *591*
Hubbard, R. *776*
Hübner, H. *355*
Hughes, C.G. *485, 489*
Hughes, Jr., R.G. *175*
Humphrey, P.E. *876*
Hunt, E. *17*
Hunt, M.B. *394*
Hupperts, A. *204*(a)
Hursthouse, M.B. *386*(a)
Husain, A. *406*
Hussain Qadri, S.M. *784, 788*
Husson, H.-P. *306*
Hutchinson, C.R. *286*
Hutchison, R.D. *450*
Hutschings, B.L. *615*
Hütter, K. *604*

Iengo, A. *199*(c), *199*(d)
Iguchi, E. *821*
Iida, T. *400*
Iimura, H. *407, 471*
Iitaka, Y. *360, 469, 471, 504, 839*(a)
Ijuin, Y. *847*
Ikeda, H. *439*
Ikegami, S. *223, 736*
Ilyin, S.G. *165, 168*
Imanaka, H. *508–512, 519, 821, 822*
Imhof, R. *37*
Imura, T. *24*(b)
Inaba, K. *130, 157*
Inagaki, T. *444*
Inami, K. *841*(a), *841*(b)
Indira, R. *806*
Inokoshi, J. *403*
Inouye, S. *466, 467, 477–479*
Irvine, D.G. *23*(a), *23*(b)
Ishibashi, F. *238*
Ishibashi, M. *180, 251, 80, 83*
Ishida, K. *83*
Ishida, N. *606*
Ishida-Okawara, A. *475*
Isogai, A. *437, 438*
Isomura, H. *736*
Isshiki, K. *552*
Issiki, K. *550*
Itō, T. *465, 466*
Ito, M. *280, 386*(a), *386*(b)
Ito, S. *88, 850*(a), *850*(b)
Ito, Y. *368, 391, 606*

Itoh, E. *557*
Itoh, J. *472*
Itoh, M. *505–507*
Itoh, S. *725*
Itoh, T. *582*
Itokawa, H. *357, 359, 360*
Iwagawa, T. *87*
Iwahashi, K. *399*
Iwamura, M. *180*
Iwao, M. *238*
Iwasaki, S. *408*
Iyengar, M.R.S. *472*
Izumida, H. *184*

Jackson, A.H. *15, 761*
Jackson, M. *876*
Jacobus, N.V. *603*
Jaffe, K. *301*
Jager, H. *837*
Jahn, T. *82*
Jain, S.C. *291*
Jakob, K. *212*
Jakupovic, J. *317, 318*
James, G.D. *833*
James, K.J. *16*
Janardhanan, K.K. *406*
Janisiewicz, W.J. *516, 517*
Jarman, W.M. *25*
Jarrah, M.Y. *8*
Jaschke, K.H. *880*
Jasinski, J.P. *27*
Jefferson, M.T. *591*
Jefford, C.W. *295*
Jeffrey, C. *320*
Jeminet, G. *671, 685, 697*
Jenden, D.J. *354*
Jerina, D.M. *281*
Jha, A. *291*
Jiang, B. *133*
Jiang, Z.D. *294*
Jiménez, C. *128*
Jimenez, D.R. *171*
Jin, Q. *204*(c), *246*
Jizba, J. *361*
Johnson, J.H. *890*
Johnson, R.E. *488, 491*
Johnson, R.K. *245*
Johnston, B.H. *667, 669*
Jomon, K. *543, 544, 849*
Jones, N.D. *656, 840*

Jones, R.C.F. *219*
Jongejan, J.A. *713, 716, 729*
Jordan, P.M. *22*
Joseph, N. *616*
Josten, I. *397*
Juillard, J. *671, 674, 685*
Julia, M. *616, 619*(a), *619*(b)
Jung, G. *650*

Kabbe, K. *266*
Kaczka, E.A. *415*
Kaib, M. *45, 46, 49, 50*
Kaiser, J. *335*
Kaji, J. *409*(a), *409*(b), *410*
Kakita, T. *504*
Kallmerten, J. *597*
Kallvy, D.M. *767*
Kaltenbronn, J.S. *498*
Kam, T.-S. *71, 307*
Kamano, Y. *262*
Kameoka, H. *275*
Kan, Y. *829*
Kanai, K. *658, 662*
Kanamaru, N. *859*(a), *859*(b), *860*
Kanazawa, S. *229*
Kanda, F. *80*
Kaneda, M. *469*
Kaneko, K. *401*
Kaneko, M. *87*
Kan-Fan, C. *306*
Kang, H. *237*(b), *243*(c), *244*
Kantoci, D. *153*
Kapadia, G.J. *293*
Kaplan, S.A. *585*
Kariyone, K. *520–523, 535, 543, 544*
Karle, I.L. *30, 35*
Karle, J. *30, 35*
Karr, A.E. *568, 570*
Karwowski, J.P. *876*
Kase, H. *399*
Kashman, Y. *95, 207, 248*(a), *248*(b)
Kassab, D.J. *97*(a)
Kasum, B. *160*
Kat, K. *274*
Katayama, H. *350, 351*
Kato, H. *90, 91, 96*
Kato, S. *389, 462*
Kato, T. *184*
Kato, Y. *210, 224*(a), *433*
Katoh, T. *557*

Katz, S. *577*
Kavai, Y. *653*
Kawaguchi, H. *104, 564–567, 886–889*
Kawai, H. *390, 462, 549*
Kawakami, K. *555*
Kawamura, N. *550–553*
Kawasaki, I. *179*
Kazlauskas, R. *259*
Keifer, P.A. *175*
Keil, J.G. *589*
Keith, D.D. *800, 804, 807–809*
Keller, O. *568, 573*
Keller-Schierlein, W. *461*
Kellner, M. *502, 813*
Kelly, R.B. *344*
Kelly-Borges, M. *138, 176, 226, 78*
Kennard, O. *118, 705*
Kenner, G.W. *761*
Kergomard, A. *682*
Kerr, R.G. *146*
Kerssebaum, R. *237*(b)
Kervagoret, J. *44*(d)
Kesztler, F. *285*
Ketcha, D.M. *44*(g)
Kettenes, D.K. *277*
Khokhar, A.R. *386*(a)
Khokhlov, A.S. *812*
Khokhlova, Yu.M. *812*
Kido, M. *152*
Kikuchi, M. *606*
Kikuchi, Y. *251*
Kim Y.-J. *444*
Kim, S. *243*(c)
King, R.M. *318*
Kinghorn, D. *97*(b)
Kinnel, R.B. *183*(a), *183*(b)
Kinney, W.A. *97*(b)
Kinoshita, K. *681*
Kinoshita, N. *550, 551*
Kinoshita, T. *843*(a)
Kinstle, T.H. *873*
Kirby, G.W. *454, 455*
Kirst, H. *840*
Kishi, Y. *666*
Kishimoto, T. *520, 535*
Kissinger, K. *647*
Kitagawa, I. *152*
Kitanaka, K. *152*
Kleschick, W.A. *659*
Klevit, R.E. *643*

Klich, M. *574, 583*
Klika, K.D. *701*
Klohr, S.E. *649*
Knoll, K.-H. *321, 322*
Knorr, L. 2
Kobaru, S. *887*
Kobayashi, J. *80, 83, 114, 116, 126, 130, 131, 141, 157–159*(a), *177, 180, 182, 251, 254*(a), *254*(b), *846*
Kobayashi, M. *152, 222*
Kobayashi, T. *423*
Köck, M. *107*(a), *122, 237*(a), *237*(b)
Koczka, I. *862*
Kodama, Y. *467, 479*
Koehn, F.E. *826*(a), *826*(b)
Kogon, I.C. *614*
Koh, Y.-H. *172*(a), *172*(b)
Kohl, H. *747*
Kohsaka, M. *821, 822*
Kojima, M. *466, 472*
Kojima, Y. *444*
Kojiri, K. *738*(a), *738*(b)
Koker, M.E.S. *175*
Kolber, M.A. *680*
Kominek, L.A. *563*
Komiyama, K. *403*
Kondo, Y. *253*(a), *253*(b)
Kong, Y.L. 105, *590*
König, G.M. *72, 82, 86*
König, W.A. *651, 652*
Koniuszy, F.R. *340*
Konno, K. *376*(a)
Konstantinova, N.V. *883–885*
Kopka, M.L. *639*
Koren-Goldschlager, G. *207*
Korenyako, A.I. *811*
Kornprobst, J.-M. *210*
Kort, M.E. *696*
Kosaka, M. *511, 512*
Kosemura, S. *358*
Koshiyama, H. *565*
Kosugi, Y. *370*
Kotake, M. *271*
Kotecha, N.R. *687*
Koul, S.K. *136*
Kousaka, M. *508–510, 519*
Kovch, A.G.B. *269*
Kowicki, K. *644*
Koyama, M. *24*(b), *466, 467, 475, 477–479*

Koyima, Y. *401*
Kozikowski, A.P. *446*
Kozuka, M. *368*
Kraft, E. 122, *742*
Krauss, D. *873*
Krowicki, K. *620, 631, 645*
Kruger, F.W.H. *452*(b)
Krumpe, K.E. *243*(a)
Kudo, Y. *367*
Küster, W. 2
Kumagai, K. *606*
Kunze, B. *836, 861*
Kuo, R.-Y. *375*
Kurihara, T. *820*
Kuroda, Y. *821, 822, 849*
Kurth, M.J. *852*(a)
Kyogoku, Y. *152*

La, T. *684, 690*
Laatsch, H. *496, 500–503, 792, 813*
Laboute, P. *105*
Labroli, M.A. *204*(c)
Lacey, R.N. *217*
Lafargue, F. *254*(b)
LaForge, F.B. *287*
Lakshmi, V. *68*
Lallemand, J.Y. *697*
Laloue, M. *302*
Lam, H.M. *842*
Lam, K.S. *648*
Lambowitz, A.M. *531*
Lange, G.L. *292*
Lapalme, R. *350, 351*
Lapper, E. *2*(a)
Lardy, H.A. *655, 670, 673*
Large, C.M. *871*
Larson, J.E. *638*
Lassaigne, P. *574, 583*
Laszlo, P. *699*
Laugal, J.A. *440*
Laurin, P. *574*
Lebrun, M.H. *422*
Lechevalier, M.P. *797*
Lee, H.-S. *213*
Lee, K.K. *242*
Lee, M. *609, 645, 647*
Lee, M.S. *649*
Lee, S.-F. *70*(a)
Lee, V.J. *874, 875*
Leet, J.E. *148*

Leimgruber, W.   571
Leitz, F.M.   495
Le Quesne, P.W.   7, 70, 296, 299(a), 299(b), 300
Leroy, S.   169
Letellier, G.   28
Leturc, D.M.   350, 351
Lewis, S.M.   746
Ley, S.V.   425, 428, 435, 439, 687, 688, 698
Li, C.   257
Li, C.-J.   138
Li, L.   842
Liao, C.-C.   350, 351
Lichte, E.   107(a), 122
Ligon, J.M.   542
Likhitwitayawuid, K.   292
Lim, D.V.   788
Lin, H.-S.   853
Linde, H.A.   71, 305
Lindel, T.   107(b), 113, 117, 119(a), 124, 127
Linden, A.   72
Lindner, E.   335
Lindquist, N.   65, 231, 256, 261
Lingens, F.   545
Lipsky, S.R.   785, 787
Lister, S.G.   688
Little, T.L.   110(b)
Liu, C.-M.   683, 684, 689–691
Liu, J.-F.   133
Liu, J.-H.   243(b)
Liu, J.R.   459
Liu, M.   690
Lively, D.H.   513, 537
Lloyd, H.A.   281
Lockley, W.J.S.   385
Loeffler, W.   377
Lombardo, L.J.   759
Longbottom, D.A.   425, 439
Longley, R.E.   73, 826(a), 826(b)
Loukaci, A.   211
Lovell, F.M.   497
Lown, J.W.   609, 620, 631, 644, 645, 647
Lynn, D.G.   70, 301, 441
Lyon, R.L.   311

Mabe, J.   538
Mabe, J.A.   513–515, 539–541
MacBeth, J.L.   648
MacConnel, O.J.   826(b)

Macdonald, D.   856, 857
MacDonald, S.F.   15
Machinek, R.   604
MacKellar, F.A.   878
MacKenzie, A.R.   710
MacLachlan, F.N.   350, 351
Maeda, H.   686
Maeda, K.   367, 834, 839(b)
Maffrand, J.-P.   350, 351
Magae, J.   818
Magdoff-Fairchild, B.   163(b)
Magerlein, B.J.   596
Magnus, P.   310(b)
Magolda, R.L.   692(b), 694
Magyar, K.   862
Mahanta, P.   320
Mahanta, P.K.   321
Mahoney, N.E.   516, 517
Mai, A.   536
Maiese, W.M.   602, 603
Majer, J.R.   23(a), 23(b)
Majumder, P.L.   303
Mak, T.C.W.   243(b)
Maksimov, O.B.   151, 165, 74
Manabe, T.   820
Manchanda, A.H.   329
Mancini, I.   89
Mancy, D.   561
Manderville, R.A.   261
Manfredi, K.P.   148
Mangoni, A.   123, 137, 79, 99
Manni, P.E.   541
Mantegani, S.   687
Marchais, S.   139(a)
Marchesini, A.   250
Marchlewski, L.P.   2
Margraff, R.   650
Maricq, J.   568
Marion, J.P.   270, 276
Marks, G.S.   780, 781
Marner, F.-J.   828, 831
Marques, S.   194
Marquez, B.   830
Marrazza, F.   350, 351
Marsh, J.J.   382
Martìn, M.V.   371
Martin, L.L.   540, 541
Martín, P.   711
Martinez, G.R.   662
Martino, R.   350, 351

Maruyama, H.   *275*
Marwood, J.F.   *259*
Mascagni, P.   *336, 337*
Mascal, M.   *254*(b)
Massa, S.   *536*
Masuda, K.   *95, 475*
Masuko, A.   *364*
Masuma, R.   *403*
Matamoros, B.   *392*
Matson, J.A.   *649*
Matsubara, T.   *527*
Matsuda, C.   *391*
Matsuda, K.   *420*
Matsuda, M.   *423*
Matsuda, N.   *504*
Matsuda, Y.   *399*
Matsukawa, S.   *24*(b)
Matsunaga, S.   *224*(a), *225, 227, 229, 258*
Matsuoka, M.   *389, 462*
Matsushima, T.   *468*
Matsushita, K.   *718, 720*
Matsuzaki, T.   *736*
Matthews, T.R.   *526, 529*
Mattia, C.A.   *155*
Matulenko, M.E.   *696*
Maugeri, T.L.   *772*
Mauvais, P.   *574, 583*
Mauzerall, D.   *11*(c), *12*(a)
Maximov, O.B.   *168*
Mazzarella, L.   *104, 155*
Mazzini, S.   *634*
Mazzocchi, P.H.   *57*
McAlpine, J.B.   *877*
McCarthy, P.   *228*
McCormick, K.D.   *52*
McDonald, E.   *16, 17*
McGrath, R.M.   *449, 453*
McInnes, A.G.   *816*
McKeon, J.E.   *754, 755, 778*
Mclean, S.   *187*
McLean, S.   *189*
McNab, H.   *326*(a), *326*(b)
McNulty, J.   *167*
Means, G.E.   *771*
Mechoulam, R.   *626, 630*
Medina-Castro, J.   *757, 764*
Méhesfalvi-Vajna, Z.   *865, 866*
Meinwald, J.   *52–55, 57, 58, 60, 62–64*
Meinwald, Y.C.   *54, 55, 57*
Meister, A.   *29*(b)

Melvin, M.S.   *261*
Menéndez, J.C.   *687*
Mendoza, L.A.   *309*
Merlin, J.   *51*
Meronuck, R.A.   *405*
Meulenberg, J.J.M.   *735*
Meyer, C.E.   *879*
Meyers, B.R.   *578, 581*
Meyers, E.   *890*
Michaeli, D.   *578, 581*
Mikami, Y.   *407*
Miller, M.W.   *818*
Miller, P.A.   *683, 684, 690*
Miller, S.L.   *189*
Milligan, K.E.   *830*
Mills, S.D.   *833*
Mimaki, Y.   *363, 364, 367*
Mimura, S.   *468*
Minale, L.   *71, 102, 190*
Minegishi, Y.   *363*
Ming-hui, D.   *44*(e)
Minoda, Y.   *493*
Minoura, K.   *391*
Minton, M.A.   *852*(b)
Miralles, J.   *210*
Mirocha, C.J.   *405*
Mishima, H.   *601*
Misiek, M.   *588*
Mito, K.   *364*
Miyadoh, S.   *472, 475*
Miyaki, T.   *104, 564, 565, 567*
Miyashita, H.   *23*(a)
Miyazaki, Y.   *238*
Miyazawa, M.   *275*
Miyazawa, T.   *141*
Mizsak, S.A.   *593*
Mizuno, K.   *681*
Mizuno, S.   *475*
Mizuoka, T.   *739*
Mochizuki, J.   *389, 462*
Mochizuki, T.   *557*
Mogi, K.   *470, 473*
Moisand, A.   *312*
Moka, W.   *142*
Molina, P.   *125*
Molinski, T.F.   *171*
Momen, A.Z.M.R.   *739*
Monaghan, R.L.   *823*
Mongelli, N.   *632, 633*
Moody, C.J.   *710*

Moore, R.E.   *825, 828, 831, 832*
Moppett, C.E.   *591, 592*
Morales, J.J.   *115, 159*(b)
Moreau, C.   *350, 351*
Moreira, D.D.O.   *42*
Morgan, E.D.   *42*
Morgan, E.N.   *763*
Morgat, M.   *312*
Mori, K.   *188*
Mori, Y.   *24*(a)
Morimoto, Y.   *520, 522, 525*
Morisaki, N.   *859*(a), *860*
Morita, H.   *357, 359, 360*
Mornet, R.   *302*
Moron, J.   *17*
Morrison, A.J.   *439*
Morrison, D.A.   *775*
Morton, G.O.   *602*
Moser, J.C.   *40*(a), *40*(b), *44*(b)
Mourabit, A.A.   *119*(b)
Mroczenski-Wildey, M.J.   *603*
Muchowski, J.M.   *191*
Müggler-Chavan, F.   *270, 276*
Mulder, A.C.   *729*
Mullaney, J.T.   *263, 264*
Müller, A.   *461*
Müller, G.   *19*(a), *19*(b)
Müller, W.E.G.   *103*
Munro, M.H.G.   *201*
Muramatsu, S.   *468*
Muramatsu, T.   *420*
Muraoka, Y.   *504*
Muratake, H.   *447*
Murayama, T.   *180*
Muro, H.   *408*
Murphy, F.X.   *610*
Murphy, P.J.   *528*
Murphy, P.T.   *259*
Musicki, B.   *574, 583*
Mutaparat, T.   *239*(b)
Muto, N.   *681*
Myasnikova, L.G.   *867*
Mynott, R.   *810*

Nabbs, B.K.   *198*
Nabney, J.   *329*
Nadelson, J.   *809*(a), *809*(b)
Naef, R.   *191*
Nagai, H.   *829*
Nagai, K.   *818*

Naganawa, H.   *471, 504, 550–553, 575, 839*(b)
Nagao, S.   *423*
Nagasaka, T.   *370*
Nagashima, S.   *357, 359, 360*
Nagatsu, J.   *798, 799*
Nagatsu, T.   *848*(a)
Nagle, D.G.   *824*
Naito, T.   *566, 886, 888, 889*
Nakagawa, S.   *886, 888, 889*
Nakahara, Y.   *664, 665*
Nakajima, S.   *738*(a), *738*(b)
Nakajima, T.   *473*
Nakamura, H.   *114, 116, 141, 158, 159*(a), *180, 471, 504, 520, 535, 553, 839*(a)
Nakamura, S.   *399, 469, 617*
Nakano, H.   *520*
Nakashima, T.   *470*
Nakatani, M.   *87*
Nakatani, S.   *557*
Nakayama, J.   *437*(b), *438*
Nakayama, Y.   *834*
Namba, K.   *81*
Namiki, M.   *403*
Namsa-aid, A.   *247*(b)
Napier, J.J.   *851, 855*
Narayanan, S.   *472*
Natsume, M.   *447*
Natu, A.A.   *321*
Naya, Y.   *271*
Nazareth, P.V.   *747*
Neethling, D.C.   *453*
Negri, D.P.   *666*
Neier, R.   *14*(a), *14*(b)
Neitz, A.W.   *457*
Nelson, A.   *676*
Nencki, M.   *2*
Nettleton, D.E.   *589*
Newmark, H.L.   *586, 587*
Ng, P.-L.   *78*
Ngo Le Van,   *321*
Niccolai, N.   *337*
Nichols, C.   *788*
Nickisch, K.   *325*
Nicolaou, K.C.   *692*(a), *692*(b), *694*
Nicolas, L.   *422*
Nicolella, V.   *622, 623, 629*
Nielsen, P.H.   *98*
Nietsche, J.A.   *474, 476*
Nigrelli, R.F.   *92*

Niida, T. *465*
Nikaido, F. *179*
Nikaido, T. *364*
Nimmesgern, H. *209*(b)
Ninet, L. *561, 618*
Nishida, M. *527, 535*
Nishijima, Y. *407*
Nishimura, N. *24*(a), *24*(b)
Niwa, T. *466, 467*
Nogami, T. *504*
Noguchi, H. *520*
Nolte, M.J. *216*
Nomoto, K. *391*
Norstrom, R.J. *25*
Northcote, P. *602*
Nose, M. *530*
Noval, J.J. *23*(c)
Novotný, L. *361*
Nowak, A. *424, 426*
Nozoe, S. *180, 251, 408, 423*
Numata, A. *391*
Nyomrkay, K.M. *268*

O'Brien, M.J. *852*(a)
O'dea, M.H. *582*
Obata, T. *471*
Oberhänsli, W.E. *221*
Occolowitz, J.L. *656, 840*
Odagawa, A. *389, 462*
Offen, P.H. *245*
Ogawa, T. *358, 664, 665*
Ohfune, Y. *81*
Ohizumi, Y. *114, 116, 141, 158, 159*(a), *177, 180, 251*
Ohkishi, H. *736*
Ohmomo, S. *451*
Ohmori, T. *493, 565, 364*
Ohmoto, T. *364*
Ohno, M. *839*(b)
Ohshiro, Y. *725*
Ohta, A. *820*
Ohta, E. *223*
Ohta, S. *179, 223*
Ohta, T. *180, 251, 423*
Ohya, N. *309*
Okamoto, M. *822*
Okamura, H. *87*
Okanishi, M. *104, 564, 565*
Okita, T. *889*
Okuda, M. *601*

Okuda, S. *408*
Okuda, T. *606*
Okuhara, M. *821, 822*
Okura, A. *738*(a)
Oliveri, M.C. *194*
Olofson, A. *112, 174*
Olsen, C.E. *291*
Olson, E.C. *878*
Omura, S. *403, 459*
Onan, K.D. *296*
Onishi, I. *272, 273*
Ono, M. *437, 438*
Ono, N. *463*
Ono, Y. *473*
Oono, J. *470*
Orezzi, P. *622, 623*
Organ, H.M. *687*
Ori, K. *364*
Ortega, M.C. *371*
Ortega, M.J. *197*
Ortiz, M. *159*(b)
Osano, Y.T. *736*
Otake, N. *653*
Ott, I. *862*
Ott, W.R. *600*
Ovcharenko, V.V. *701*
Overman, L.E. *185*
Ozawa, N. *370*

Paal, C.L. *2*
Padua, S. *76*
Padwa, A. *209*(a), *209*(b)
Pai, N.N. *299*(a), *300*
Palermo, J.A. *205*
Palleroni, N.J. *684, 690*
Palmer, B.D. *688*
Pang, G.-L. *820*
Papahatjis, D.P. *692*(a), *692*(b), *694*
Paquette, L.A. *853, 856, 857*
Paredes, M.C. *371*
Paris, R.-R. *332, 333*
Parker, K.A. *209*(a)
Parker, M.A. *696*
Parmar, V.S. *291*
Pasteels, J.M. *46, 51*
Patel, D.J. *642*
Patel, M. *758*
Pattenden, G. *833*
Pattisina, L.A. *142, 178*
Paul, V.J. *824*

Pavone, P.   *772*
Pawlik, J.R.   *107*(a), *107*(b)
Pearce, C.J.   *882*
Pecherer, B.   *568, 573*
Pelizaeus, B.   *503*
Pelloux-Léon, N.   *48*
Peltier, M.   *332*
Penco, S.   *622–624, 628, 629*
Pendrak, I.   *460*
Pennella, G.   *633*
Perni, R.B.   *855*
Perry, T.L.   *176*
Peschko, C.   *241, 247*(a)
Peseckis, S.M.   *693*
Pestka, S.   *843*(b)
Peter, S.R.   *187*
Petersen, P.J.   *603*
Peterson, G.E.   *219*
Pettit, G.R.   *148, 153, 167, 262–264*
Petty, R.L.   *62*(a), *62*(b)
Peverada, P.   *787*
Pfäffli, P.   *378, 383*
Pfeiffer, D.R.   *670, 672, 673, 675, 677*
Pfister, R.M.   *495*
Pham, A.T.   *214*
Phan, G.   *143*
Phillips, N.J.   *441*
Phillipson, J.D.   *336, 337*
Pia, I.C.   *194*
Pierce, N.J.   *245*
Pierens, G.K.   *234*
Pietra, F.   *169, 170, 173, 89*
Pihlaja, K.   *701*
Pinchin, R.   *133, 838*
Pini, H.   *794*
Pinnert-Sindico, S.   *618*
Pirelli, A.   *623*
Piscopio, A.D.   *696*
Pjura, P.   *639*
Plata, D.J.   *597*
Pliske, T.E.   *56*
Plowman, T.   *281*
Pohlan, S.   *266*
Polborn, K.   *249*
Polz, L.   *309*
Pomerantz, M.W.   *150*
Pomilio, A.B.   *282*
Pomponi, S.A.   *146*
Pon, R.T.   *645, 647*
Ponasik, J.A.   *97*(a), *97*(b)

Poncet, J.   *265*
Postma, P.W.   *735*
Potatova, N.P.   *883–885*
Potgieter, D.J.J.   *457, 458*
Potier, P.   *105, 110*(a), *111, 118, 119*(b), *139*(a), *306*
Poupat, C.   *105, 110*(a), *111, 118, 139*(a)
Pouteau-Thouvenot, M.   *702*
Powell, A.D.G.   *614*
Prager, R.H.   *160, 161*
Prasad, A.K.   *291*
Préau-Joseph, N.   *619*(a), *619*(b)
Preud'Homme, J.   *561, 618*
Price, K.E.   *588, 589*
Probert, C.L.   *657*
Probst, G.W.   *646*
Procita, L.   *353*
Proksch, P.   *103, 142, 178*
Prosser, B.   *684, 690*
Prudhomme, M.   *685*
Przybylska, M.   *348*
Pudleiner, H.   *496, 500, 503*
Puliti, R.   *104, 155*
Puri, A.K.   *675*
Pusset, J.   *105, 169*
Pusset, M.   *105*

Qadri, S.M.H.   *782*
Qiu, F.   *226*
Quinn, R.J.   *186, 234*
Quirion, J.-C.   *306*

Radema, M.H.   *334, 338*
Radhakrishnan, A.N.   *29*(b)
Raffauf, R.F.   *296, 297*
Ragg, E.   *634*
Ragnarsson, U.   *627*
Rahalkar, P.W.   *806*
Rainey, T.   *310*(b)
Raistrick, H.   *89, 430*
Rajwanshi, V.K.   *291*
Rama Rao, M.   *195*
Ramesh, P.   *69*
Rangel, H.R.   *194*
Ranomenjanahary, S.   *422*
Rao, J.V.   *69*
Rao, M.R.   *193, 235, 236*
Rao, T.P.   *69*
Rapoport, H.   *43, 123, 756*
Ratcliffe, A.H.   *304, 310*(a)

Ravi, B.N.   76, *196*
Rdl, S.   *584*
Redaelli, S.   *624, 629*
Reddy, M.V.R.   *235, 236*
Reddy, N.S.   69, *134, 135*
Reed, J.K.   *826*(a), *826*(b)
Reed, P.W.   *655, 673*
Rees, A.H.   *485, 486, 489, 499*
Rees, C.W.   *710*
Reißig, H.-U.   *44*(f)
Reichenbach, H.   *836*
Reid, B.R.   *643*
Reid, R.J.   *596*
Reif, B.   *237*(a), *237*(b)
Reil, S.   *12*(b)
Reilly, M.   *94*
Reinecke, M.G.   *339*
Reiter, M.   *355*
Reitner, J.   *82*
Rengaraju, S.   *472*
Renneberg, B.   *502*
Reshetnyak, M.V.   *165, 168*
Retico, A.   *536*
Reusser, F.   *869*
Reymond, D.   *270, 276*
Reyna, S.   *392*
Reynolds, W.F.   *187, 189*
Rho, J.-R.   *213*
Rhodes, D.   *236*
Ribes, O.   *169*
Riccio, R.   *71*
Richardson, C.H.   *288*
Rickards, R.W.   *483, 70*(b)
Riley, R.G.   *41*
Rimington, C.   *3, 6, 776*
Rinehart, Jr., K.L.   *138, 254*(a), *254*(b), *599, 870, 872–875, 878, 881, 882*
Rinehart, K.L.   *106, 175*
Rinkes, I.J.   *75*
Ripamonti, M.   *173*
Rittschof, D.   *175, 97*(b)
Roberts, M.A.   *824*
Robertson, A.V.   *790*
Robertson, J.   *559*(a), *559*(b)
Robins, D.J.   *315*
Robinson, H.   *318*
Roder, E.   *356*
Röder, E.   *61*
Rodgers, G.C.   *804, 807, 808*
Rodgers, Jr., G.C.   *800*

Rodriguez Brasco, M.F.   *205*
Rodriguez, A.D.   *115, 159*(b)
Rogers, B.N.   *185*
Rogers, E.F.   *340*
Rohr, J.   *604, 605*
Rohwedder, W.K.   *440*
Roitman, J.N.   *516, 517*
Romine, J.L.   *853*
Rosa, R.   *159*(b)
Rosen, T.   *220*
Rosenstein, R.D.   *311*
Rosset, T.   *404*
Rössner, E.   *651, 652*
Roth, M.M.   *44*(f), *769*
Rothhaas, A.   *122, 751, 752*
Roush, W.R.   *693, 854*
Roussakis, C.   *89*
Rowan, D.D.   *393, 394, 396*
Royles, B.J.L.   *215*
Rozynov, B.V.   *885*
Ruangrungsi, N.   *292*
Rubasheva, L.M.   *885*
Rubins, K.   *236*
Ruchirawat, S.   *239*(b), *247*(b)
Rudd, B.A.M.   *801*
Rudi, A.   *248*(a), *248*(b), *95*
Ruest, L.   *343, 350, 351*
Runge, F.F.   *2*
Ruth, J.M.   *40*(a), *40*(b)
Rützler, K.   *148*

Sachs, P.   *4*
Sacks, C.E.   *659*
Saint-Laurent, L.   *350, 351*
Saintonge, R.   *350, 351*
Sakaguchi, N.   *179*
Sakai, S.   *309, 479*
Sakakibara, H.   *839*(a), *839*(b)
Sakakibara, T.   *444*
Sakamoto, B.   *829*
Sakamoto, T.   *253*(a), *253*(b)
Sakemi, S.   *444*
Sakuda, S.   *437, 438*
Salcher, O.   *545*
Salisbury, S.A.   *705*
Salv, J.   *197*
Salvador, U.J.   *491*
Samek, Z.   *361*
Sancovich, H.A.   *18*
Sankawa, U.   *432, 436*

Sankhala, R.H.   *404*
Sannai, A.   *274*
Sano, H.   *400*
Santacrone, C.   *199*
Santer, U.V.   *777, 778*
Santo, L.   *880*
Sanz, M.A.   *125*
Sargent, M.V.   *484*
Sasaki, H.   *849*
Sasaki, T.   *83, 846*
Sashida, Y.   *363, 364, 367*
Sata, N.U.   *225, 227*
Satari, R.   *222*
Sato, H.   *157*
Sato, R.I.   *851*
Sato, S.   *253*(a), *253*(b)
Sato, Z.   *408*
Satomi, T.   *848*(a)
Sawa, R.   *550–553*
Sawa, T.   *468, 550–553*
Sayah, B.   *48*
Scannell, J.   105, *590*
Schabacher, K.   *861, 863*
Schabort, J.C.   93, *457, 458*
Schacht, U.   *335*
Schaufelberger, D.E.   *148*
Schaufelberger, R.   *37*
Scheer, M.   *880*
Scheuer, P.J.   *183*(a), *183*(b), *200, 214, 242*
Schleyer, M.   *95, 207, 248*(a)
Schmalz, H.-G.   *858*
Schmidt, B.   *397*
Schmidt, E.W.   *224*(b)
Schmidt, J.M.   *148, 167, 262*
Schmitz, F.J.   *68, 78, 138, 149*(a), *149*(b), *226*
Schmitz, H.   *589*
Schneider, D.   *60, 62*(a), *62*(b), *64*(a)
Schneider, H.   *62*(a)
Schneider, U.   *461*
Schocher, A.J.   *568, 573*
Schönewolf, M.   *604, 605*
Schröder, D.R.   *649*
Schröder, F.   *45–47, 49, 50*
Schröder, K.   *541*
Schramm, H.W.   *209*(c)
Schreier, J.A.   *44*(g)
Schubert-Zsilavecz, M.   *209*(c)
Schultes, R.E.   *281*
Schumacher, D.   *266*

Schwartz, R.E.   *175*
Schwarz, B.   *103*
Schwarz, O.   *858*
Scott, J.J.   *21*
Sebek, O.K.   *563*
Seff, K.   *831*
Sehon, C.A.   *204*(c)
Seino, A.   *848*(a)
Seki, M.   *188*
Seldes, A.M.   *205*
Sellink, E.   *735*
Sello, L.H.   *683, 691*
Semper, A.T.J.   *277*
Sennett, S.H.   *826*(b)
Sensi, P.   133, *835*
Seo, Y.   *213*
Séquin, U.   *654*
Sergeeva, L.N.   *812*
Seto, H.   *555, 653, 848*(b), *859, 860*
Sévenet, T.   *306, 312*
Seybold, P.G.   *44*(g)
Sezaki, M.   *472*
Shang, M.Y.   *172*(a)
Shankaran, K.   *696*
Shao, Y.   *356*
Shapiro, L.   *642*
Sharma, G.M.   *150, 162, 163*
Shavel, Jr., J.   *340*
Shaw, C.-K.   *786*
Shaw, C.K.   *787, 817*
Shea, R.G.   *647*
Shearer, G.M.   *818*
Sheichenko, V.I.   *812*
Sheldon, B.   *823*
Sheldrick, W.S.   *381*
Shell, J.W.   *871*
Shemin, D.   20
Shen, X.   *176*
Shenin, Yu.D.   137, *867, 868*
Shepard, H.H.   *288*
Sheuer, P.J.   *76*
Shi, X.   *53*
Shiba, T.   *841*(a), *841*(b)
Shibakawa, R.   *591*
Shibanuma, Y.   *432*
Shibata, N.   *557*
Shibata, S.   *431, 432, 436*
Shieh, T.L.   *541*
Shield, L.S.   *254*(b)
Shigemori, H.   *80, 83, 846*

Shigeura, H.T. *417*
Shigihara, Y. *504*
Shimazu, H. *859*(b)
Shimoi, K. *459*
Shimomura, H. *363, 367*
Shimura, H. *280*
Shin, J. *213*
Shinada, T. *81*
Shindo, K. *389, 390, 462, 549*
Shiomi, K. *403, 471*
Shirahama, H. *376*(a), *376*(b)
Shirahata, K. *400*
Shizuri, Y. *184*
Shoji, J. *89, 431, 432*
Shoji, N. *85*
Shomura, T. *465, 466, 472, 477, 479*
Shrimpton, D.M. *781*
Shuford, J.S. *243*(a)
Sica, D. *199*(a)
Siegel, M.M. *476, 602*
Sienkiewicz, K. *295*
Sierankiewicz, J. *413*
Sigg, H.P. *377*
Sikorski, J.A. *243*(a)
Silva, W. *159*(b)
Silverstein, R.M. *40*(a), *40*(b), *41*
Sim, C.J. *213*
Simmons, C.J. *831*
Simon, M. *25*
Simonetti, G. *536*
Singh, M.P. *443, 603*
Singh, P.D. *890*
Singh, S.B. *263*
Singh, S.-B. *264*
Sinnwell, V. *45, 49, 50, 651*
Sipma, G. *277*
Skorobogaty, A. *631, 644*
Skoropowski, G. *183*(b)
Slayman, C.W. *531*
Slusarchyk, D.S. *890*
Smith, D.G. *816*
Smith, G.D. *678*
Smith, G.F. *304, 310*(a)
Smith, G.N. *304, 310*(a)
Smith, J. *186*
Smith, Jr., J.M. *612*
Smith, L. *754, 755*
Smith, L.W. *59, 313, 316*
Smith, M.C. *415*
Smith, S.C. *428*

Snatzke, G. *323*
Snchez-Ziga, A.A. *766*
Sndor, P. *269*
Snipes, C. *384*
Snyder, W.C. *599*
Sobin, B.A. *610*
Sodano, G. *102, 104*
Soenen, D.R. *246*
Sohler, A. *23*(c)
Son, S. *243*(c)
Sondheimer, F. *657*
Sonnet, P.E. *44*(a), *44*(b)
Sosnick, T.R. *734*
Soucy, P. *350, 351*
Späht, E. *285*
Spande, T.F. *34*
Speigelberg, H. *568*
Spicer, L.D. *870*
Spiegelberg, H. *573*
Springer, J.P. *446*
Srinivasan, C.V. *662*
Srivastava, S.N. *348*
Srkny, S. *268*
Stähelin, H. *377*
Stahly, D.P. *803*
Stanfield, M.K. *837*
Steele, J.A. *405*
Steffan, B. *397, 424, 426, 427*
Steglich, W. *206, 239*(a), *241, 247*(a), *249, 397, 427, 429*
Stein, W.J. *615*
Stein, Z. *248*(a), *95*
Steinberg, D.A. *603*
Steiner, E. *711*
Stempien, Jr., M.F. *92*
Steube, K. *142*
Steyn, P.S. *216, 449, 453–456*
Stickings, C.E. *404, 411, 412*
Stien, D. *172*(a), *172*(b)
Stierle, D.B. *45, 192*
Still, J.L. *790*
Stipanovic, R.D. *494*
Stöckigt, J. *309*
Stoessi, A. *494*
Stoll, C. *377*
Stoll, M. *279*
Stoller, C. *120*
Stothers, J.B. *494*
Struchkov, Yu.T. *168*
Stüpfel, J. *819*

Su, J. *149*(a), *149*(b)
Subramanian, G. *307*
Suda, H. *738*(a), *738*(b)
Sudarsono *142, 178*
Sugawara, Y. *607*
Sugie, Y. *444*
Sugita, M. *451*
Sugiura, A. *444*
Suhl, K. *380*
Sumaria, S. *219*
Sumaryono, W. *103*
Supriyono, A. *103*
Surez, A.I. *194*
Sutcliffe, J.A. *444*
Suwita, A. *320, 321*
Suzuki, A. *407, 437, 438*
Suzuki, H. *738*(a), *843*(a)
Suzuki, K. *475*
Suzuki, M. *182, 607, 608*
Suzuki, R. *368, 369*
Svatos, A. *52*
Sveshnikova, M.A. *750*
Swali, R. *183*(b)
Sykes, R.J. *785–787, 817*
Szentpétery, G.B. *269, 268*
Szillat, H. *810*
Szolcsanyi, P. *435*
Szymanski, E.F. *840*

Taber, T.R. *659*
Tackett, L.P. *148, 167*
Tada, H. *65*
Taglialatela-Scafati, O. *79, 99–101, 123, 129, 137*
Taguchi, H. *432, 436*
Taguchi, R. *848*(a)
Takahashi, A. *280*
Takahashi, C. *391*
Takahashi, M. *400, 554*
Takahashi, Y. *504, 550–553*
Takaishi, T. *607, 608*
Takaoka, S. *398*
Takayama, H. *309*
Takeda, K. *834*
Takeda, R. *96, 480–483*
Takeshima, H. *403*
Takeuchi, T. *468, 471, 504, 550–553, 834*
Takeuchi, Y. *471, 472*
Takeya, K. *357, 359, 360*
Takiguchi, Y. *601*

Takita, T. *575*
Tamai, M. *280*
Tamari, K. *409*(a), *409*(b), *410, 420*
Tamm, C. *82, 377, 378, 383, 384*
Tamura, G. *508–512, 519*
Tamura, S. *407*
Tan, W. *600*
Tanabe, M. *848*(b)
Tanaka, H. *459, 660*
Tanaka, K. *520–523, 535*
Tanaka, N. *554*
Tanaka, T. *799*
Tanaka, W.K. *779*
Tanaka, Y. *280*
Taneja, S.C. *136*
Tanner, E.M. *763*
Tapiolas, D.M. *196*
Tashiro, K. *479*
Tatematsu, A. *608*
Tatsuoka, T. *154*
Tatsuta, K. *505–507, 554*
Tawara, K. *834*
Taylor, D.R. *343*
Taylor, M.E.U. *404*
Taylor, R.W. *670, 672, 675*
Tee, Y.-M. *307*
Tellew, J.E. *185*
Tempesta, M.S. *434*
Teramoto, T. *81*
Teranaka, M. *85*
Terao, M. *601*
Terashima, S. *557*
Terpin, A. *206, 239*(a), *249*
Terrasi, T.C. *772*
Tertzakian, G. *771*
Testa, R.T. *474*
Thaller, V. *8*
Theander, O. *837*
Theilleux, J. *561*
Theriault, R.J. *876*
Thirumalachar, M.J. *806*
Thoison, O. *110*(a), *306, 312*
Thomas, E.W. *851*
Thomas, R. *404*
Thomas, T.P. *672*
Thompson, W.R. *58, 62*(a)
Thomson, R.H. *792*
Thornley, C. *326*(a), *326*(b)
Thornton, S.R. *295*
Thornton, T.J. *263, 264*

Threlfall, T.L. *561*
Thrum, H. *613*
Thureen, D.R. *245*
Tillekeratne, L.M.V. *68*
Timmermans, M. *51*
Timyk, O.E. *750*
Tinto, W.F. *187, 189*
Tissier, C. *671, 674, 685*
Tito, A. *371*
Titsworth, E. *580*
Tittlemier, S.A. *25*
Tius, M.A. *560*
Tjarks, L.W. *440*
Toda, S. *888, 889*
Todorov, T.A. *765*
Tokuyama, T. *30, 31, 33*
Tollon, Y. *312*
Tolstykh, I.V. *883–885*
Tomiie, Y. *492*
Tomita, H. *272*
Tomita, K. *648, 887*
Tomizawa, J. *582*
Tomko, J. *365, 372–374*
Tomono, Y. *847*
Tone, J. *591, 686*
Toogood, P.L. *842*
Tori, M. *398*
Torrey, M.J. *474, 476*
Totsuka, K. *358*
Toube, T.P. *385–388*
Townsend, R.J. *412*
Toyama, H. *720*
Tozyo, T. *65*
Tramontano, A.J. *707*
Tramontano, W. *301*
Tran Huu Dau, E. *105*
Treibs, A. *212, 753*
Trejo, W.H. *890*
Trenkle, W.C. *185*
Trigos, A. *392*
Tripathi, R.K. *532*
Troupe, N. *683, 691*
Trujillo, J. *314*
Trumbull, E.R. *362*
Tsang, J.C. *767*
Tsao, R. *602*
Tsao, S.-W. *801*
Tsopelas, C. *160, 161*
Tsuchida, N. *798, 799*
Tsuchimori, N. *736*

Tsuda, M. *126, 130, 131, 157, 177, 180, 182*
Tsuji, T. *847*
Tsukamoto, S. *90, 91, 96*
Tsukida, K. *386*(b)
Tsukiura, H. *565–567, 887, 889*
Tsunakawa, M. *887, 889*
Tsuruoka, T. *467, 477–479*
Tumlinson, J.H. *40*(a), *40*(b)
Turos, E. *442*
Tweeddale, H.J. *316*
Tyler, P.C. *661*

Ubik, K. *366, 372, 374*
Uchida, N. *309*
Uchiyama, T. *401*
Ueda, A. *493*
Ueda, I. *520, 522*
Uemoto, H. *126, 131*
Uemura, D. *847*
Ueno, Y. *89, 433*
Uhlig, G. *266*
Uhrín, D. *365, 366, 372–374*
Umetsu, N. *409*(a), *409*(b), *410, 420*
Umeyama, A. *85, 88*
Umezawa, H. *133, 471, 575, 834, 839*(a), *839*(b), *843*(a)
Umezawa, K. *468*
Umezawa, S. *617*
Umio, S. *520–523, 535*
Unkefer, C.J. *732–734*
Uno, H. *463*
Ura, A. *24*(b)
Urano, S. *848*(b)
Urban, S. *48, 60, 186, 201–203, 233*
Utkina, N.K. *151, 165, 168, 74*
Utsumi, Y. *492*

Vahlquist, B. *11*(a)
Valadon, L.R.G. *385*
Valasinas, A. *13*(b)
Valenta, Z. *345, 347*
Valeriote, F. *93*
Vallée, Y. *48*
Vallefucco, T. *101*
van der Baan, J.L. *134, 844, 845*
van der Helm, D. *78, 149*(a), *149*(b)
Van der Wal, B. *277*
Van Duyne, G.D. *230, 231*
van Eijk, J.L. *334, 338*

Van Engen, D. *109*
van Krimpen, S.H. *729*
van Pée K.-H. *542, 503, 545*
van Rooyen, P.H. *448*
van Soest, R. *103*
van Soest, R.W.M. *87, 120, 122, 142, 178, 225, 227*
van Tamelen, E.E. 109, *614*
Van Zeeland, J.K. *706*
Vander Meer, R.K. *63*
Variyar, P.S. *740*
Varley, M.J. *454*
Vaterlaus, B.P. *568, 573*
Vazquez, D. *418, 419, 421*
Veen, G. *327*
Velterop, J.S. *735*
Venkatachalam, S.R. *740*
Venkatesham, U. *195*
Venkateswarlu, Y. *69, 134, 193, 195, 235, 236*
Verley, M.J. *455*
Verwiel, P.E.J. *722–724, 729*
Vesnaver, G. *647*
Vesonder, R.F. *440*
Viani, R. *270, 276*
Vierling, W. *352, 355*
Vincent, L.P.D. *318*
Vining, L.C. *816*
Viscontini, M. *703*(a)
Vleggaar, R. *454, 455*
Vogel, H.J. *777*
von Baeyer, A. *1, 2*
Vuillemin, B. *561*
Vul'fson, N.S. *812*
Vuorela, P.M. *701*

Wada, C.K. *854*
Wada, S. *227*
Wähnert, U. *641*
Waikedre, J. *173*
Wakamatsu, K. *141*
Waldenström, J. *11*(a)
Walizei, G.H. *44*(c)
Walker, R.P. *109*
Waller, C.W. 109, *615*
Walter, F. *45*
Walter, J.A. *816*
Walts, A.E. *693*
Ward, B. *631, 644*
Ward, F.B. *148*

Ware, R.S. *592*
Warner, J.F. *576*
Wartell, R.M. *638*
Wasserman, H.H. 123, 124, *754, 755, 759, 775, 778, 785–787, 793, 800, 804, 807–809, 817*
Watabe, S. *227*
Watanabe, K. *611*
Watanabe, N. *280, 527*
Waterhouse, A.L. *341, 342*
Watkin, D. *559*(b)
Watson, T.R. *483*
Watts, P.S. *592*
Waxham, F.J. *760*
Webb, J.S. *612*
Webb, T.M. *243*(a)
Webber, S.E. *110*(b)
Weber, K. *66*
Weedon, B.C.L. *385, 386*(a)
Wehrli, H. *11, 37*
Wei, Y. *252*
Weidner, C.H. *855*
Weinreb, S.M. *172*(a), *172*(b), *708*
Weinstein, L. *578, 581*
Weintritt, H. *204*(a), *556*(a), *556*(b), *558*
Weiss, M.J. *612*
Wells, R.E. *638*
Wells, R.J. *259*
Wemmer, D.E. *643*
Wengel, J. *291*
Wenkert, E. *540*
Wessels, P.L. *216*
Westall, R.G. *3, 5*
Westerling, J. *723*
Westley, J.W. *245, 683, 684, 689–691, 700*
Weyland, H. *66, 813*
Whaley, H.A. *593, 594*
Wheeler, J.W. *54*
Whipple, E.B. *592*
Whitaker, W.D. *572*
White, A.D. *687*
White, D.M. *614*
White, E.P. *328*
Whitesell, J.K. *852*(b)
Whittern, D. *877*
Whittingham, D.J. *344*
Wick, A.E. *571*
Wiedenfeld, H. *61*
Wiedhopf, R.M. *362*

Wiese, K.J. *166*
Wiesinger, D. *377*
Wiesner, K. *344–347, 349*
Wildfeuer, M.E. *621*
Wilkens, D.C. *457*
Wilkins, D.C. *450*
Willhalm, B. *279*
Williams, D.H. *156, 181*
Williams, D.J. *688*
Williams, D.R. *602*
Williams, E. *658, 660, 662*
Williams, R.H. *757*
Williams, R.P. *743, 782–784, 788, 789*
Williams, T.L. *528*
Williamson, R.T. *830*
Willis, A.C. *140*
Willoughby, R. *146*
Wilson, J.M. *761*
Windisch, W.W. *490*
Wink, J. *604*
Winkler, J. *790*
Winkler, T. *711*
Winklhofer, C. *241*
Winter, M. *279*
Wiryowidagdo, S. *142*
Witkop, B. *30–32, 39*
Witte, L. *103, 327*
Wnuk, R.J. *593, 878*
Wolf, C.F. *615*
Wolf, D. *226*
Wolf, G. *29*(a)
Wolf, H. *651*
Wolfe, M.S. *464*
Wong, D.-J. *140*
Wong, D.T. *533, 534*
Wong, H.N.C. *243*(b)
Wood, W.F. *60*
Woodruff, H.B. *415*
Woods, B.L. *646*
Woodward, P.R. *428*
Worthington, R.E. *789*
Wörheide, G. *82*
Wray, V. *103, 142, 178, 327*
Wrede, F. *122, 745, 751, 752*
Wright, A.D. *72, 82, 86*
Wright, A.E. *121*
Wright, D.E. *561*
Wright, J. *853, 856*
Wright, J.L.C. *816*
Wright, M. *312*
Wu, Y.-C. *375*
Wurziger, H.K.W. *16*

Xiao, D. *44*(g)
Xu, Q. *403*
Xu, S.-C. *52*
Xu, Y. *143, 144*(a), *147*

Yagen, B. *626, 630*
Yaginuma, S. *681*
Yakushijin, K. *112, 132, 139*(b), *143, 144*(a), *144*(b), *147, 166, 174, 368, 369*
Yamada, K. *493, 847*
Yamada, T. *88, 391*
Yamagishi, Y. *389, 390, 549*
Yamaguchi, H. *834*
Yamaguchi, M. *24*(a), *24*(b)
Yamaguchi, T. *606*
Yamamoto, K. *273*
Yamamura, S. *251*
Yamanaka, H. *253*(a), *253*(b)
Yamano, K. *376*(a), *376*(b)
Yamasaki, K. *436*
Yamashita, M. *179, 821, 822*
Yang, C.-C. *595, 598*
Yang, H. *403*
Yang, J.-P. *189*
Yang, Q.-C. *243*(b)
Yano, K. *470, 473*
Yasuda, A. *398*
Yasuzawa, T. *400*
Yazaw, H. *544*
Yazawa, H. *543*
Yazawa, K. *846*
Yazawea, K. *847*
Ye, Y. *222*
Yenkateswarlu, Y. *135*
Yokoyama, A. *184*
Yonehara, H. *617, 848*(b)
Yoon, C. *639*
Yoshida, J. *465*
Yoshida, M. *400*
Yoshida, S. *768*
Yoshida, W.Y. *214, 242*
Young, D.W. *329*
Yu, S.-X. *299*(a), *299*(b)
Yuasa, E. *88*
Yugami, T. *736*
Yuhara, T. *820*

Zaccardi, J. *443*
Zaehner, H. *880*
Zaghloul, A.M. *289*
Zähner, H. *461, 651, 819, 861*
Zaikin, V.G. *812*
Zard, S.Z. *44*(d)
Zaretskii, V.I. *812*
Zatman, L.J. *704*(a), *704*(b)
Zdero, C. *318–321, 323, 324*
Zeeck, A. *604, 651, 652, 819, 861, 863, 864*
Zeng, L. *149*(a), *149*(b)
Zennie, T.M. *296–298*
Zhao, K. *356*
Zhao, S.-Y. *133*
Zhao, Y.-Y. *339*
Zhuze, A.L. *625*
Ziegler, F.E. *668*
Zimmer, C. *637, 641*
Zimmer, R. *44*(f)
Zimmer-Galler, R. *753*
Zimmerman, S.B. *823*
Zubia, E. *197*
Zurita, M.B. *118*
Zweig, J. *64*(a)

Subject Index

Aburatubolactam A   135
Acetyltetramic acids, synthesis   54
2-(2-Acetoxyacetyl)-1-(2-hydroxy-
  ethyl)pyrrole   69
2-(2-Acetoxyacetyl)-1-(2-hydroxy-
  propyl)pyrrole   69
2-Acetylpyrrole   67
Acanthella aurantiaca   24, 33
A. carteri   21, 29, 31, 33, 37
Acetobacter aceti   118
A. pasteurianus   118
Acremomium loliae   84
Acromyrmex subterranous   12
Actinomadura sp.   101, 127, 128
A. madurae   128
A. pelletieri   128
A. spiralis   102
Actinomyces aureoverticillatus   127
A. pneumonicus   137
Actinosporangium vitaminophilum   94, 95, 97
Adenosine, 6-(3-Methylpyrrol-1-yl)-   71
Aflastatin A   89
  biosynthesis   90
Aflastatin B   89
  biosynthesis   90
Agelas sp.   17, 38, 42
A. cf. nemoechinata   41
A. clathrodes   23, 24, 26
A. conifera   24, 26, 41, 42
A. ceylonica   29
A. dendromorpha   39
A. dispar   23, 24, 26, 29
A. flabelliformis   18
A. longissima   19, 24, 26, 29
A. mauritiana   18, 27, 40, 42
A. nakamurai   20, 21, 26, 28, 29, 41
A. oroides   18, 20, 23, 27, 30
A. sceptrum   41
A. wiedenmayeri   25
Agelasine G   20

Agelastatin A   39
  biosynthesis   39
Agelastatin B   39
Agelastatin C   39
Agelastatin D   39
Ageliferins   41
Ageliferin, 12,12'-dimethyl   41
Ageliferins, biosynthesis   42
Ageline B   20
Agelongine   19
δ-ALA   8
Aldisin   35
Altamycin A   137
Alteramide A   135
Alteramide B   135
Alternaria alternata   86
A. longipes   86
A. tenuis   86
Alteromonas sp.   135
A. luteoviolaceus   97
A. rubra   126
Althiomycin   133
  biosynthesis   133
5-Aminolaevulinic acid   8
Aminopyrrolnitrin   100
Ancorina sp.   55
Ancorinoside A   55
Anhydroproferrorosamine B   117
Anhydroryanodine   79
Anthelvencin A   111
Anthelvencin B   111
Antibiotic
  2814A (netropsin)   109
  Al-R2081 (pyrroxamycin)   95
  A 23187 (calcimycin)   114
  AC 7230   115
  Bu-2313A   139
  Bu-2313B (nocamycin I)   139
  CJ-17,572   91
  CP-47,444 (nargenicin A_1)   106
  CP-61,405 (routiennocin)   115
  FR-900148   130

HS3 (pyrroxamycin)   95
K16 (malonomycin)   134
LL-F42248α (pyrroxamycin)   95
LL-49F233α   91
N-461 (capsimycin)   135
RP 18,631 (chlorobiocin)   102
S 5185 RP   113
SS 46506A (pyrroxamycin)   95
T 1384 (netropsin)   109
U-61,732 (18-deoxynargenicin)   106
X-14885A   115
X-14547A (indanomycin)   116
Aspergillus flavus   92
A. oryzae   92
A. parasiticus   89
A. versicolor   92
Aspidosperma quebracho blanco   71
Aster tataricus   79
Asterinin A   79
Asterinin B   79
Asterinin C   79
Astin J   79
Astrosclera willeyana   20, 42
Atapozoa sp.   64
Atta cephalotes   12
A. texana   12
Aurantoside A   55, 56
Aurantoside B   55, 56
Aurantoside C   57
Aurantoside D   57
Aurantoside E   57
Aurantoside F   57
Auxarthron umbrinum   83
Axinella sp.   33, 44
A. carteri   24, 34, 35
A. damicornis   24
A. proliferans   29
A. tenuidigitata   17, 29
A. verrucosa   24
Axinellamide   45
Axinellamine A   44, 45
Axinellamine B   44, 45
Axinellamine C   44
Axinellamine D   44
Axinellidae sp.   18, 34–36
Axinohydantoin   33

Bacillus prodigiosus   122
Balanus amphitrite   22, 40
Batrachotoxin   11
Batrachotoxinin A   11
Beneckea gazogenes   122, 126
BE-18591   121
2,2′-Bipyrrole
  Hexabromo-   94
  Hexabromo-1,1′-dimethyl-   10
  Tetrabromo-5,5′-dichloro-1,1′-dimethyl-   10
Botanpi   67
Bromoageliferin   41
2-Bromoaldisin   35
3-Bromohymenialdisine   34
Bromonitrin A   101
Bromopalau'amine   42, 43
Bromophakellin   36
N-(4-Bromopyrrol-2-yl) carbonylguanidine   21
Bromostyloguanidine   43
Brunfelsamidine   68
Brunfelsia grandiflora   68
Bugula dentata   65
Bugula pyrrole   65
Butylcycloheptylprodiginine   128
  biosynthesis   183

Cacospongia mollior   47
Cadia ellisiana   77
C. purpurea   78
Calcimycin   114
  biosynthesis   115
Calpaurine   77
Calpurnia aurea   77
C. subdecandra   77
Calpurnine   77
Camellia thea   67
Capsimycin   135
  biosynthesis   136
Catacandin A   139
Catacandin B   139
Cezomycin   115
  biosynthesis   115
Chilocorine   15
Chilocorine A   15
Chilocorine B   15
Chlorobiocin   102
  biosynthesis   104
Cholocorus cacti   15
Chromobacterium sp.   10, 94, 97

Subject Index

C. violaceum 85, 121
Chromopyrrolic acid 85
Chromoviridan 121
  biosynthesis 122
Chromoxymycin 113
Ciona savignyi 21
Cladophora sp. 122
N-Cinnamoylpyrrole 68
  2′,3′-Dihydro- 68
Clathramide A 23
Clathramide B 23
Clathramide C 23
Clathramide D 23
Clathrodin 24
Cliona sp. 50
Clitocybe acromelalga 81
Cocoa 67
Coffee 67
Congocidin 109
Coumermycin A_1 104
Coumermycin A_2 104
Creatonotos gangis 16
C. transiens 16
Crotalaria 16
Cyclocinamide A 22
Cyclononylprodiginine 128
Cyclooroidin 30
α-Cyclopiazonic acid 92
  biosynthesis 92
β-Cyclopiazonic acid 92
Cycloprodigiosin 126
Cylindramide 57
Cymbastela sp. 39
Cystamidin A 93
Cystobacter fuscus 133

Dactylosporangium sp. 115
Danaidal 16
Danaidone 16
Danainae 16
Danaus gilippus 16
Deazainosine 120
Debromoaxinohydantoin 34
Debromohymenialdisine 33, 34
  synthesis 34
2-Debromohymenin 31
2-Debromolongamide 29
2-Debromomidpacamide 27
Debromosceptrin 41
2-Debromostevensine 32

4′-Dechloropyrrolnitrin 101
Dehydroheliotridine 73
Dehydroisosenaetnine 75
  biosynthesis 76
Dehydrorhazinilam, 3-Oxo- 71
Dendrilla cactos 48
18-Deoxynargenicin 106
  biosynthesis 106
Deoxychromoviridan 121
  biosynthesis 122
Desmapsamma anchorata 45
Desmethylaminocalcimycin 115
Deoxymajusculamide D 130
Desoxyverrucarin E 82
  biosynthesis 82
Dibromoagelaspongin 38
Dibromoageliferin 41
Dibromocantharellin 37
Dibromoisophakellin 37
Dibromopalau'amine 42, 43
Dibromophakellin 36
  synthesis 37
Dibromophakellstatin 37
Dibromostyloguanidine 43
5,7a-Didehydroheliotridin-3-one,
  synthesis 75
Didehydroryanodine 78
Didemnum sp. 62
D. chartaceum 58
4,7-Dihydroisoindole-4,7-dione,
  6-methoxy-2,5-dimethyl- 51
  synthesis 51
Dihydrokikumycin B 109
Dihydropyrrolizine alkaloids 73
  biosynthesis 73
  metabolism 73
Dihydropyrrolizin-3-carboxylic acid,
  1-Oxo- 121
Dihydrorhazilinam 71
Dioxapyrrolomycin 95
  biosynthesis 95
Discoderma dissoluta 57
Discodermide 57
Dispacamide A 26
Dispacamide B 26
Dispacamide C 26
Dispacamide D 26
Distamycin A 111
  synthesis 110

Dolabella auriculaira 66
Dolastatin 15 66
Dysidea herbacea 55
Dysidin 55

Enteromorpha intestinalis 83
Epiryanodine 78
Equisetin 91
Eremophilene lactam 80
Erythroskyrine 89
  biosynthesis 89
Ethylcyclononylprodiginine 128
Eudistoma cf. rigida 63
E. olivaceum 64
Eudistomin A 64
Eudistomin M 64
Eupenicillium hirayamae 82
Exochomine 15
Exochomus quadripustulatus 15

Frankia 117
Frankiamide 117
Fuligorubin A 88
Fuligo septica 88
Funebral 70
Funebradiol 70
Funebrine 70
  biosynthesis 70
1-Furfurylpyrrole 68
3-Furfuryl pyrrole-2-carboxylate 78
Fusarium equiseti 91
Fuscin 33

Glycerinopyrin 108
  biosynthesis 108
Goreauiella sp. 24
Gracilariopsis lemaneiformis 69

Halichondria cylindrata 57
H. okadai 135
Haliclona tulearensis 50
Halitulin 50
  biosynthesis 51
Halocynthia roretzi 21
Hanishin 29
  biosynthesis 30
Haumanamide 53
Hedamycin 113
heliotrope 73

Heliotropium alkaloid 73
Heliotropium europaeum 73
Hippodamine 15
Homaxinella sp. 21, 29
Homobatrachotoxin 11
Homophymia conferta 57
Hormaomycin 113
5-(4-Hydroxybenzyl)-3-acetyltetramic acids 89
Hydroxydanaidal 16
Hymeniacidon sp. 19, 27, 31, 33, 34, 42
H. aldis 35
Hymenialdisine 33
  3-bromo- 34
Hymenidin 24
  biosynthesis 25
Hymenin 31
  biosynthesis 31
  synthesis 31, 32

Ikarugamycin 135
  biosynthesis 136
Inaequidenine 75
  biosynthesis 76
Indanomycin 116
Iotrochota sp. 62
Isobatrachotoxin 11
Isomalyngamide A 131
Isopterophorine 75
  biosynthesis 76
Isopyrrolnitrin 101
Isosenaetnin 75
  biosynthesis 76

Jacmaia alkaloid 74
Jacmaia incana 74
Japanese hops 67
Jathropham 80
Jatropha macrorhiza 80
Jatropham dimer 81

Karebakitsunetake 87
Keramadine 24
Keronopsins A 66
Keronopsins B 66
Kikumycin A 108
Kikumycin B 108
Konbu'acidin A 42
  biosynthesis 43

Kopsia singapurensis 71
Kryptopyrrole 10
Kukoshi 67

Laccaria vinaceoavellanea 87
Laccarin 87
Lamellaria sp. 49
Lamellarin α, 20-sulfate 59
Lamellarin A 60
Lamellarin B 59
Lamellarin C 59
Lamellarin D 59
Lamellarin E 60
Lamellarin F 59
Lamellarin G 59
Lamellarin H 59
Lamellarin I 59
Lamellarin J 59
Lamellarin K 59
Lamellarin L 59
Lamellarin M 59
Lamellarin N 58
Lamellarin O 48
Lamellarin P 48
Lamellarin Q 48
Lamellarin R 48, 50
Lamellarin S 59, 60
Lamellarin T 59
Lamellarin U 59
Lamellarin V 59
Lamellarin W 59
Lamellarin X 59
Lamellarin Y 59
Lamellarin Z 59
Lamellarins 48, 58, 59
  biosynthesis 61
Laxosuberites sp. 45
Leocarpus fragilis 88
Leuconotis eugenifolia 71
Licogarubin A 85
α-Lipomycin 137
β-Lipomycin 137
Lilium sp. 80
L. candidum 81
L. hansonii 80
Lissodendoryx sp. 17, 35
Lolium perenne 84
Longamide 29
  biosynthesis 30

Longamide B 30
  biosynthesis 30
  methyl ester 29
Lophophora williamsii 69
Loroco 73
Loroquin 73
Lukianol A 49, 61
  synthesis 49
Lukianol B 61
Lycium chinense 67
Lycogala epidendrum 85, 86
Lycogalic acid A 85
  dimethyl ester 85
Lycogalic acid B 85
Lycogalic acid C 85
  dimethyl ester 85
Lycogarubin A 85
Lycogarubin B 85
Lycogarubin C 85
Lydicamycin 137
Lyngbya majuscula 130–132
Lysobacter gummosus 139

Macrophomina phaseolina 84
Macrophominol 84
  biosynthesis 84
Majusculamide D 130
Makaluvic acid A 18
Makaluvic acid B 18
Malonomycin 134
  biosynthesis 134
  synthesis 134
Malyngamide A 131
Malyngamide Q 131
Malyngamide R 131
Manzacidin A 19
Manzacidin B 19
Manzacidin C 19
Manzacidin D 20
Marasmiellus ramealis 6
Matamycin 133
Mauritamide A 27
Mauritiamine 40
  biosynthesis 40
mauve factor 10
Melodinus australis 71
Melophlin A 55
Melophlin B 55
Melophlus sarassinorum 55

Metacycloprodigiosin 127
  biosynthesis 129
  synthesis 128
Methoxatin 117
  biosynthesis 119
  catalytic activity 119
  synthesis 119
Methylcycloctylprodiginine 128
Methylcyclodecylprodiginine 128
Micrococcus sp. 122
Micromonospora sp. 112
Microtetraspora spiralis 102
Microcolin A 130
Microcolin B 130
Midpacamide 27
Milbemycin α_9 107
Milbemycin α_{10} 107
Molliorins A–E 47
  synthesis 47
Monanchora sp. 34
Monobromodispacamide 26
Mukanadin A 26
Mukanadin B 26
  10,11-Dihydro- 27
Mukanadin C 29
Mycalazal 1 46
Mycalazal 2 46
Mycalazol 11 46
Mycalazols 47
Mycale micracanthoxea 47
M. microsigmatosa 45
M. mytilorum 45
M. tenuispiculata 45
Mycalecarmia monanchrorata 45
Myrmicaria sp. 12
M. eumenoides 13
M. opaciventris 14
M. striata 15
Mycaleoxime 45
Myrmicarin 213A 14
Myrmicarin 213B 14
Myrmicarin 215A 14
Myrmicarin 215B 14
Myrmicarin 215C 14
Myrmicarin 217 14
  synthesis 14
Myrmicarin 237A 13
Myrmicarin 237B 13
Myrmicarin 430A 14

Myrmicarin 663 14
Myrmicarins 12
  biosynthesis 13
Myrmicyrin 645 15
Myrothecium verrucaria 82
Myxococcus sp. 133

Nakamuric acid 41
Nakamurol D 20
Nargenicin A_1 106
  biosynthesis 105
Nembrotha kubaryana 65
Neopyrrolomycin 98
Netropsin 109
  biosynthesis 109
  synthesis 110
β-Nicotyrine 68
Nicotiana tabacum 68
Nierenbergia hippomanica 68
Ningalin A 62
Ningalin B 61, 62
Ningalin C 62
Ningalin D 62
3′-Nitropyroluteorin 96
Nocamycin 139
Nocamycin I 139
Nocamycin II 139
Nocardia argentinensis 106
Nocardiopsis sp. K-290 85
N. syringae 139
Nodusmicin 106
Nonylprodiginine 127
bis-Norfunebral 70
Norprodigiosin 126
Novobiocin 104
  biosynthesis 104

3-Octadecyl-2-
  pyrrolecarboxaldehyde 45
Odiline 32
Oleficin 137
Oroboidine 77
Oroidin 23
  biosynthesis 25
  synthesis 24, 25
5,5′-Oxydi(3-methyl-3-pyrrolin-2-one) 81
Oscarella lobularis 45
Oxypyrrolnitrin 101
Oxysceptrins 41

Paeonia moutan 67
Palau'amine 42
Pea 70
Penicillium sp. 83, 86
P. camemberti 92
P. cyclopium 92
P. islandicum 89
P. viridicatum 92
Penochalasin A 83
Penochalasin B 83
Penochalasin C 83
Pentabromopseudilin 97
Peramine 84
   biosynthesis 85
Petasites hybridus 80
Peyoglunal 69
Peyonine 69
   synthesis 69
Peyote 69
Pezicula 92
Phacellia fusca 33
P. flabellata 33, 36
P. mauritiana 37
Phenopyrrocin 86
Phorba aff. clathrata 22
Phorbazole A 22
Phorbazole B 22
Phorbazole C 22
Phorbazole D 22
Phyllobates aurotaenia 11
P. terribilis 11
Physarorubinic acid 88
Physarorubinic acid B 88
Physarum polycephalum 88, 90
Piper argyrophyllum 68
P. sarmentosum 68
Pisum sativum 70
Pitohui 11
Polycephalin B 90
   biosynthesis 91
Polycephalin C 90
   biosynthesis 91
Polycitone A 62
Polycitone B 63
Polycitor sp. 62
P. africanus 63
Polycitrin A 62
   synthesis 63
Polycitrin B 62
Polyphysia crassa 17

Porphobilinogen 6
   biosynthesis 8
   synthesis 6
Porphobilinogen lactam 6
PQQ (Pyrroloquinoline Quinone) 117
Prodiginine 124
Prodigiosene 124
Prodigiosin 122
   biosynthesis 124
   synthesis 123
Prodigiosin 25 C 127
Prodigiosins 122
Proferrorosamine A 117
   biosynthesis 117
Proferrosamine B 117
Propionibacterium shermanii 8
Psammocinia sp. 22
Pseudaxinyssa cantharella 24, 33, 35, 37
Pseudoceratidine 21
Pseudoceratina purpurea 21, 22
Pseudokeronopsis rubra 66
Pseudomonas sp. 101, 118
P. acidula 101
P. aeruginosa 96, 98
P. aureofaciens 98, 99, 100, 137
P. bromoutilis 97
P. cepacia 99, 100
P. fluorescens 99
P. magnesiorubra 122
P. multivorans 99
P. pyrrocinia 98
P. pyrrolnitrica 98, 101
P. roseus fluorescens 117
Pseudostellaria heterophylla 78
Pterophorin 75
   biosynthesis 76
Ptilocaulis walpersi 25
Pukeleimide A 132
Pukeleimide B 132
Pukeleimide C 132
Pukeleimide D 132
Pukeleimide E 132
Pukeleimide F 132
Pukeleimide G 132
Purpurone 62
Pyoluteorin 96
   biosynthesis 96, 97
   synthesis 96
Pyralomycins 1 102
Pyralomycins 2 102

Pyricularia oryzae 86
Pyrroindomycin A 107
Pyrroindomycin B 107
Pyrrolam A 129
Pyrrole 2
  2,3-Dibromo 17
  2,3-Dibromo-5-methoxymethyl- 17
  3-Ethyl-2.4-dimethyl 10
  2,3,4-Tribromo- 17
  Tetrabromo 94
Pyrrole-3-acrylamide 93
Pyrrole-3-acrylic acid 93
Pyrrole-3-carbamidine 68
Pyrrole-2-carbonitrile
  −4,5-Dibromo 18
Pyrrole-2-carboxaldehyde
  5-Alkyl- 45
  5-(12′-cyano-6′-dodecenyl)- 45
  5-(23′-Cyano-23′-hydroxy-
    6′-tricosenyl)- 45
  5-Nonyl- 46
Pyrrole-2-carboxamide 18, 20
  4-bromo- 21
  −4-Bromo-,N'-[Amino(imino)methyl]- 21
  −5-Bromo- 21
  −5-Bromo-N'-methoxymethyl- 21
  −4,5-Dibromo- 18, 21
  −N'-Methoxymethyl- 21
  −N'-Formyl- 20
Pyrrole-2-carboxylic acid 5, 10
  1-Allyl-4,5-dibromo- 18
  4-Bromo-, methyl ester 17
  5-Bromo-, methyl ester 18
  4,5-Dibromo 18
  4,5-Dibromo, methyl ester 17, 18
  4,5-Dibromo-1-ethyl- 18
  4,5-Dibromo-1-methyl- 18
  4,5-Dibromo-1-methyl, methyl ester 18
  4-methyl-, methyl ester 12
  3,4,5.Tribromo- 18
  3,4,5-Tribromo-1-methyl- 18
Pyrrole-3-carboxylic acid
  2,4-dimethyl- 11
  2-ethyl-4-methyl- 11
Pyrrole-1,2-dicarboxamide 10
Pyrrole-3-propanamide 93
Pyrrolnitrin 98
  biosynthesis 100
  synthesis 99

Pyrrolomycin A 94
Pyrrolomycin B 94
Pyrrolomycin C 95
Pyrrolomycin D 95
Pyrrolomycin E 97
Pyrrolomycins 94
Pyrrolomycins F 95
Pyrrolosine 120
Pyrrolosporin A 112
Pyrrolostatin 93
Pyrrolo[2.1.5-*cd*]indolizine 13
Pyrrolylalanine 81
  biosynthesis 82
O-(Pyrrol-2-ylcarbonyl)-4-
  hydroxyepilupinine 76
O-(Pyrrol-2-ylcarbonyl)virgiline 77
O^8-(Pyrrol-2-ylcarbonyl) cadiamin 78
Pyrroxamycin 95
  biosynthesis 95

Quararibea funebris 70

Readea membranaceae 77
Reniera sp. 51
Rhazinal 71
Rhazinilam 71
  synthesis 71
Rhazya stricta 71
Rigidin 63
Roboastra tigris 65
Rollipyrrole 81
Rollinia mucosa 81
Roseophilin 102
  synthesis 103
Routiennocin 115
Rubrosides A–H 57
Rumbrin 83
Ryania speciosa 78, 79
Ryanodine 78
Ryanodol 79
Ryegrass 84

Saccharopolyspora hirsuta 106
Sandal-wood 68
Santalum album 68
Sarcotragin A 52
Sarcotragin B 52
Sarcotragus sp. 52
Scalaradial 47
Sceptrin 40, 41

Subject Index

bis-Secodehydrocyclopiazonic acid 92
1,10-Seco-10-hydroxycalpurnine 78
(15-*E*)-Senaetnine 75
Senaetnine 75
  biosynthesis 76
Senampeline A 74
Senampeline B 74
Senampeline C 75
Senampeline D 75
Senampeline E 75
Senampeline F 75
Senampeline G 75
Senecio sp. 73, 75
S. cissampelinus 74
S. mikanoides 74
S. stapeliaeformis 74
Serratia marcescens 122–125, 129
Sessibugula translucens 64
Siliquariaspongia japonica 57
Sinamomycin 109
Slagenin A 28
Slagenin B 28
Slagenin C 28
Solanum sodomaeum 68
Solsodomine A 68
Solsodomine B 68
Spiganthine 79
Spigelia anthelmia 79
Spongia sp. 53
Spongiacidin A 34
Spongiacidin B 34
Spongiacidin C 34
Spongiacidin D 34
Staurosporinone 85
Stevensine 32
  biosynthesis 32
  synthesis 31, 32
Storniamide A 50
Storniamide B 50
Storniamide C 50
Storniamide D 50
Streptolic acid 138
Streptolydigin 138
  biosynthesis 139
Streptomyces sp. 65, 67, 95, 113–115, 121, 126, 128, 135, 138
S. albocinerescens 102
S. albus 120
S. althioticus 133
S. antibioticus 116

S. aureofaciens 98
S. avermitilis 107
S. chartreusensis 114, 115
S. chrestomyceticus 93
S. coelicolor 127, 128
S. distallicus 111
S. flaveolus 138
S. fragilis 95
S. fumanus 95
S. griseoflavus 113
S. griseoviridis 102
S. hazeliensis 104
S. hygroscopicus 102, 107
S. KP-1241 93
S. longisporus ruber 127, 129
S. lydicus 137, 138
S. matensis 133
S. netropsis 109
S. olivaceus 121, 129
S. parvulus 93, 137
S. phaeochromogenes 108, 135
S. rimosus 134
S. rishiriensis 104
S. roseochromogenus 102
S. routiennii 116
S. rubropurpureus 113
S. rugisporus 107
S. spinichromogenes 105
S. spinicoumarensis 105
S. tirandis 138
S. variegatus 122
S. venezuelae 111
S. violaceus 108
S. viridogenes 130
S. xanthocidicus 130
Streptorubrin A 127
Streptorubrin B 128
Streptoverticillium sp. 127, 128
S. rubrireticuli 127, 128
Stylissa carteri 31, 35, 148
Styloguanidine 43
Stylotella agminata 43
S. aurantium 34, 43
Sugordomycins 104

Tai-zi-shen 78
Tambjamine A 64
  N''-Dodecyl- 65
Tambjamine B 64
Tambjamine C 64

Tambjamine D   64
Tambjamine E   64
Tambjamine F   64
Tambjamine G   65
Tambjamine H   65
Tambjamine I   65
Tambjamine J   65
Tambje abdere   64, 65
T. eliora   64, 65
Tauroacidin A   27
Tauroacidin B   27
Tea   67
Teichaxinella morchella   25, 32
Telesto sp.   46
Taurodispacamide A   27
Tenuazonic acid   86
  biosynthesis   87
  synthesis   86
Tetramic acid   53
Theonella swinhoei   55
Thiazohalostatin   101
Tirandalydigin   138
Tirandamycic acid   139
Tirandamycin A   138
Tirandamycin B   138
Tobacco   67
Tobacco crown gall   71
Trikendiol   52
Trikentramine   52
Trikentrion loeve   52

Undecylprodigine   127
  biosynthesis   129
Undecylprodigiosin   127
Urechites kaerwinsky   73
Uroporphyrinogen   6
Utetheisa ornatrix   16

Valerian   67
Valeriana officinalis   67
Verrucarin E   82
  biosynthesis   82
  synthesis   82
Vibrio psychoerythreus   122
Violacein   122
Virgilia capensis   77
V. divaricata   76
V. oroboides   76, 77
Vitamycin A   127

Wallemia A   83
Wallemia C   83
Wallemia E   83
Wallemia F   83
Wallemia sebi.   83

Ypaoamide   130

Zyzzya fuliginosus   18

SpringerChemistry

Fortschritte der Chemie organischer
Naturstoffe / Progress in the Chemistry
of Organic Natural Products

Edited by W. Herz, H. Falk, G. W. Kirby

Volume 84

F.-P. Montforts, M. Glasenapp-Breiling
Naturally Occurring Cyclic Tetrapyrroles

D. G. I. Kingston, P. G. Jagtap, H. Yuan, L. Samala
The Chemistry of Taxol and Related Taxoids

2002. VIII, 253 pages. 12 figures.
Hardcover EUR 142,–*⁾
Reduced price for subscribers to the series: EUR 128,–*⁾
ISBN 3-211-83707-8

Volume 85

K. Krohn
**Natural Products Derived from Naphthalenoid Precursors
by Oxidative Dimerization**

P. Messner, C. Schäffer
Prokaryotic Glycoproteins

D. P. Chakraborty, S. Roy
Carbazole Alkaloids IV

2003. X, 257 pages. 12 figures.
Hardcover EUR 160,–*⁾
Reduced price for subscribers to the series: EUR 144,–*⁾
ISBN 3-211-83783-3

*⁾ Recommended retail prices.
All prices are net-prices subject to local VAT.

SpringerWienNewYork

A-1201 Wien, Sachsenplatz 4–6, P.O. Box 89, Fax +43.1.330 24 26, e-mail: books@springer.at, **www.springer.at**
D-69126 Heidelberg, Haberstraße 7, Fax +49.6221.345-229, e-mail: orders@springer.de
USA, Secaucus, NJ 07096-2485, P.O. Box 2485, Fax +1.201.348-4505, e-mail: orders@springer-ny.com
EBS, Japan, Tokyo 113, 3–13, Hongo 3-chome, Bunkyo-ku, Fax +81.3.38 18 08 64, e-mail: orders@svt-ebs.co.jp

SpringerChemistry

Monatshefte für Chemie/Chemical Monthly

An International Journal of Chemistry

Österreichische Akademie der Wissenschaften (Mathematisch-Naturwissenschaftliche Klasse) und Gesellschaft Österreichischer Chemiker

Editorial Board
H. Falk, Linz (Managing Editor)
H. Gamsjäger, Leoben
B. Kräutler, Innsbruck
F. Pittner, Wien
U. Schubert, Wien
P. Schuster, Wien

Regional Editors
P. Braunstein, Strasbourg
H. Brunner, Regensburg
K. Sawada, Niigata
E. Vogel, Köln
O. Vogl, Amherst, MA

and an **International Advisory Board**

Monatshefte für Chemie/Chemical Monthly was conceived in its very beginnings as an Austrian journal of chemistry. However, during recent times it was gradually transformed into an international journal including all branches of chemistry. It features the most recent research in analytical, inorganic, medicinal, organic, physical, structural, and theoretical chemistry, including the chemically oriented areas of biochemistry.

Monatshefte für Chemie/Chemical Monthly publishes refereed original papers and emphasizes a rapid publication section entitled "Short Communication". Invited reviews, symposia in print, and issues devoted to special fields will also be included.

Subscription Information
2003. Vol. 134 (12 issues). Title No. 706
ISSN 0026-9247 (print), ISSN 1434-4475 (electronic)
EUR 1.288,– plus carriage charges

View table of contents and abstracts online at: **www.springer.at/mochem**

A-1201 Wien, Sachsenplatz 4–6, P.O. Box 89, Fax +43.1.330 24 26, e-mail: books@springer.at, **www.springer.at**
USA, Secaucus, NJ 07096-2485, P.O. Box 2485, Fax +1.201.348-4505, e-mail: orders@springer-ny.com
EBS, Japan, Tokyo 113, 3–13, Hongo 3-chome, Bunkyo-ku, Fax +81.3.38 18 08 64, e-mail: orders@svt-ebs.co.jp